THE OSKAR KLEIN MEMORIAL LECTURES

1988–1999

THE OSKAR KLEIN MEMORIAL LECTURES

1988–1999

Editor

Gösta Ekspong

Stockholm University, Sweden

 **World Scientific**

NEW JERSEY • LONDON • SINGAPORE • BEIJING • SHANGHAI • HONG KONG • TAIPEI • CHENNAI

Published by

World Scientific Publishing Co. Pte. Ltd.

5 Toh Tuck Link, Singapore 596224

USA office: 27 Warren Street, Suite 401-402, Hackensack, NJ 07601

UK office: 57 Shelton Street, Covent Garden, London WC2H 9HE

Library of Congress Cataloging-in-Publication Data

The Oskar Klein memorial lectures, 1988–1999 / editor Gösta Ekspong, Stockholm University, Sweden.

   pages cm

  Includes bibliographical references and index.

   ISBN 978-9814571609 (hardcover : alk. paper) -- ISBN 9814571601 (hardcover : alk. paper)

  1. Klein, Oskar. 2. Physics. 3. Quantum theory. 4. Physicists--Sweden--Biography.

I. Ekspong, Gösta, editor of compilation.

  QC71.O85 2014

  530--dc23

                                      2013049080

British Library Cataloguing-in-Publication Data

A catalogue record for this book is available from the British Library.

Typeset by Stallion Press

Email: enquiries@stallionpress.com

Printed in Singapore

Oskar Klein (1894–1977)
Professor in Theoretical Physics, 1931–1962

Oil Painting by Anna Klein

FOREWORD

This volume is a compilation of lectures and articles published earlier in the series of books entitled *The Oskar Klein Memorial Lectures*. Some of the world renowned physicists were invited to lecture at the Department of Physics of Stockholm University, sponsored by the Royal Swedish Academy of Sciences through its Nobel Committee for Physics. These lectures were at extraordinarily high level and continue to be of great value to students and teachers alike.

Oskar Klein was one of the pioneers of quantum mechanics. After some years at the Niels Bohr Institute in Copenhagen, he transferred in 1930 to Stockholm University as Professor of Mathematical Physics (later renamed Theoretical Physics). His name is coupled to phenomena such as Klein's paradox, the Klein-Nishina formula for X-ray scattering, Klein-Gordon equation, as well as with Jordan second quantization. Independently of Kaluza, he introduced a fifth dimension, which Klein proposed as compactified. Some pages in the volume are devoted to commemorate Klein and his achievements. Important papers by Klein, otherwise not easily available, have been translated to English and are reproduced here.

Gösta Ekspong
Professor Emeritus Stockholm University
Member the Royal Swedish Academy of Sciences

Stockholm September 2013

CONTENTS

PART I

PREFACE

The present volume contains the first and second Oskar Klein Memorial Lectures, given in 1988 and 1989 at the Department of Physics at the University of Stockholm. This series of lectures, held once a year by a world renowned physicist, has been made possible through a grant from the Nobel Committee for Physics at the Royal Swedish Academy of Sciences.

In 1988, Chen Ning Yang, professor at the State University of New York at Stony Brook, USA, and Nobel prize winner in Physics 1957, gave the first lectures. The lectures of the second year, 1989, were given by Steven Weinberg, professor at the University of Texas, Austin, USA, Nobel prize winner in Physics 1979. The main lecture every year is aimed towards an audience of gifted students and their teachers, whereas the second lecture contains material of a somewhat more advanced nature. Thanks to an initiative from World Scientific in Singapore it is now possible to present these highly stimulating and enjoyable lectures to a wider audience.

Oskar Klein was 23 years old when he, in 1918, came to Niels Bohr in Copenhagen. He stayed there with some interruptions until 1931, at which time he became professor at Stockholm University, devoting himself to theoretical physics. Before his arrival in Copenhagen, Klein had published a few scientific papers, the first at the age of 18.

When he was 16, his father, the chief rabbi in Stockholm, arranged for him to work in the chemistry laboratory of Svante Arrhenius, famous for his dissociation theory of electrolytes and his Nobel prize in Chemistry in 1903.

Klein's professorship lasted until his retirement in 1962. A biography was written after his death in 1977 jointly by two scientists who knew Klein well. The authors were Professor Inga Fischer-Hjalmars, a specialist in molecular theory who had been a student of Klein and who later became his successor, and Professor Bertel Laurent, also a student of Klein, specializing in general relativity and field theory. Their article, first published in the yearbook of the Swedish Physical Society, *Kosmos*, is here presented in an English translation. The biography clearly shows Klein's broad interest in various fields of physics. This is also evident from the list of his scientific publications, included at the end of this volume.

Klein's attempt to unify general relativity and electromagnetism dates back to the 1920s. One of his key papers, originally published in German, introduces five-dimensional space–time. It is here reproduced for the first time in an English translation. Klein has told the story of how he struggled for years with his attempt to reach a deeper understanding. He was ready in 1926, only to be told by Wolfgang Pauli that Th. Kaluza had done similar work already in 1921. Typical of Klein was that he never hesitated to give Kaluza the honour. He did that in his short paper in *Nature* in 1926, in which he proposed that the fifth dimension was rolled up, i.e. compactified. As a service to present day "compactifiers" this paper is also included here.

Klein was ready for Schrödinger's wave mechanics when it appeared in 1926. He had already for several years considered matter wave propagation as a natural way to introduce whole numbers through interference. When Dirac's paper on the quantization of the radiation field appeared, Klein immediately sat down to generalize it; a work he published jointly with Jordan. Dirac's paper on the electron led Niels Bohr to send Oskar Klein to Cambridge to learn more. Soon after Klein and Nishina did their joint work on the Compton effect.

The third scientific paper by Klein included here was originally inspired by Yukawa's theory of mesons, which led Klein to apply his five-dimensional ideas to the problem of forces between the neutron–proton pair and the electron–neutrino pair through intermediary charged fields. The complete Lagrangian would entail couplings between the heavy and light fermions, i.e. beta-decay processes. This famous paper was published in French in connection with a conference in Warsaw in 1938. It is included here in a new English translation done by Dr. Lars Bergström, Stockholm, who also helped the editor to correct some misprints in the original as well as in the German paper.

Both Yang and Weinberg refer in interesting ways to Klein's contributions to physics in their main lectures. The changed role of symmetry in modern physics is emphasized by Professor Yang in his main lecture. The earlier passive role is now an active one, so that *symmetry dictates interaction.* This will certainly remain as a lasting contribution to science from the second half of this century. Along this road the Yang–Mills gauge theory paper from 1954 represents a milestone.

In his second lecture Yang asks a question regarding the Yang–Baxter equation, namely, why is this equation involved in so many different areas of physics and mathematics? His own answer lies in his proof of its deep relation to the permutation group, being a generalization of its structure.

Professor Weinberg's seminal work on the unification of weak and electromagnetic forces was based on symmetry considerations. From this grew the successful standard model of elementary particle theory. In his first lecture Professor Weinberg brings the reader to the fore-front of research in this field as well as in that of cosmology. He is convinced that there must be something new beyond. The unanswered questions are addressed by Weinberg who emphasizes that one must not be satisfied with what the standard models offer. To quote him: "we understand that we know less than we thought we did."

Quantum mechanics is a linear theory. Weinberg deals in his second lecture with the question of how precisely the absence of nonlinearities is known. His recent proposal to test this by experiment has led to impressive and successful results. This paper would in all likelihood have appealed to Oskar Klein who, in the pre-Schrödinger days, worried about possible nonlinearities.

The volume begins with the biography of Klein and ends with his autobiography in the form of a lecture once held at the International Centre for Theoretical Physics at Trieste, Italy. It makes interesting reading for those who want to know more about the roots of modern physics. Klein's kindness, as experienced by those who met him in person, can also be noticed in his writings.

Finally, I wish to thank my colleagues in the Organizing Committee for the Oskar Klein Memorial Lectures, Professor Tor Ragnar Gerholm, Professor Cecilia Jarlskog and Professor Bertel Laurent. To Dr. Lars Bergström I owe special thanks for his expert help with the translations from French and from German of two of Klein's papers.

Holmsveden, Sweden, August 1990
Gösta Ekspong

OSKAR KLEIN*

Inga Fischer-Hjalmars and Bertel Laurent

Department of Physics, Stockholm University, Stockholm, Sweden

Oskar Klein died on the fifth of February 1977 at the age of eighty-two. One of the most prominent Swedish physicists ever and an outstanding personality in the field of culture had passed away. He was a man whose interests knew no limits and as a scientist he greatly enriched our understanding of Nature. All those who knew him were astounded by his profound thinking, wealth of ideas, extensive insight and humanism, qualities that obviously had been stimulated by the spirit in his parents' home. His father, rabbi and professor in Stockholm, was deeply engaged in theological and humanitarian issues.

Oskar Klein has described how his interest in natural science arose from popular science magazines. Already at the age of sixteen did he get the opportunity to start working in the laboratory of the famous Svante Arrhenius, and the following year his first scientific publication appeared. It discussed the solubility of zinc hydroxide. After studies at Stockholm University Oskar Klein became assistant to Arrhenius. His important investigation on the dielectric constant of certain solutions dates from this time. This branch of research gradually led to his inaugural dissertation in 1921 on the statistical theory of suspensions and solutions.

Even before the defence of his doctor's thesis Oskar Klein's interest had been seized by a completely new complex of problems. In the year 1918

*Translated from the Swedish by I. Fischer-Hjalmars; from *Oskar Klein* by I. Fischer-Hjalmars and B. Laurent. Originally published in *Kosmos* **55** (1978) 19–29.

he was granted a fellowship for studies abroad. On his way to the continent Oskar Klein intended to visit Niels Bohr in Copenhagen. The visit became a long one. Except for certain intermissions it lasted until 1931. During this period Copenhagen developed into the focal point for a revolution in our conception of atoms and molecules. H. A. Kramers was the first in the row of Niels Bohr's assistants and collaborators, and Oskar Klein became the second. In a few years the group of collaborators grew to one of the greatest concentrations of physical geniuses of all time. Here, Oskar Klein received decisive stimuli for his continued research, and he himself also presented a number of essential contributions to the development.

Oskar Klein's stay in Copenhagen was interrupted for almost two years (1923–1925) owing to appointments at the University of Michigan, first as instructor and later as assistant professor. In 1928, after his return to Copenhagen, Oskar Klein received a position as reader in theoretical physics at Bohr's institute. In 1930 he was appointed full professor at Stockholm University, a chair that he held until 1962.

Molecular Physics

Oskar Klein's research was with few exceptions concerned with scientific problems of a deep, fundamental nature. His attitude towards principles and theories was, however, characterized by an endeavour to find how these were manifested and based upon physical experience. He was therefore especially eager to study the operation of the new theories in genuine applications. It is perhaps such investigations that have made him most renowned.

An early example is Oskar Klein's work in 1920 with the Norwegian astronomer S. Rosseland. The year before, Franck and Hertz had made an experiment showing that when a beam of slow electrons is sent into a gas of atoms (or molecules), an atom can absorb energy when colliding with an electron and thus get excited or ionized. Klein and Rosseland pointed out that such processes would be incompatible with the second law of thermodynamics, unless the effect was counterbalanced by other processes which they called *collisions of the second kind*. During such collisions an already excited atom would collide with and transfer energy to an electron, thus increasing the velocity of the electron. Klein and Rosseland predicted that this kind of collision would lead to extinction of radiation. Such an effect was found experimentally some years later.

Another example is Klein's quantum theoretical treatment (in 1929) of a molecule without rotational symmetry. The classical energy calculation is already complicated. However, Oskar Klein made an elegant derivation of the quantum mechanical expression using Bohr's correspondence principle, i.e. the fact that at the limit of high quantum numbers the quantum mechanical expressions must be transformed into the corresponding classical expressions.

In 1932, Oskar Klein performed yet another important piece of research directly applicable to molecular spectroscopy. Several scientists had tried to find a method to derive the distance dependence of the force between the atoms of a molecule from the observed molecular energy levels. R. Rydberg in Stockholm had worked out a method to solve this problem. Oskar Klein developed the theoretical foundation of the problem and the resulting solution allowed him to improve Rydberg's method, thus increasing its applicability considerably. Today, the method is called the RK or RKR method, as A. L. G. Rees later introduced further modifications. The method is still widely used in spectral analysis of diatomic molecules.

Quantum and Relativistic Theory

During his stay in Michigan, Oskar Klein, like many others, became impressed by the similarities between. the electromagnetism, as formulated by Maxwell, and the gravitational theory of Einstein. By employing the idea of the existence of a fifth dimension, in addition to the three space dimensions and the time dimension, he also succeeded (in 1927) in formulating a theory in which the two above-mentioned theories merged harmoniously. The same thought, although less completely developed, *was* independently formulated by Th. Kaluza.

After the discovery of new phenomena in addition to the electromagnetic and gravitational ones, the Klein-Kaluza theory was less used.[1]

---

[1] Since 1977, when the original Swedish version of this article was written, there has been an upsurge of interest in the theory. It is now generalized to still more dimensions and contains much more than the five-dimensional theory. The generalization of the theory is a result of the attempts to harmonize it with quantum theory. (Author's note during translation.)

However, it played an important role for Oskar Klein himself as a kind of general starting point for much of his later work. This was for instance the case with the so-called Klein–Gordon equation which is a relativistic counterpart to the well-known Schrödinger equation. "Relativistic" means that the equation is in agreement with the special theory of relativity. As a matter of fact, Oskar Klein discovered in this way the Schrödinger equation independently of Schrödinger, but the publication was delayed on account of a lingering disease.

The name of Oskar Klein is strongly associated with the earliest achievements in relativistic quantum theory. Every physicist is familiar with both the Klein–Gordon equation and the Klein–Nishina formula, the Klein paradox and the Klein–Jordan second quantization. These accomplishments date back to the period between 1927 and 1929.

The Klein–Nishina formula shows the probability of different scattering angles after collisions between photons and electrons. The calculation, which has the Dirac radiation theory as the starting point, is even today arduous for anyone who embarks upon it, and when it was carried out for the first time it must have been extremely exacting. Only by carrying out all the manipulations of the formulae independently were Klein and Nishina finally successful in accomplishing a satisfactory result.

The Klein paradox is the name given to the discovery that the Dirac electron theory predicts that an electric current will start flowing *in vacuum* when the electric field strength is increased above a certain value. This seemingly absurd and much debated result was eventually interpreted as a creation of electron–positron pairs.

The Klein–Jordan second quantization inaugurated a new epoch in quantum theory. Klein and Jordan proposed that the *wavefunction* of a single particle should be considered as a *field* (cf. the electromagnetic field), which, in turn, should be subjected to the laws of quantum mechanics. With regard to the electromagnetic field, Dirac had already proposed a formalism requiring a relation (similar to the well-known Heisenberg uncertainty relation) to be valid between the value of the field in any space point and its time derivative in that point. Klein and Jordan now introduced corresponding rules for their wavefunction. The second quantization presented a deep and integrated picture of the connection between particles and fields, and serious difficulties earlier encountered in relativistic quantum theory were elegantly dissolved by this idea.

In fact, Klein and Jordan showed that the quantum states of such a "field" in a box are the same as those of particles in the same box when the number of particles is unspecified. This implies that the list of quantum states has to contain all states of *one* particle in the box, of *two* particles in the box, etc. Thus, the second quantization is equivalent to a many-particle theory. Every kind of field corresponds with one kind of particle, and vice versa. As an example, the electromagnetic field and the photons correspond with each other. The advantage of this point of view was soon revealed, e.g. through the proposal by H. Yukawa that a new particle (later discovered and called the pi meson) ought to correspond with a field responsible for the nuclear forces. Nowadays, Klein–Jordan's concept of treating many-particle problems by the use of field quantization not only provides the foundation of elementary particle theories but is also used as an extremely important tool in solid state physics.

Statistical Physics

The development of classical statistical mechanics encountered the difficulty that the mean value, assumed to represent the increasing entropy, turned out to be constant in time as a result of the conservative microdynamics. When the first attempts to frame a statistical quantum mechanics were made at the end of the 1920s it was hoped that the "quantum mechanical uncertainty" would eliminate this difficulty. Yet, the problem seemed to remain until Klein, in 1931, due to his deep understanding of quantum mechanics as well as statistical mechanics, could resolve the paradox by modifying the entropy expression in accordance with the above-mentioned uncertainty. He showed that in this new representation the entropy increases irreversibly with time, precisely as required by the second law of thermodynamics. This elegant proof is nowadays called Klein's lemma.

Gauge Theory

A less well-known work, carried out by Oskar Klein in 1939, points far ahead in time. The essential idea in this work was repeated by C. N. Yang and R. L. Mills in 1954, obviously without any knowledge of Oskar Klein's achievements fifteen years earlier. The work of Yang and Mills has later become the model of methods that are at present mainly employed for theoretical approaches to elementary particles as well as gravitation. Presumably, there are several reasons why so little attention has been paid to the above-mentioned work by Oskar Klein: It is included in a publication that is difficult to obtain (and written in French) and it appeared perhaps *too* early. Typically, it was inspired by the five-dimensional theory; a theory with which most elementary particle physicists were not very well acquainted at that time.

Astrophysics

In 1945, Oskar Klein started on a new line of research when he, together with G. Beskow and L. Treffenberg, began to study the relative abundance distribution of chemical elements as measured on earth, in meteorites, in stars' atmospheres, etc. They compared this distribution with the theoretical equilibrium distribution of a heavy star and found a high level of agreement in several respects. At about the same time, Gamow, Alpher and Hermann developed an alternative theory, based on the idea that the elements had been formed through successive capture of neutrons during the first few minutes of the history of the universe. At present, the prevalent opinion about the distribution of elements is that it must be considered as a result both of early neutron capture (this applies to the occurrence of deuterium and helium) and of processes in heavy stars that have exploded as supernovae and dispersed their content in the interstellar space.

His investigations into the origin of elements directed Oskar Klein further towards cosmological studies and to the proposal of the so-called metagalactic model (1954). He took exception to the idea that the universe had been "created" at a definite time ten billion years ago by a process principally concealed from us. This hypothesis is often called the Big-Bang theory. Oskar Klein suggested that the part of the universe that we can observe (containing for instance the well-known expansion) is not necessarily representative of the universe as a whole. He believed that we now live in a part of the universe (our metagalaxy) that, much more than ten billion years ago, was a thin gas cloud which first contracted, then reversed its motion due to radiation pressure, and, consequently, is now in the process of expanding. He showed that this hypothesis in a natural way could explain the Eddington relations which had aroused much wonder and speculation. In general, these relations had been presented in the form: *The relation between the electrical and gravitational attraction between a proton and an electron ≈ the relation between the radii of the universe and the electron ≈ the square root of the number of particles in the universe.*

Somewhat later, Oskar Klein suggested that the ultimate origin of the expansion of the metagalaxy was the matter–antimatter annihilation. The one who in particular has adopted and developed this idea is Hannes Alfvén.

We have neglected to mention the work that Oskar Klein carried out within meson theory, on the divergence problem of quantum field theory, on Mach's principle, on star models in general relativity theory, on superconductivity, on the Dirac equation in general relativity theory, on the systematics of elementary particles, on the extension of the equivalence principle to quantum field theory, and many other problems. Our selection has by necessity been guided by the present consensus of opinion regarding which part of his work that is most valuable. It would not be surprising if, sometime in the future, equally important thoughts will be found in that which has been left out here.

Teaching Activity

Oskar Klein was not only an impressive character as a scientist but also as a teacher. When he arrived at Stockholm University in 1930 he was appointed professor of "Mechanics and Mathematical Physics." It was to these fields his predecessors had devoted themselves, in Stockholm as well as at the other Swedish universities and institutes of advanced studies. For Oskar Klein it was natural to expand the introductory lessons to cover a larger part of theoretical physics. Little by little, this change was generally accepted. During the last few years of Oskar Klein's term as professor even the name of the subject was changed to "Theoretical Physics."

Through his fascinating manner of teaching Oskar Klein succeeded in inspiring many students to continued studies in the subject. During his period as professor he had around 30 research students. The wide range of topics that characterized the content of his own research was also reflected in his students' various fields of activity. He had an eminent power to stimulate the students to active contributions of their own. He always had important views on the research problems but left much of the details to the initiative of his students.

Oskar Klein did not limit his teaching activities only to those who pursued academic studies. He was an exceedingly active representative of the old tradition, originating from the very first years of the existence of Stockholm University — that is that the professors should also participate in the education of common people. He delivered series of lectures in popular science and wrote a large number of highly esteemed popular works about all areas of physics, in particular, however, about quantum theory and relativistic theory. His wish to disseminate information about physics also included the history of physics. He had an intimate knowledge of the world of ideas of such principal characters in physics as Galileo, Newton and Pascal. Some of these works were published as booklets, others as articles, e.g. in *Kosmos*, the yearbook of the Swedish Physical Society, where he published many papers during four decades.

Cultural Activities

Oskar Klein contributed frequently to public discussions, and not only when the subject was a scientific one. As a student of Niels Bohr he had had ample opportunity to develop his interests in philosophy and the foundation of knowledge. But his cultural activities were even wider.

He was keenly interested in all kinds of current topics. We who had the privilege of being with him every day are aware of how frequently he took up a definite position and how intensely he discussed all issues — cultural, social and political. His contributions were marked not only by knowledge but also by his warm human perspective. In his relations to the people surrounding him he was modest, sometimes even shy, full of natural kindness, humour and sympathy. He was always ready to help, in minor as in major matters. In particular, during the Second World War he assisted a great number of expelled Jews. From his parents' home he had acquired a rather negative attitude towards Zionism, but under the pressure of events it changed to a serious commitment to the reborn State of Israel. During his last few years he took a strong interest in the complex problems of the Judaism–Christianity confrontation and harmonization.

Physicists of today are often forced to restrict their research to a narrow field of speciality. With his achievements, Oskar Klein has demonstrated the opposite kind of activity — he was always at the forefront, irrespective of which field of physics was being discussed. In many fields he made lasting contributions. He never tried to be efficient, nevertheless, he accomplished a lot. His perpetual effort was to strive towards *understanding* — scientific and human.

SYMMETRY AND PHYSICS

Chen Ning Yang

The Chinese University of Hong Kong

Early Beginnings

The concept of symmetry is as ancient as human civilization. How it was born will probably remain for ever a mystery. But early men must have been deeply impressed with the amazingly symmetrical structures in both the biological world (Fig. 1) and the physical world (Fig. 2). And of course, the bilateral symmetry of the human body could not have failed to inspire the creative instincts of early men. Out of such beginnings, it is easy to imagine, the concept of symmetry was abstracted, first perhaps subconsciously, later in more explicit forms. As civilization developed, symmetry gradually permeated into all areas of human activities: painting, sculpture, music, architecture, literature, etc. Figure 3 exhibits a bronze vessel called *Gu* (or *ku* in the Wade–Giles system) dating back some 3200 years to the Shang Dynasty in China. Its elegant shape reveals a mature grasp, on the part of the artist, of the appeal of symmetrical forms. Figure 4 exhibits a poem written by the great poet of the Sung Dynasty, Su Dongpo (1036–1101), which consists of eight vertical lines of seven characters each, a standard Chinese poetic form. The poem is read vertically downwards, starting from the first column on the right, followed by the second column, etc. But it can also be read *backwards*, starting upwards from the bottom of the last (eighth) column, followed by the seventh column, etc. Both ways the poem reads beautifully, with correct meter and correct rhyme.

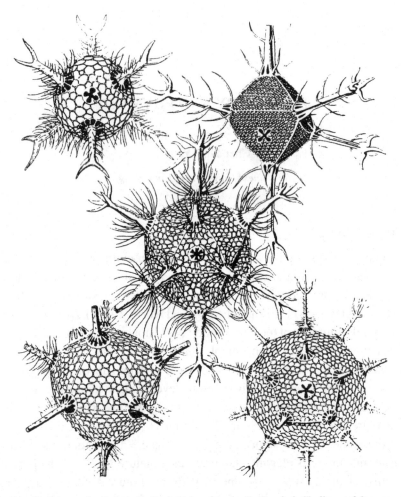

Fig. 1.   Skeletons of radiolarians. Orginal drawings by E. Haeckel, Challenger Monograph, *Report on the Scientific Results of the Voyage of H. M. S. Challenger*, Vol. XVIII, Pl. 117, H.M.S.O. (1887). (From H. Weyl, *Symmetry*. Copyright 1952 Princeton University Press. ©1980 renewed by Princeton University Press. Reprinted here with permission from Princeton University Press.)

Figure 5 is the *"Crab Canon"* of J. S. Bach (1685–1750) representing a violin duet in which each violin's music is the time-reversed version of that of the other violin. It is hard to judge which of the two, Su's poem or Bach's music, was the more difficult to compose. It is certain that both

Fig. 2. Snow flakes. U.S. Weather Bureau, photograph by W. A. Bentley. (From H. Weyl, *Symmetry*. Copyright 1952 Princeton University Press. ©1980 renewed by Princeton University Press. Reprinted here with permission from Princeton University Press.)

resulted from a deep appreciation, on the part of the artist, of the appeal of the concept of symmetry.

Perhaps the earliest entry of the concept of symmetry into science dates back to the Greek mathematicians and philosophers. It is well-known that the Greeks discovered the five *regular solids* which were highly symmetrical (Fig. 6). It has even been suggested by some authorities that the compilation by Euclid (∼300 B.C.) of the *Elements* was in fact to prove the theorem that these five are the only regular solids. That suggestion may or may not be correct, but we do know that the Greeks were so impressed with this discovery that they tried to associate the basic elements of the

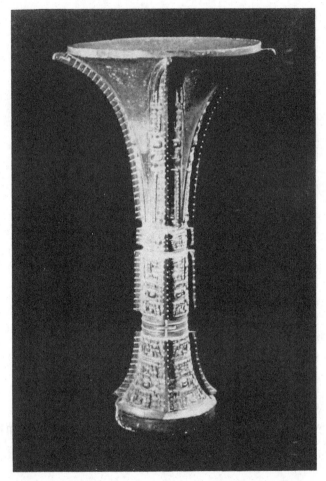

Fig. 3.   Bronze vessel *Gu*. This beautiful vessel is in the collection of the Freer Gallery, Washington D.C., USA.

structure of the universe with these five symmetrical solids. We shall return later to the efforts of Kepler (1571–1630), at the beginning of the scientific age, to fit these regular solids into the orbits of the planets.

The Greeks were so obsessed with the concept of symmetry that they perpetuated the idea of the *Harmony of the Spheres* and the *Dogma of the Circles*. According to this idea, the heavenly bodies must observe the most symmetrical rules, and the circle and the sphere are the most symmetrical

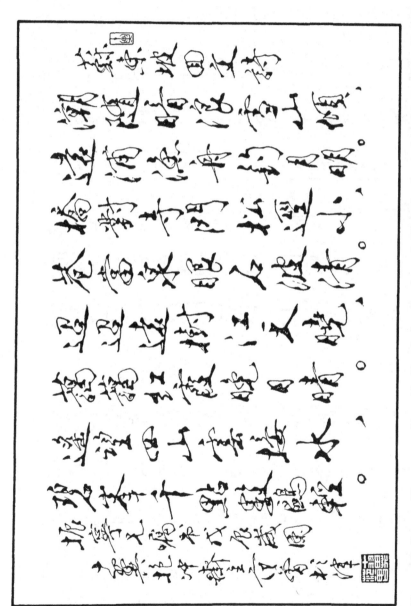

Fig. 4. Poem of Su Dongpo. The poem can be read backwards as well as forwards. The author is grateful to Mr. Fan Zeng (范曾) for the calligraphy reproduced here.

Fig. 5.   Crab canon from *The Musical Offering* by J. S. Bach. Revised, edited and in part arranged by Hans T. David. Copyright (c) 1944 (Renewed) G. Schirmer, Inc., international copyright secured. All rights reserved. Used by permission.

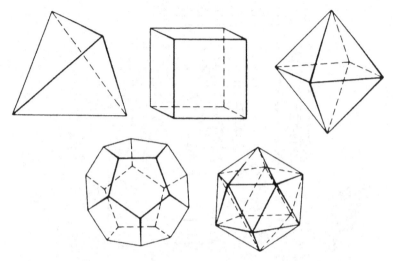

Fig. 6.   The five regular solids with maximum symmetry. Reprinted from A. V. Shubnikov and V. A. Koptsik, *Symmetry in Science and Art* (Plenum, 1974). (Reprinted with permission from Plenum Publishing Corporation.)

forms. But the heavenly bodies do not make simple circular motions. So they tried to fit their motions with circular ones superposed on circular ones. When that did not work either, they tried circular ones on circular ones on circular ones, etc.

The *Harmony of the Spheres* retarded the progress of astronomy for at least 1500 years. But the legacy of the idea was not entirely negative. When Kepler started his career as an astronomer, he inherited the Greeks' obsession with symmetrical forms and tried to construct a theory of the ratios of the diameters of the planetary orbits based on the five regular solids (Fig. 7). He had already been convinced of the heliocentric system of Copernicus (1473–1543). There were six known planets at that time: Saturn, Jupiter, Mars, Earth, Venus and Mercury. Kepler imagined Saturn to be represented by a large sphere, inscribed in which he constructed a cube. Inside this cube, and tangent to it, he placed the next sphere representing Jupiter, inscribed in which he constructed a regular tetrahedron. Inside this tetrahedron, and tangent to it, he placed the next sphere representing Mars, etc. The five regular solids thus provided the interpolation between the six spheres. He then computed the ratios of the diameters of these spheres and

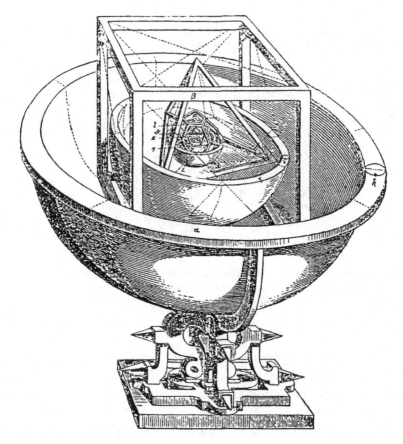

Fig. 7.   Kepler's construction. From *Mysterium Cosmographicum* (Tubingen, 1596).

compared them with the ratios of the observed diameters of the orbits of the six planets. This procedure depended on the order of the five solids: cube, tetrahedron, dodecahedron, icosahedron and octahedron. There are $5! = 120$ permutations of this order. Kepler tried them all and found the one illustrated in Fig. 7 to best approximate astronomical data.

Of course, Kepler's theory was totally wrong: we know today that there are other planets than his six, and there are only five regular solids available for interpolations. But his later discovery of the three famous Kepler laws were, according to himself, initiated by this first effort as a research astronomer. As is well-known, these Kepler laws were in turn the foundation on which Newton (1642–1727) later erected the whole edifice of modern physics.

We should also emphasize that although the idea of Kepler, as embodied in Fig. 7, was wrong, his *method of approach* is nevertheless entirely similar to a method in use today in elementary particle physics: to explain some observed regularities in physics, theorists try to match them with mathematical regularities resulting from symmetry considerations. If there are several ways of matching, theorists try them one by one. The effort usually fails. But once in a while a new facet is found in the meaning or the type of symmetry used and progress is made. Occasionally that progress did lead to profound new conceptual advances in fundamental physics.

19th Century: Group and Crystals

In the 19th century, a mathematical idea was formulated which later became one of the most profound concepts in mathematics. This is the idea of the *group*. Although several mathematicians had touched on this idea earlier, it was Galois (1811–1832) who, in 1830, showed the power of this concept with his brilliant resolution of the insolubility problem of fifth degree polynomial equations. It is thus commonly said that Galois originated the concept of the *group*, which was extensively developed in the latter part of the 19th century. In the 1880s, Sophus Lie (1842–1899) generalized the idea of the group and created the theory of *continuous groups*, or *Lie groups*. It turns out that *the concept of the group and of the continuous group are the proper mathematical representations of the concept of symmetry.*

On the physics side, crystallography was an important area of investigation. It was natural to classify crystals by *crystal classes*, with crystals belonging to the same class possessing much the same mechanical, thermal, and electrical properties. That crystal classes were related to the mathematical theory of groups was not an obvious idea, and took decades to develop, culminating around 1890 in the conclusion of Fedorov, a crystallographer, and Schonflies and Barlow, mathematicians, that each crystal class is associated with a *space group* and that there are exactly 230 different space groups in three dimensions. Therefore, they concluded, *there are 230 different crystal classes.*

C. N. Yang

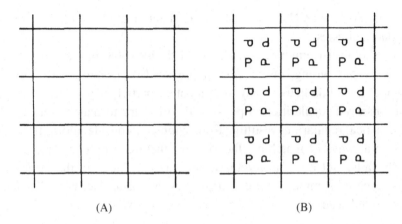

(A)   (B)

Fig. 8.   (A) Simple square lattice. (B) Simple square lattice decorated with the letter *d*.

It would be too much to describe in detail this profoundly beautiful
and extremely useful development. But we can illustrate the spirit of it
with the same mathematical problem in two dimensions. Figure 8(A),
extended in both dimensions, represents a simple square lattice. It has
many symmetries: if one displaces the diagram by one unit to the right,
or three units upwards, or by one unit downwards followed by two units
to the left, etc., the lattice remains unchanged, or remains *invariant*. These
displacements, or combinations thereof, are called *symmetry elements* of
the lattice. There are other symmetry elements: a rotation around one of
the corners through 90 degrees, or 180 degrees, etc., a rotation around the
center of any square through 90 degrees or 180 degrees, etc., all leave the
lattice invariant. These are also symmetry elements. One can also reflect
the lattice with respect to any of the vertical lines or any of the horizontal
lines, or any line midway between two vertical lines, etc. All of these are
also symmetry elements. In addition, if one reflects the lattice with respect
to a 45 degrees line passing through many lattice points, one also obtains a
symmetry element. All these symmetry elements together generate a *group*,
a *two-dimensional space group*. We say that pattern 8(A) *belongs* to this
space group and vice versa.

Now let us turn to the lattice B in Fig. 8(B). This lattice, extended
both ways to infinity, also has symmetry elements. In fact, all the symmetry
elements of lattice (A), which do not involve a reflection, are also symmetry

elements of lattice (B), as can be easily verified by inspection. But any reflection would *not* leave lattice (B) unchanged because it would turn the letter *d* into the letter *b* which is nowhere to be found in lattice (B). The reflections therefore are not symmetry elements of lattice (B). Thus the space group of lattice (B) is different from, and smaller than, the space group of lattice (A).

Figure 9 exhibits 17 different patterns, each of which should be imagined to be extended both ways to infinity. They form patterns like tiles in bathrooms. Each pattern has it own space group. The one in the center in the fourth row has the same space group as that of lattice (A) of Fig. 8. The pattern to its right has the same space group as that of lattice (B) of Fig. 8. It is easy to verify that the 17 different space groups belonging to these 17 patterns are all different. It can be proved, but that is not so easy, that these 17 are the *only* space groups in two dimensions. This statement is the generalization of the theorem mentioned earlier that there are 230 space

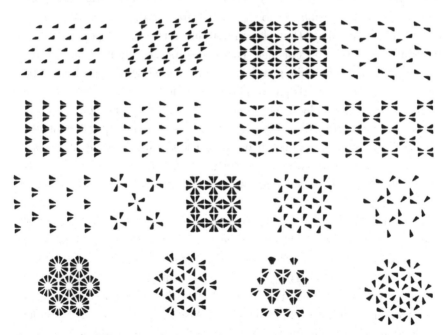

Fig. 9.   The 17 different two-dimensional patterns. Reprinted from A. V. Shubnikov and V. A. Koptsik, *Symmetry in Science and Art* (Plenum, 1974). (Reprinted with permission from Plenum Publishing Corporation.)

groups in three dimensions. (Historically, the problem in three dimensions was solved first. The generalization to the easier problem in two dimensions came later, betraying the *physical* origin of these investigations.)

The application of group theory to analyze the concept of symmetry in crystallography provided physicists with the first example of the elegance and power of the abstract mathematical concept of groups. Additional examples were to follow in the 20th century, *affecting the course of development of fundamental physics in a profound way*, as we shall outline in what follows.

20th Century: Enlarged Role of Symmetry

When Einstein (1879–1955) created special relativity in 1905, he also paved the way for the concept that *space* and *time* are symmetrical in an abstract mathematical sense. Years later, in 1982, in a conversation in Erice, Italy, Dirac (1902–1984) asked me what I thought was Einstein's most important contribution to physics. I answered, "General relativity of 1916." Dirac said, "That was important, but not as important as his introduction of the concept that space and time are symmetrical." What Dirac meant was that while general relativity was singularly profound and original, the symmetry of space and time had more pervading influence on later developments. Indeed, the symmetry of space and time, so contrary to the archetypical perception of the human race, is today incorporated into the primordial frame of reference of physics. The symmetry is called *Lorentz invariance* or *Lorentz symmetry*, after the great Dutch physicist Lorentz (1853–1928) who had stumbled on the mathematics, but not the physics, of this symmetry.

Another important development concerning symmetry was the gradual appreciation in the first 20 years of the century that conservation laws are related to symmetry. Conservation laws had been known, of course, since Newton's times. But it was only in the 20th century that it was realized that the conservation of momentum is related to the displacement invariance, or displacement symmetry, of physical laws. Also the conservation of angular momentum is related to the rotational symmetry of physical laws.

Why was it that the relationship between conservation laws and symmetry was not discovered for two hundred years before the beginning of this century? The answer lies in the fact that in classical physics this relationship, though present, was nevertheless not very useful. When quantum mechanics was developed in 1925–1927, the importance of this relationship really came to the fore. In quantum mechanics, the state of a dynamic system is labeled by quantum numbers which designate the symmetry properties of the state. Together with quantum numbers there arise selection rules which govern the change of quantum numbers in transitions between states. Quantum numbers and selection rules had been empirically discovered before quantum mechanics, but their meaning in terms of symmetry became apparent only after the development of quantum mechanics (Fig. 10). Thus, after 1925, symmetry began to permeate the very language of atomic physics. Later, as physics moved into nuclear phenomena and elementary particle phenomena, symmetry permeated also the languages of these new areas of physics.

The greatly enlarged role of symmetry in quantum physics derives[1] from the fact that the mathematics of quantum mechanics (Hilbert space) is linear. Because of this linearity there exists in quantum mechanics the *superposition principle.* In classical physics, an elliptical orbit is less symmetrical than a circular orbit. In quantum physics, because of the

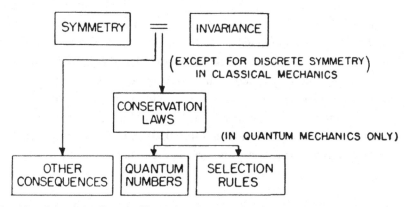

Fig. 10. Schematic diagram illustrating the use of symmetry in fundamental physics. From C. N. Yang, in *Les Prix Nobel*, 1957 (Nobel Foundation), p. 95. Also in *Science* **127** (1958) 565.

---

[1]C. N. Yang, *Phys. Teach.* **5** (October 1967) 311.

superposition principle, one discusses the symmetry of an elliptical orbit (*p* state) on an equal footing with the symmetry of a circular orbit (*s* state). Indeed, the rotational symmetry of *all* orbits are analyzed in quantum physics *at once* through the *representation theory* of the rotation group, a profoundly beautiful branch of mathematics.

As an example of the deep insight that these developments gave, we can mention the structure of the periodic table which was discovered in the 19th century. It was a great discovery, but the periods two, eight, eighteen, etc., were empirical numbers found heuristically by comparing the chemical properties of various elements. There was no deep understanding of their origin. After the development of quantum mechanics, it became gradually clear that these numbers were not just random numbers. They follow directly from the rotational symmetry of the Coulomb force. The elegance and perfection of the mathematical reasoning, and the depth and complexity of the physical consequences of this new insight, were deeply inspiring to physicists, and enhanced their respect for the usefulness of the concept of symmetry.

Another example of the far-reaching consequences of symmetry considerations in quantum physics was the prediction by Dirac of the existence of the antiparticle. I have likened this bold and original prediction of Dirac's to the first introduction of negative numbers,[2] an introduction that enlarged and completed our understanding of the integers which lie at the foundation of all mathematics.

An unexpected development occurred in 1956–1957 with the discovery[3] that right–left symmetry, though very accurately observed, is nevertheless not valid in *weak interactions*. This was a great surprise to all physicists at the time and was one of the most exciting discoveries since the end of World War II. It highlighted the fundamental importance of the concept of symmetry (and of asymmetry) in elementary particle physics.

---

[2]C. N. Yang, *Symmetry Principles in Modern Physics*, a lecture given at Bryn Mawr College, November 6, 1959. Printed in C. N. Yang, *Selected Papers 1945–1980 with Commentary* (Freeman, 1983), p. 276.

[3]C. S. Wu, E. Ambler, R. W. Hayward, D. D. Hoppes and R. P. Hudson, *Phys. Rev.* **105** (1957) 1413.

Caught[4] in this excitement, Werner Heisenberg (1901–1976) and Wolfgang Pauli (1900–1958) began to collaborate on *a field equation with a very high degree of symmetry*. Heisenberg later wrote:

> With every step Wolfgang took in this direction, he became more enthusiastic — never before or afterward have I seen him so excited about physics.[4]

For a few months they were extremely optimistic, but eventually the effort failed, ending in a bitter and sarcastic attack that Pauli launched against Heisenberg during a conference at CERN in 1958 in front of a stunned audience of mostly physicists of my generation.

Preoccupation with symmetry and asymmetry did not abate with this fiasco. If anything, it increased and became one of the dominant themes of elementary particle physics in the late 1950s and early 1960s, especially after the discovery of many resonances. Figure 11 is reprinted from an article published in 1964, illustrating a symmetry based on the Lie group SU_3 which was exuberantly dubbed the *eightfold way*.

In 1964 another great excitement swept through physics with the discovery[5] of CP nonconservation, and of the concurrent asymmetry with respect to time reversal. This tiny, subtle effect has led to much important physics during the last 20 odd years.

20th Century: Gauge Symmetry

Figure 10 is a schematic diagram made in 1957 to summarize the role of symmetry considerations in fundamental physics at that time. Subsequent developments have greatly enlarged this role so that today, in 1988, that diagram has to be greatly modified, as illustrated in Fig. 12. As a matter of fact, there is altogether a basic change of character of the role of symmetry considerations in fundamental physics: from a *passive* role in which symmetry is the property of interactions, to an *active* role in which symmetry serves to determine the interactions themselves — a role that I have called[6] *symmetry dictates interaction*.

---

[4]W. Heisenberg, *Physics and Beyond* (Harper and Row, 1971), Chap. 19.

[5]J. H. Christenson, J. W. Cronin, V. L. Fitch and R. Turlay, *Phys. Rev. Lett.* **13** (1964) 138.

[6]C. N. Yang, *Phys. Today* **33** (June 1980) 42.

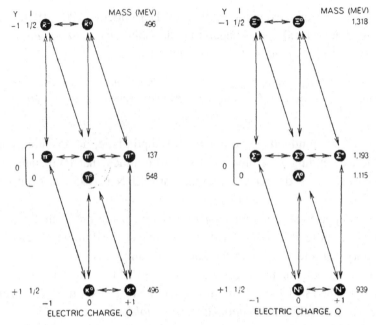

Fig. 11. Systematics of particles according to SU_3 symmetry. Adapted from G. F. Chew, M. Gell-Mann and A. H. Rosefeld, *Scientific American* **210** (1964) 74, February issue. (Copyright ©(1964) by Scientific American, Inc. All rights reserved.)

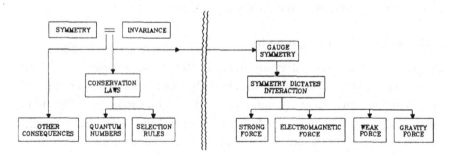

Fig. 12. Schematic diagram illustrating the role of symmetry in fundamental physics, 1988.

In 1954 it was observed[7] that the electromagnetic interactions of a particle are solely determined by its charge, a conserved quantity. The

---

[7]C. N. Yang and R. L. Mills, *Phys. Rev.* **95** (1954) 631, **96** (1954) 191. See also C. N. Yang, *Selected Papers 1945–1980 with Commentary* (Freeman, 1983), p. 19.

question was raised: may other conserved quantities not also determine interactions? The answer is yes and is beautifully unique. It involves a generalization of the gauge symmetry in electromagnetism already known from the work of Weyl (1885–1955), F. London (1900–1954) and Fock (1898–1974) in the period 1918–1929. Gauge symmetry of electromagnetism is based on a very simple Lie group, U(1). The generalization requires considerations of more complicated Lie groups, with the consequence that the generalized equations have strange-looking nonlinear[8] terms, as illustrated below:

U(1) Gauge Theory
(Maxwell Theory)

General Gauge Theory

$$f_{\mu\nu} = A_{\mu,\nu} - A_{\nu,\mu} \qquad f^i_{\mu\nu} = b^i_{\mu,\nu} - b^i_{\nu,\mu} - c^i_{jk} b^j_\mu b^k_\nu$$

$$f_{\mu\nu,\nu} = -J_\mu \qquad f^i_{\mu\nu,\nu} + c^i_{jk} b^j_\nu f^k_{\mu\nu} = -J^i_\mu. \tag{1}$$

In this illustration commas denote derivatives. The first equation on the left side is the familiar combined covariant form of Gauss' law and Faraday's law. The second equation on the left side is the familiar combined covariant form of Coulomb's law and Ampere's law. The two equations on the right side are their respective generalizations. We notice the replacement of the electromagnetic potential A_μ by b^i_μ. The superscript i runs from 1 to n where n is the *dimension* of the Lie group. The symbols c^i_{jk} denote the important *structure constants* that characterize the Lie group. They are integers, positive, negative or zero.

For electromagnetism, the Lie group is U(1). For this group $n = 1$, and we can drop the superscript i. Furthermore, $c^i_{jk} = 0$ and all equations reduce to the Maxwellian form.

For the general case, $c^i_{jk} \neq 0$, and the gauge theory equation contains nonlinear[8] terms; We shall come back to this point later.

Developments in the 1960s and 1970s have converged to the view, now generally accepted, that the strong, electromagnetic and weak interactions

---

[8]The nonlinearity here has nothing to do with the linearity of the Hilbert space mentioned earlier. Quantum physics, as we understand it now, is based on the mathematics of the Hilbert space which is always linear. The nonlinearity here refers to equations such as the gauge theory equations exhibited in the text, which are *superstructures* on the Hilbert space.

are all due to gauge fields based on different Lie groups. There is also general agreement that the gravitational interaction is due to a kind of gauge field, but the additional subtleties,[9] because of (i) the mixing of the gauge degree of freedom (i.e. index i in b^i_μ) with the coordinate degrees of freedom (i.e. index μ in b^i_μ), and (ii) the spin, have prevented a clear resolution of the precise manner in which the gravitational interaction is to be identified as a gauge field. This has remained an outstanding fundamental problem of physics.

It was shown[10] in 1975 that the concept of gauge fields in physics is related to the concept of fiber bundles in mathematics. This relationship clarifies the *intrinsic geometrical meaning of electromagnetism*, and the topological aspect of the Aharonov-Böhm experiment and that of the Dirac monopole.

A table of translation of terms was drawn up in this 1975 paper and is reproduced in Fig. 13. The table aroused great interest among mathematicians. In particular, they began to concentrate on the concept of the *source J* (see the ? in Fig. 13), which they had not studied before, but which had been so *fundamental* and *natural* to the physicists, both in Maxwell equations and in general gauge theory. The source J is defined in physics as the divergence of the field strength. In the mathematician's notation today, this definition becomes

$$*\partial * f = J.$$

A sourceless case would satisfy the equation

$$*\partial * f = 0.$$

The study of this equation turned out to be extremely fruitful, leading to a surprising breakthrough in topology, for which Donaldson received a Fields medal in 1986.

---

[9]It is interesting to observe that also in mathematics, the tangent bundle is more subtle than other bundles. See my paper in *Herman Weyl 1885–1985*, ed. K. Chandrasekharan (Springer–Verlag, 1986), especially footnote 29.

[10]T. T. Wu and C. N. Yang, *Phys. Rev.* **12** (1975) 3845.

Translation of Terminology.

Gauge field terminology	Bundle terminology
gauge (or global gauge)	principal coordinate bundle
gauge type	principal fiber bundle
gauge potential b_μ^k	connection on a principal fiber bundle
S_{b_a} (see Sec. V)	transition function
phase factor Φ_{QP}	parallel displacement
field strength $f_{\mu\nu}^k$	curvature
source[a] J_μ^K	?
electromagnetism	connection on a $U_1(1)$ bundle
isotopic spin gauge field	connection on a SU_2 bundle
Dirac's monopole quantization	classification of $U_1(1)$ bundle according to first Chern class
electromagnetism without monopole	connection on a trivial $U_1(1)$ bundle
electromagnetism with monopole	connection on a nontrivial $U_1(1)$ bundle

[a] I.e., electric source.

Fig. 13. Translation of terminology in gauge field theory (Physics) and in fiber bundle theory (Mathematics). Reprinted from T. T. Wu and C. N. Yang, *Phys. Rev.* **12** (1975) 3845, with permission from the American Physical Society.

About gauge symmetry we want to make several additional remarks:

(1) Einstein's general relativity was the first example where symmetry was used[11] *actively* to determine gravitational interactions. In today's language, Einstein had used the tangent bundle for his symmetry. The tangent bundle, being more subtle than other bundles, did not easily lend itself to generalization to other bundles. That is the reason why the works in the 1920s and 1930s by Schrödinger (1887–1961),[12] by O. Klein (1894–1977)[13] and many others, all grounded in general relativity, did not lead to general gauge theory. (Compare footnote 9 above.)

[11]A. Einstein, *Autobiographical Notes*, in *Albert Einstein, Philosopher-Scientist*, ed. P. A. Schilpp (Open Court, Evanstan, Ill. 1949). See also footnote 6 above.
[12]E. Schrödinger, *Sitz. der Preuss. Akad. der Wiss.* (1932), p. 105.
[13]O. Klein, in *New Theories in Physics*, Warsaw Conference, May 30–June 3, 1938 (Nyhoff, Hague, 1939).

(2) The fiber bundle is a sophisticated *geometrical* concept. That it turns out to be an essential element of the structure of fundamental fields would have pleased Einstein, who had repeatedly emphasized (see footnote 6) that fundamental fields must be *geometrical* in nature. He would also have been pleased by the natural nonlinearity of Eq. (1), as he had emphasized[11] that *"true laws cannot be linear nor can they be derived from such."*

(3) I have mentioned above the paper[13] by O. Klein which was his report at a 1938 conference in Warsaw. It is appropriate at this Oskar Klein Memorial Lecture to pay tribute to this very remarkable paper which presented a theory of fields satisfying equations that contain nonlinear terms very similar to those of Eq. (1) above. How did Klein arrive at these terms? The answer is: he had started from the Kaluza–Klein theory which, being based on general relativity, had nonlinear terms. Unfortunately, as already mentioned in remark (1), general relativity (i.e. tangent bundle) does not easily lend itself to generalizations to other gauge fields. Thus Klein did not discover non-Abelian gauge symmetry and his remarkable paper did not produce strong impact.

(4) As was emphasized[14] by Wigner, the word symmetry means quite different things in gauge symmetry as compared with its meaning in usual symmetries. This difference is illustrated in Fig. 14. We shall take

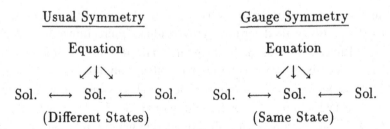

Fig. 14.   Schematic diagram illustrating the difference between usual symmetry and gauge symmetry. The horizontal arrows represent symmetry transformations which relate the solutions (sol. in the diagram). For the left column, these solutions represent different physical states. For the right column, they represent the same physical state.

---

[14]E. Wigner, in *Gauge Interactions, Theory and Experiment*, ed. A. Zichichi, Proceedings of the Erice School, August 3–14, 1982 (Plenum, 1984), pp. 725–733.

the rotational symmetry of the equation for the hydrogen atom as an example of a usual symmetry, in either classical or quantum mechanics. Out of *one* solution of the equation describing an elliptical orbit one can obtain *other* solutions by rotational transformations indicated by horizontal arrows in Fig. 14. These solutions represent *different* physical states. For gauge symmetry, out of one solution of the equation one can also obtain other solutions by gauge transformations indicated by horizontal arrows in Fig. 14. However, all these solutions related by gauge transformations represent *one and the same physical state.*

(5) One of the deep mysteries of contemporary fundamental physics is the concept of *renormalization*. This is not the place to discuss this concept in detail. Suffice it to say that on the one hand it has produced unimagined accuracy in predicting fundamental properties of elementary particles (to one part in 10^{11}), but on the other hand it has remained ungrounded on solid mathematics.

One of the lessons we have learned in the past forty years is that *symmetry is good for renormalization*: (a) The need for renormalization was already evident in the 1930s but renormalization as a program was only understood in the late 1940s. This is because physicists had not, until the 1940s, made *sufficient use of Lorentz symmetry*. (b) General gauge theory, being highly nonlinear, had appeared difficult to renormalize. Through the beautiful work[15] of the early 1970s it became clear that general gauge theory is renormalizable because *it contains a high degree of symmetry*.

To demystify renormalization we may need additional insight into the concept of symmetry.

(6) Through the work of many physicists, the concept of *broken symmetry* was introduced into elementary particle physics in the 1960s and 1970s. The idea was, in the simplest language, to keep the mathematical forms symmetrical, but the physical consequences unsymmetrical. The *standard model*, for which Glashow, Salam and Weinberg shared the Nobel prize in 1979, was based on a gauge theory with broken symmetry. It has been extremely successful. But the manner in which the broken symmetry was

---

[15]G. 't Hooft, *Nucl. Phys.* **B33** (1971) 173, **B35** (1971) 167; G. 't Hooft and M. Veltman, *Nucl. Phys.* **B50** (1972) 318.

introduced in this model has left much to be desired. It is the opinion of most physicists that the last word has not been said on this subject.

(7) Recent developments of supersymmetry, supergravity and superstring theories are all efforts at exploring and exploiting various new aspects of symmetry in field theory and in generalizations of field theory.

21st Century: New Facets of Symmetry?

Reviewing the progress of the concept of symmetry in physics through the centuries one cannot fail to be impressed by how the instincts of ancient Greek philosophers seem to have been in the right direction, by how the metamorphoses of this concept in mathematics and in physics have led to deep penetrations in both disciplines, and by how the deepest mysteries of fundamental physics, which remain unresolved at the present time, seem to all entangle with various aspects of this concept.

Will the next century witness new facets of the concept of symmetry in our deepening understanding of the physical world? I would answer, yes, very probably.

FROM THE BETHE–HULTHÉN HYPOTHESIS TO THE YANG–BAXTER EQUATION

Chen Ning Yang

Institute for Theoretical Physics
State University of New York, Stony Brook,
New York 11794-3840, USA

1. Introduction

In 1931, when solving the one-dimensional Heisenberg model of a ferromagnet, Bethe[1] formulated a hypothesis about the wave function of the model. His idea was later generalized to the antiferromagnetic case by Hulthén.[2] Bethe's hypothesis, or the Bethe-Hulthén hypothesis, can be stated in any of the following three equivalent forms:

1. If the coordinate space of the system is divided into a number of regions, then, in each of these, the wave function is a *finite* sum of pure exponentials.
2. The dynamical system exhibits no diffraction, only reflection.
3. The dynamical system has a large number of conservation laws.

The Bethe–Hulthén hypothesis was later successfully used for many one-dimensional quantum mechanical problems, especially in the 1960s, by many physicists. Of particular interest was the problem of the one-dimensional Bose gas with a delta function interaction solved by Lieb and Liniger.[3] In 1967 I generalized this work to a Fermion gas. While in

---

[1] H. A. Bethe, *Z. Phys.* **71** (1931) 205.
[2] L. Hulthén, *Arkiv. Mat. Astro. Fys.* **26A** (1938) No. 11.
[3] E. H. Lieb and W. Liniger, *Phys. Rev.* **130** (1963) 1605.

previous applications, the self-consistency of the Bethe–Hulthén hypothesis was obvious, for the Fermion problem this was not so. Instead, it was found that matrices $A(u)$ and $B(v)$, which occurred in the development of the hypothesis, had to satisfy[4] the following equation:

$$A(u)B(u + v)A(v) = B(v)A(u + v)B(u),\qquad(1)$$

in order for the Bethe–Hulthén hypothesis to be self-consistent. I showed that indeed this equation was satisfied.

We now know many one-dimensional quantum mechanical problems for which the Bethe–Hulthén hypothesis is valid. In each case, the consistency condition is Eq. (1), with $A(u)$ and $B(v)$ taking different forms in different problems.

Quite independently of these developments, the ice problem — a two-dimensional classical statistical mechanics problem — was solved by Lieb,[5] who showed that the transfer matrix of the problem could be diagonalized by using the Bethe–Hulthén hypothesis. Rapidly a number of other two-dimensional classical statistical mechanics problems[6] were solved, all using the Bethe–Hulthén hypothesis. The next step in this direction was taken by Baxter[7] in his solution of a generalization of the ice problem — the 8-vertex model. A crucial step in Baxter's work is again Eq. (1), written in terms of matrix elements.

Both lines of development were pursued in several centers of research, especially in the USSR, where the largest efforts were concentrated. In 1980, Faddeev coined[8] the term the Yang–Baxter relation or Yang–Baxter equation, and that became the generally accepted name for Eq. (1).

During the last five or six years, a number of exciting developments in physics and mathematics have led to the conclusion that the Yang–Baxter

---

[4]C. N. Yang, *Phys. Rev. Lett.* **19** (1967) 1312, *Phys. Rev.* **168** (1968) 1920.
[5]E. H. Lieb, *Phys. Rev. Lett.* **18** (1967) 1046.
[6]B. Sutherland, *Phys. Rev. Lett.* **19** (1967) 103; E. H. Lieb, *Phys. Rev. Lett.* **19** (1967) 108; C. P. Yang, *Phys. Rev. Lett.* **19** (1967) 586.
[7]R. J. Baxter, *Ann. Phys.* **70** (1972) 193.
[8]L. D. Faddeev, *Physica Scripta* **24** (1981) 832.

equation is a fundamental mathematical structure with connections to various subfields of physics and mathematics such as:

Physics:

One-dimensional quantum mechanical problems
Two-dimensional classical statistical mechanical problems
Conformal field theory

Mathematics:

Knot theory, braid theory
Operator theory
Hopf algebra
"Quantum groups"
Topology of 3-manifold
Monodromy of differential equations

There is an explosion of literature on these subjects. In order to find these, one could consult the three recent review volumes and reprint collections listed in footnote.[9]

Why does the Yang–Baxter equation enter into so many different areas of mathematics and physics? I believe the answer is that the equation is a kind of generalization of the structure of the permutation group. To illustrate this we discuss a simple Fermion problem in the rest of this article.

2. A Simple Fermion Problem

Consider the simple one-dimensional Fermion problem which originally[4] led to the Yang–Baxter equation (1). Our discussion here follows that of Gu and Yang[10] which is easier to understand than Ref. 4, because no familiarity with group theory is required.

---

[9] *Braid Group, Knot Theory and Statistical Mechanics*, eds. C. N. Yang and M. L. Ge (World Scientific, Singapore, 1989); *Yang–Baxter Equation* in *Integrable Systems*, ed. M. Jimbo (World Scientific, Singapore, 1990); *Yang–Baxter Equations, Conformal Invariance and Integrability* in *Statistical Mechanics and Field Theory*, eds. M. Barber and P. Pearce (World Scientific, Singapore, 1990).

[10] C. H. Gu and C. N. Yang, *Commun. Math. Phys.* **122** (1989) 105.

We take N fermions and consider the Hamiltonian:

$$H = -\sum_i \frac{\partial^2}{\partial x_i^2} + 2c \sum_{i<j} \delta(x_i - x_j),\tag{2}$$

where $c =$ real. Each particle has m "spin" states designated by $s_1, s_2 \ldots s_N$ where

$$1 \leq s_i \leq m.\tag{3}$$

The Schrödinger equation is $H\psi = E\psi$ where

$$\psi = m^N \times 1 \quad \text{column.}\tag{4}$$

For the Fermion problem we are only interested in wave functions that are antisymmetrical with respect to the interchange:

$$Q^{ij} : (x_i, s_i) \leftrightarrow (x_j, s_j).\tag{5}$$

3. The Bethe–Hulthén Hypothesis

Consider the scattering of two particles with initial momenta k_1 and k_2 into states with final momenta k_1' and k_2'. Momentum and energy conservation give

$$k_1' + k_2' = k_1 + k_2,$$
$$k_1'^2 + k_2'^2 = k_1^2 + k_2^2,\tag{6}$$

which has two, and only two, solutions:

$$(k_1'k_2') = (k_1, k_2) \quad \text{or} \quad (k_2, k_1).$$

The two solutions are *reflections* of each other in the (k_1, k_2) plane with respect to the mirror $k_1 = k_2$.

For the scattering of three particles, momentum and energy conservation still give two equations

$$k_1' + k_2' + k_3' = k_1 + k_2 + k_3,$$
$$k_1'^2 + k_2'^2 + k_3'^2 = k_1^2 + k_2^2 + k_3^2,\tag{7}$$

which obviously has the following six special solutions:

$$(k'_1, k'_2, k'_3) = (k_1, k_2, k_3), \quad \text{or five other permutations.} \tag{8}$$

The six solutions exhibited in (8) represent *reflections* of each other in the (k_1, k_2, k_3) space. But there are many other solutions of (7) which represent *diffractions* in the (k_1, k_2, k_3) space. The quantum mechanical three-body scattering problem will in general yield outgoing states including both *diffracted* and *reflected* waves, and is therefore difficult to solve.

However, in some special cases, the outgoing waves consist of *only* *reflected waves*. This is the Bethe–Hulthén hypothesis. If the hypothesis works, the solution of the Schrödinger equation becomes an *algebraic* problem, as we shall now see.

4. N = 2

For two particles, the Schrödinger equation describes two free particles, except on the line $x_1 = x_2$ in Fig. 1. The Bethe–Hulthén hypothesis states that in region I $(x_1 < x_2)$:

$$\psi = \alpha_{12}e^{i(k_1x_1+k_2x_2)} + \alpha_{21}e^{i(k_2x_1+k_1x_2)}, \tag{9}$$

where α_{12} and α_{21} are $m^2 \times 1$ column matrices. The antisymmetrization requirement for ψ says that in region II $(x_2 < x_1)$:

$$\psi = -(P^{12}\alpha_{12})e^{i(k_1x_2+k_2x_1)} - (P^{12}\alpha_{21})e^{i(k_2x_2+k_1x_1)}, \tag{10}$$

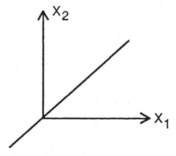

Fig. 1. Coordinate space for two particles. The potential energy is nonvanishing only along the line $x_1 = x_2$.

where $P^{12} =$ operator on the $m^2 \times 1$ column that interchanges

$$s_1 \leftrightarrow s_2. \tag{11}$$

Now we transform to the center of mass coordinate X, and the relative coordinate y:

$$X = (1/2)(x_1 + x_2), \quad y = x_2 - x_1. \tag{12}$$

The Schrödinger equation becomes

$$H\psi = \left[-\frac{1}{2}\frac{\partial^2}{\partial X^2} - 2\frac{\partial^2}{\partial y^2} + 2c\delta(y) \right]\psi = E\psi.$$

H commutes with $i\frac{\partial}{\partial X}$. Hence we can put $\frac{\partial}{\partial X} = 0$, obtaining

$$\left[-2\frac{\partial^2}{\partial y^2} + 2c\delta(y) \right]\psi = E\psi. \tag{13}$$

Thus ψ is continuous at $y = 0$:

$$\alpha_{12} + \alpha_{21} = -P^{12}(\alpha_{12} + \alpha_{21}), \tag{14}$$

and $\frac{\partial^2}{\partial y^2}\psi$ has a δ function singularity at $y = 0$. Integrating (13) from $y = 0$ to $y = 0+$ we obtain

$$\left.\frac{\partial\psi}{\partial y}\right|_{y=0+} - \left.\frac{\partial\psi}{\partial y}\right|_{y=0-} = c\psi|_{y=0}.$$

Substituting (9) and (10) into this equation results in

$$(i/2)(1 - P^{12})(k_1 - k_2)(-\alpha_{12} + \alpha_{21}) = c(\alpha_{12} + \alpha_{21}). \tag{15}$$

Eliminating the term $P^{12}\alpha_{21}$ from (14) and (15) leads to

$$[i(k_1 - k_2) - c]\alpha_{21} = [-i(k_1 - k_2)P^{12} + c]\alpha_{12}. \tag{16}$$

Thus

$$\alpha_{21} = Y\alpha_{12}, \tag{17}$$

where

$$Y = \frac{-i(k_1 - k_2)P^{12} + c}{i(k_1 - k_2) - c}, \tag{18}$$

if the denominator is nonvanishing. It is easy to check that if (17) is satisfied, then (14) and (15) are satisfied.

Thus if $k_1 \neq k_2$ is real, (9) and (10) will indeed give a solution of the Schrödinger equation satisfying the Bethe–Hulthén hypothesis.

5. $N \geq 3$

For $N \geq 3$, the Bethe–Hulthén hypothesis states that in one of the regions, where $x_1 < x_2 < \cdots < x_N$, we have a generalization of (9):

$$\psi = \alpha_{12\ldots N}e^{i(k_1x_1+k_2x_2+\cdots+k_Nx_N)}$$
$$+ \alpha_{2134\ldots N}e^{i(k_2x_1+k_1x_2+\cdots+k_Nx_N)}$$
$$+ (N! - 2) \quad \text{other terms.} \tag{19}$$

The column α has the dimension $m^N \times 1$. In other regions, the wave function ψ is determined from (19) by the requirement of antisymmetrization. The energy is given by

$$E = k_1^2 + k_2^2 + \cdots + k_N^2. \tag{20}$$

We will now examine the Schrödinger equation along the plane $x_3 = x_4$. The $N!$ terms in (19) form pairs, with the terms containing $\alpha_{1234\ldots N}$ and $\alpha_{1243\ldots N}$ forming one pair, and the terms $\alpha_{37154\ldots2}$ and $\alpha_{37514\ldots2}$ forming another pair, etc. In each pair only the third and fourth subscripts are different. Exactly the same procedures apply to the two terms in each pair as to the two terms in (9) of the last section.

Thus we obtain in the same way that we obtained (17) and (18):

$$\alpha_{\ldots lj\ldots} = Y_{jl}^{34}\alpha_{\ldots jl\ldots}, \tag{21}$$

where

$$Y_{jl}^{34} = \frac{-i(k_j - k_l)P^{34} + c}{i(k_j - k_l) - c}, \tag{22}$$

provided the denominator does not vanish.

6. The Yang–Baxter Equation

The subscripts in Eq. (21) can be chosen in $N!$ different ways. The superscript 34 can be replaced by $12, 23, \ldots (N-1)N$. Thus Eq. (21) is actually representative of $(N-1)(N!)$ different linear equations between the $N!$ columns $\alpha \ldots$ Are these equations mutually consistent?

For instance for $N = 3$, we have, by using (21) three times,

$$\alpha_{123} = Y_{21}^{12}\alpha_{213} = Y_{21}^{12}Y_{31}^{23}\alpha_{231} = Y_{21}^{12}Y_{31}^{23}Y_{32}^{12}\alpha_{321}. \tag{23}$$

We can use (21) three times in a different way, obtaining,

$$\alpha_{123} = Y_{32}^{23}\alpha_{132} = Y_{32}^{23}Y_{31}^{12}\alpha_{312} = Y_{32}^{23}Y_{31}^{12}Y_{21}^{23}\alpha_{321}. \tag{24}$$

Thus for the Bethe–Hulthén hypothesis to be consistent, we require

$$Y_{21}^{12}Y_{31}^{23}Y_{32}^{12} = Y_{32}^{23}Y_{31}^{12}Y_{21}^{23}. \tag{25}$$

It is easy to show, with the explicit form (22) of Y, that (25) is indeed satisfied.

If we write $k_2 - k_1 = u$, $k_3 - k_2 = v$, then $k_3 - k_1 = u + v$. Further, if we write $Y_{21}^{12} = A(u)$, $Y_{31}^{23} = B(v)$, then (25) reduces to the Yang–Baxter equation (1).

We can schematically illustrate the order of the subscripts for α in (23) and (24) as

$$
\begin{array}{ccc}
 & 213 \longrightarrow 231 & \\
\nearrow & & \searrow \\
123 & & 321 \\
\searrow & & \nearrow \\
 & 132 \longrightarrow 312 &
\end{array}
$$

Thus the Yang–Baxter equation is deeply related to the permutation group. In fact, putting $c = 0$ in Eq. (22) gives

$$A(u) = -P^{12}, \quad B(v) = -P^{23},$$

and the Yang–Baxter equation reduces to

$$P^{12}P^{23}P^{12} = P^{23}P^{l2}P^{23},$$

which is a fundamental property of the permutation group. Thus the Yang–Baxter equation is a generalization of the structure of the permutation group, and that is the reason why it has found its way into so many subfields of mathematics and physics.

Note Added in Proofreading

At the recent International Congress of Mathematicians in Kyoto in August 1990, there were four winners of the prestigious Fields Medals: Drinfeld, Jones, Mori and Witten. The work of three of them was related to areas of mathematics mentioned in this paper: Drinfeld on Hopf algebra and quantum groups, Jones on knot theory and operator theory and Witten on topology of 3-manifolds. The re-emergence of close cross fertilization between physics and mathematics is causing great excitement in both fields.

Acknowledgment

This work is supported in part by the National Science Foundation under grant PHY 89–08495.

BEYOND THE STANDARD MODELS

OSKAR KLEIN MEMORIAL LECTURE
UNIVERSITY OF STOCKHOLM, JUNE 6, 1989

Steven Weinberg

Theory Group, University of Texas Austin, Texas 78712, USA

I am grateful for this chance to return to Stockholm and speak in honor of a great theoretical physicist, Oskar Klein. All physicists know of Klein's famous contributions to quantum mechanics, recalled to us when we speak of the Klein–Nishina formula, the Klein paradox, and the Klein–Gordon equation. More than that, Klein seems to have had the gift of prophecy — he could see farther into the future of physics than is given to most of us.

In 1938, reasoning from analogies with general relativity, Klein anticipated the Yang–Mills structure of non-Abelian gauge theories, identifying the unit spin particles of the theory as photons and what were then called mesotrons. Earlier, in 1926, he showed how electromagnetism could arise from pure gravity in five space–time dimensions when the extra spatial dimension is curled up in a tiny circle. In this case the five-dimensional Einstein field equations turn into the equations of four-dimensional general relativity plus Maxwell's electrodynamics, with the electric current four-vector given by $J^\mu = (16\pi G)^{1/2} T^{\mu 5}/c$, so that the electric charge of any system is related to the five-component of its momentum vector by $q = (16\pi G)^{1/2} p^5/c$. The quantization of momentum circulating in a circle of circumference L requires that $p^5 = 2\pi \hbar n/L$, when n is an integer. All charges here are evidently quantized in units of the charge for $n = 1$, so Klein was led to identify the charge for $n = 1$ with the electronic charge e, and hence was able to calculate the circumference of the rolled-up fifth dimension as

$$L = 2\pi\hbar(16\pi G)^{1/2}/ec = 0.8 \times 10^{-30} \, \text{cm}.$$

(Incidentally, I can't resist mentioning here that almost everyone quotes Klein's 1926 paper in *Zeitschrift für Physik* as the reference for this work, but this paper is the wrong reference; it is devoted to tidying up the rather formal work of Kaluza published in 1921. The correct reference, in which the idea of a compactified fifth dimension first appears in physics, is another 1926 paper by Klein, "The Atomicity of Electricity as a Quantum Theory Law," published in *Nature*.[1]) It was this work on gravity in five space–time dimensions that led Klein to his 1938 paper linking electromagnetic and strong interactions (though we now understand that in order to get a non-Abelian gauge theory in four space–time dimensions at least two extra spatial dimensions must be compactified). The ideas of Klein and Kaluza were revived in the early 1980s, and are still actively pursued, especially in the context of superstring theory. What will be their significance for physics? When asked about the significance of the French Revolution of 1789, the Chinese premier Zhou En-Lai is said to have answered, "It's too soon to tell." About higher-dimensional space–times, it's also too soon to tell.

I am going to talk here about particle physics and cosmology, with special attention to their historical context, but also with an eye to their future. The subject is an appropriate one for a talk in honor of Oskar Klein — his last paper, published in 1971, bears the title "Arguments Concerning Relativity and Cosmology."

It has become almost a cliché to say that there are profound connections between particle physics and cosmology. We see the connection even institutionally, when a high energy physics laboratory like Fermilab supports a research group in theoretical cosmology. (On reflection this seems only fair; high energy physicists make a practice of explaining new accelerator projects to the public by emphasizing how the accelerator will recreate the conditions in the early universe.) Unfortunately, at the present moment the connection between particle physics and cosmology

---

[1] O. Klein, "The Atomicity of Electricity as a Quantum Theory Law," *Nature* **118** (1926) 516.

that is strongest may be a shared sense of frustration. Particle physicists and cosmologists are today victims of their own previous successes. In both fields there are *standard models* that incorporate a vast range of experimental data. These standard models are clearly not final or complete, but they have been so successful that it is hard to see how to go beyond them.

The standard model of cosmology was more or less complete by the late 1960s. Its theoretical foundations are general relativity and some well established principles of atomic and nuclear physics. To these, one adds the working assumption that the universe is homogeneous and isotropic. This implies that the large-scale space–time structure of the universe is entirely specified by one real scale factor $R(t)$, plus a curvature parameter k that can take only the values $+1$, -1, or 0. Seventy years of measurements of redshifts and distances of galaxies and supernovas give us the information that the Hubble constant $H_0 \equiv (R^{-1}dR/dt)_0$ at the present time (indicated by the subscript zero) has the value $(\frac{1}{2}$ to $1) \times 10^{-10}$/yr, and that the deceleration parameter $q_0 \equiv -(R^{-1}d^2R/dt^2)_0/H_0^2$ is at most of order unity. The greatest qualitative support for the standard model was provided by the discovery in 1965 of the cosmic microwave radiation background. Its thermal spectrum confirms that the universe was once so highly compressed as to be in a state of thermal equilibrium, as required by the solution of the Einstein field equations for the scale factor $FR(t)$, and its impressive isotropy confirms our assumption about the isotropy and (by inference if we are not in a privileged position) the homogeneity of the universe at large scales.

Beyond this, we can point to just two solid quantitative successes of the standard model of cosmology. One success is that mass observed in clusters of galaxies is not very different from the density required by the Einstein equations for plausible values of H_0 and q_0. Specifically, it provides some 10% to 20% of the critical density $3H_0^2/8\pi G$ required for the widely popular value $q_0 = \frac{1}{2}$ for which $k = 0$. This is often referred to as the cosmological missing mass problem, but the rough agreement between the observed mass density and that inferred from the Hubble constant is really quite gratifying.

The other quantitative success is more precise. The production of light elements at the end of the first three minutes depends on just one unknown environmental parameter, the ratio of photons to baryons. It is striking that if we take this ratio to be 2.5×10^9, then the abundances of ^{1}H, ^{2}H, ^{3}He, ^{4}He, and ^{7}Li all turn out to agree with those that are inferred from observations for the material from which the first generation of stars formed.

The ratio of 2.5×10^9 photons per baryon, together with an observed present cosmic microwave background temperature of $2.7°$K, leads to a rather firm estimate of 2×10^{-31} g/cc for the present cosmic baryon mass density, which (depending on the Hubble constant) is some 1% to 4% of the critical density $3H_0^2/8\pi G$. Indeed, on this basis the present cosmic baryon mass density is considerably smaller even than the mass density observed gravitationally in clusters of galaxies, thus confronting us with a second missing mass problem. It is somewhat ironic that the progress of theoretical cosmology might have been seriously impeded if the modern calculations of the cosmological production of the light elements had been carried out before the discovery of the cosmic microwave radiation background. The theorists doing these calculations would probably have taken the present baryonic mass density to be of the order of the critical value $3H_0^2/8\pi G$, in which case calculated value of 2.5×10^9 for the photon to baryon ratio would have led to an estimated present cosmic microwave background temperature of $8°$K to $12°$K. A search for a cosmic microwave radiation background with such a temperature would of course have failed, discrediting what we now call our standard cosmological model.

The standard model of elementary particles was completed in essentially its present form by the mid-1970s. It theoretical foundations are quantum mechanics and special relativity, joined together in the principles of quantum field theory. The added assumption of renormalizability (about which more later) acts as a requirement of simplicity, sharply limiting the numbers of field factors and space–time derivatives that can appear in the Lagrangian of such a theory. Within this general context, we assume the existence of a specific gauge invariance group, SU(3) × SU(2) × U(1), spontaneously broken to the SU(3) × U(1) of quantum chromodynamics and electrodynamics. The Lagrangian of the theory is constructed from the 12 gauge fields, plus the fields of six species each of leptons and quarks, plus something else to break the gauge symmetries. Assuming the simplest possible symmetry-breaking mechanism (SU(2) doublets of scalar fields), the structure of the whole Lagrangian is tightly constrained by the requirements of Lorentz invariance, gauge invariance, and renormalizability. The successes of this theory can be summarized by saying that wherever it can be used to draw general conclusions or to do specific calculations, its predictions correctly account for all data at energies accessible to present accelerators.

A word about the expression *standard model*. When I wrote the chapter on cosmology in my 1972 book on general relativity, I decided to give it the title *Cosmology: The Standard Model.* I chose the adjective *standard* in order to emphasize that the value of this cosmological model was to serve as a useful reference point for calculations and observations, and not to establish some sort of orthodoxy. To quote that chapter, "Of course, the standard model may be partly or wholly wrong. However, its importance lies not in its certain truth, but in the common meeting ground that it provides for an enormous variety of observational data. By discussing these data in the context of a standard cosmological model, we can begin to appreciate their cosmological relevance, whatever model ultimately proves to be correct." When the new understanding of elementary particle physics was coming together in the early 1970s, I and others found it natural to speak in the same spirit of a *standard model* of elementary particles. Of course, some skepticism survives about the standard model in cosmology, and perhaps also in elementary particle physics. Such skepticism is healthy, but not when it hinders communication among scientists by denying us a common language, as was so often the case in particle physics and cosmology when we lacked a standard model. Anyway, the real problem with the standard models is not some possibility that they might not be correct, but the clear certainty that they are incomplete.

First, cosmology. We don't know why the universe is homogeneous and isotropic. This is especially puzzling, because points in opposite directions in the sky at a distance corresponding to a redshift greater than $1 + (2/q_0)^{1/2}$ are so far apart that there has not yet been time during the present matter-dominated expansion of the universe for any signal to have reached these points from a common source. We also don't know why the photon to baryon ratio has the inferred value of 2.5×10^9. Most urgently, we do not understand the origin of the observed distribution of matter in galaxies and clusters of galaxies, perhaps because we do not know the nature of the dark matter of which these structures are presumably mostly composed.

Next, particle physics. We don't know why the gauge group of the standard model is SU(3) × SU(2) × U(1). We don't know the details of the mechanism by which this gauge group is spontaneously broken. There may be scalar fields in the Lagrangian, or there may be extra strong forces that break the gauge symmetries dynamically, in somewhat the same way that electromagnetic gauge invariance is spontaneously broken in a superconductor. We also don't know why the quarks and leptons transform as they do under the gauge group, or why the quarks and leptons form 12 separate irreducible representations (the three generations) of this group. Most annoyingly, we don't know why the 18+ parameters of the standard model take their observed values. If we like we can add gravitation to the standard model, but then renormalizability is lost, a problem to which I will return later.

Another incompleteness that affects both cosmology and elementary particle physics has to do with the long-standing problem of the cosmological constant. Any realistic quantum field theory gives an infinite vacuum energy density, and with a plausible cutoff we find an energy density ranging roughly from a GeV per cubic fermi in quantum chromodynamics to a Planck mass per cubic Planck length in quantum gravity. This is larger than allowed by observations of the cosmic expansion rate by a ridiculous number of powers of ten. Of course, the gravitational effects of this vacuum energy density can be cancelled by a suitable adjustment of the cosmological constant term in the field equations of general relativity, but this cancellation then has to be accurate to a ridiculous number of decimal places.

These incompletenesses of the standard models have naturally led to a wide range of speculations. First, in particle physics it has been speculated that the SU(3) and SU(2) × U(l) gauge groups of the strong and electroweak interactions are combined in a simple *grand unified* gauge group of some sort. This has had a single quantitative success, that in order for the three strong and electroweak couplings to become equal at some single high energy, in a wide variety of grand unified gauge models it is necessary that the electroweak mixing angle Θ should have $\sin^2 \Theta$ equal to about 0.22, in good agreement with the observed value. Beyond grand unification, there have been increasingly innovative speculations about supersymmetry, about the higher-dimensional *Kaluza–Klein* theories mentioned earlier, and most recently about superstrings. This work has been marked with great mathematical brilliance and physical ingenuity, and it may provide the theoretical foundation for the physics of the 21st century, but so far it has had no solid quantitative successes beyond the prediction of $\sin^2 \Theta$. For superstrings, the obstacle seems to be that no one knows how to do calculations involving interactions that are strong at the string scale of 10^{18} GeV or so.

This has also been a time of active speculations in cosmology, based in part on the speculations going on in elementary particle physics. The baryon nonconservation that is endemic to theories that go beyond the standard model has been considered as a source of the baryon–antibaryon imbalance observed in the present universe. The photon to baryon ratio predicted in these theories does turn out typically to be rather large, but it has been impossible to make any detailed quantitative predictions. Also, the spontaneous symmetry breaking built into all modern theories of elementary particles has suggested that the universe may get hung up in a supercooled *false-vacuum* state before it finally executes some phase transition. In such a state the universe would undergo a period of exponential expansion known as inflation. Such inflationary models offer the possibility of solving some of the outstanding problems of cosmology. Particles like monopoles that are predicted to be produced in the early universe but are not observed in our present universe may be diluted to acceptable levels in the inflation.

Also, the long period of exponential expansion provides an opportunity for all parts of the observable universe to get into contact with each other, thus solving the problem posed by the isotropy of the universe at large redshifts that I mentioned earlier. Inflationary cosmologies have their own problems, to solve which various elaborations have been introduced, but the general idea seems likely to survive.

It should be evident from this survey of particle physics and cosmology that theory and experiment have gotten woefully out of touch with each other. I don't think that this can be blamed on anyone's poor judgment, either theorists or experimentalists. It is just that the standard models have been so successful that the frontiers of particle physics and cosmology have been moved out beyond the reach of direct observation. In particle physics we now have to deal with energies in the range of 10^{15} GeV to 10^{18} GeV where we suppose that strong and electroweak interactions become unified, higher string or compactification modes are excited, and gravitation becomes a strong interaction. In cosmology the problem is that the universe was in a state of approximate thermal equilibrium from the time when the temperature was of order $e/\sqrt{G} \approx 10^{17}$ GeV, when collision rates of order $(kT)^3 e/(kT)^2$ became larger than the expansion rate of order $(G(kT)^4)^{1/2}$, down to a temperature of order 1 MeV, when weak interaction processes began to lag behind the expansion. There is not much that could survive that period of thermal equilibrium and be observed today, except the baryon–photon ratio itself, plus possibly small inhomogeneities and perhaps some particles like monopoles, axions, or photinos that went out of equilibrium before other particles.

The progress made in theoretical particle physics and cosmology over the last decade has been real, but largely negative: We now understand that we know less than we thought we did.

In cosmology, we now take seriously the possibility that the homogeneous isotropic Robertson–Walker–Friedmann universe in which we live is just one bubble in a megauniverse of many such bubbles. On an even grander scale, we have recently learned to think of the quantum state of the whole universe as a superposition of many terms, in which supposed constants of nature like the cosmological constant may take different values. There has been much argument recently about whether the statistical distribution of the cosmological constant in this state is sharply peaked about a value that would cancel the vacuum energy density produced by quantum fluctuations, but even if it isn't, the existence of a distribution of values allows a solution of the cosmological constant problem on anthropic grounds: We can only be living in that term in the state vector of the universe that allows for the appearance of living beings capable of worrying about the cosmological constant. Such anthropic arguments have a bad reputation in physics, and I am not entirely happy about them myself, but here is an analogy that may help to explain in what cases they do make sense. Textbooks on elementary physics frequently have tables of fundamental physical constants in the back of the book, and in such tables one often finds a constant g, the acceleration due to gravity, equal to 980 cm/sec^2. A student with nothing else to do might reflect that this value is well suited to the appearance of intelligent life; a much smaller value would (for a given planetary radius) allow the gases of the atmosphere to escape into space, and a much larger value might prevent the appearance of large land animals. Would it be stupid for the student to offer this as an explanation of the order of magnitude of g? Not at all, if the student recognized that g is an environmental parameter that takes different values on the surfaces of different planets. Naturally, any physics student who worries about the value of g can only be native to a planet where g is in a range that allows for the evolution of intelligent life. In the same way, *if* the vacuum energy density can vary from term to term in the state vector of the universe, or from subuniverse to subuniverse in a larger megauniverse, then scientists who worry about its value can only be living in terms or subuniverses in which the total vacuum energy (including the cosmological constant) is rather small, for otherwise perturbations in the early universe could never have grown into galaxies or stars.

In particle physics too we have learned to have reservations about what we thought we knew. In the 1970s we might have outlined our understanding of elementary particle physics in something like the following terms:

> "The world is described by a quantum field theory, because this is the only way of reconciling the principles of quantum mechanics and special relativity. Elementary particles are just those whose fields appear in the Lagrangian of this quantum field theory. In order to provide for the cancellation of ultraviolet divergences, the theory must be renormalizable, which forces the structure of the Lagrangian to be very simple. Assuming a group of symmetries and a menu of elementary particles that transform under given representations of this group, the whole theory is uniquely determined in terms of a finite number of fundamental constants."

Our point of view today is rather different. We now know that there are theories like superstring theories that are not quantum field theories in the usual sense but that successfully combine relativity (even general relativity) and quantum mechanics. However, such theories, and other exotic theories such as the higher-dimensional Kaluza–Klein theories, inevitably look like ordinary quantum field theories at relatively low energies. Furthermore, the nonrenormalizable terms in such effective quantum field theories will be suppressed by negative powers of the string or compactification or Planck mass scale M: anomalous magnetic moments and neutrino masses by a factor M^{-1}, baryon nonconserving four-fermion interactions by a factor M^{-2}, and so on. Thus, as long as we are limited to energies much smaller than M, we can expect to go on describing elementary particles in terms of renormalizable quantum field theories, but the success of these theories should not convince us that we know anything about physics at the fundamental scale M.

Something like this shift in point of view has happened before in particle physics. In the 1930s and 1940s physicists speculated about quantum field theories of nucleons and mesons, and it was widely believed that the observed pion exchange term in nuclear forces provided some sort of evidence for these theories. Later, in the 1950s, it became understood that the force between nucleons produced by pion exchange just depends on the existence of a pseudoscalar particle that couples to nucleons, and has nothing to do with whether this pion is an *elementary* particle. In the 1960s we developed effective field theories of pions and nucleons that could be used to calculate soft pion processes on the assumption of a spontaneously broken approximate symmetry, chiral $SU(2) \times SU(2)$, without having any idea of what underlying theory it is that has this symmetry. Today, of course, we know that this underlying theory is quantum chromodyamics, and in fact it involves fields for neither the nucleon nor the pion.

I don't want to be entirely gloomy in this talk. There are many ways in which experimental discoveries may open up exciting new possibilities for progress in the near future, some of which involve both particle physics and cosmology. Over the past decade it has been hoped that someone might discover tiny effects like proton decay or neutrino masses or oscillations, produced by nonrenormalizable terms in the effective Lagrangian that describes physics at ordinary energies, and which would give a clue to the new physics at inaccessibly high energies. There are also rare kaon decays, anomalous magnetic moments, and anomalous terms in quark–quark scattering that could arise from new physics at somewhat lower energies. These hopes have so far been unfulfilled, but we may yet get lucky. In this connection it is useful to recall that there is one tiny effect produced by nonrenormalizable terms in the low energy effective interaction that has already been observed — it is of course gravitation. The only reason that we can detect gravitation at all is that it has the unique property of coherence — the gravitational effects of N particles are proportional to N. But there is no reason to think that gravitation is the only nonrenormalizable remnant of high energy physics in our low energy effective field theory.

The nonrenormalizability of these extra terms in the effective Lagrangian does not mean that ultraviolet divergences cannot be absorbed into a redefinition of physical constants, as in quantum electrodynamics, but only that unlike quantum electrodynamics we need an unlimited number of terms in the effective Lagrangian to provide counter-terms to the ultraviolet divergences in each order of perturbation theory. This infinite number of free parameters is a symptom of our ignorance of the underlying theory that describes physics at the fundamental scale M. However, such "nonrenormalizable" theories still have plenty of predictive power: they provide an expansion in powers of energy divided by the fundamental scale M, with only a finite number of unknown parameters in each order, just as in the old theory of low-energy pions.

Our hopes for progress rest largely on the new instruments that are just now becoming available. The new accelerators such as the Tevatron collider, LEP, and SLC may have enough energy to produce superpartners of known particles. Also, our speculations in particle physics have provided a number of candidates for the particles that make up the supposed dark matter of the universe, and these might be discovered in the new accelerators. My favorite candidate is the photino of supersymmetry theories, because it would be quite natural for it to have a mass that would allow just enough photinos to survive to provide the needed dark matter. Nothing that experimental physics could do for cosmology in the near future would be as important as the discovery of the photino or whatever other particle makes up the dark matter.

On the astronomical side, there is the COBE satellite which will greatly improve our knowledge of the angular and wavelength distribution of the cosmic microwave radiation background; the various new gravitational wave detectors; axion detectors; and the Hubble Space Telescope which will open up a new chapter in astronomy, not least by finally providing a precise value of the Hubble constant. Also, surveys of galaxy redshifts (such as those that would be carried out at the proposed SST telescope in Texas) can provide better information about the behavior of the cosmic scale factor $R(t)$. Among other things, this can help us decide whether an anthropic solution to the cosmological constant problem makes sense. (This is because there is no anthropic reason why the effective vacuum energy density needs to be any smaller than the matter density at the time that gravitational condensations were becoming strongly nonlinear, a density that we know is at least hundreds of times larger than the present mass density. A vacuum energy density this large would also help with the discrepancy between the Hubble age of the universe and the estimated ages of globular clusters. Redshift surveys can determine whether the cosmological expansion is in fact dominated by a large vacuum energy density; if not, as seems so far to be the case, then we will have to conclude that anthropic considerations do not explain why the effective vacuum energy is as small as it is.) Finally, if funding is continued, we will eventually also have the SSC, where among other things the problem of the mechanism of electroweak symmetry breaking will be settled.

In closing, I want to mention one other kind of experiment that can also throw some light on the mechanism of electroweak symmetry breaking. It is the measurement or bounding of the electric dipole moment of the neutron. So far, an experiment at Leningrad in 1986 has provided a value of $(-14 \pm 6) \times 10^{-26}$ e cm, while an experiment at Grenoble gives a new value of $(-3 \pm 5) \times 10^{-26}$ e cm. The standard model with only one doublet of scalars allows CP violation only in the $K - M$ angle δ in W exchange, and this predicts a neutron electric dipole moment less than 10^{-30} e cm. However, a more complicated scalar sector allows a richer variety of mechanisms for CP nonconservation and larger values for the neutron electric dipole moment. At the opposite extreme from the straight $K - M$ picture, one might suppose that all CP nonconservation is due to scalar exchange, but this picture seems less plausible today than when it was suggested in 1976. It tends to give too large a value for the ϵ'/ϵ ratio in K_L^0 decay; it requires uncomfortably light Higgs scalars, which leads to too large a neutron electric dipole moment; and anyway with three generations it is rather unnatural for CP violation not also to be violated in W exchange. It seems to me that a more reasonable point of view is that CP conservation is violated wherever it can be. It is violated in W exchange, and this presumably makes the dominant contribution to K_L^0 decay into two pions. With a sufficiently complicated scalar sector, CP conservation is also violated in scalar exchange, and this would then make the dominant contribution to the neutron electric dipole moment. I have recently begun a survey of the various CP (and P) violating operators that can be produced by charged and neutral scalar exchange, to see if bounds on the neutron electric dipole moment can be used to constrain the scalar sector of the standard model, while we are waiting for the SSC.

PRECISION TESTS OF QUANTUM MECHANICS*

Steven Weinberg

*Theory Group, Department of Physics,
University of Texas, Austin, Texas 78712, USA*

It is proposed to set stringent limits on possible nonlinear corrections to ordinary quantum mechanics by searching for the detuning of resonant transitions. A suggested nonlinear generalization of quantum mechanics is used to show that such detuning would be expected in the rf transition in $^9\text{Be}^+$ ions that is used to set frequency standards. Measurements at the National Bureau of Standards already set limits of order 10^{-21} on the fraction of the energy of the ^9Be nucleus that could be due to nonlinear corrections to quantum mechanics, with good prospects of improving this by 2–3 orders of magnitude.

It seems long overdue to find precision tests of quantum mechanics itself, that would be more stringent than existing tests of specific quantum-mechanical theories. One sensitive way to test the linearity of quantum mechanics is to look for the detuning phenomenon characteristic of any system of nonlinear oscillators. The resonant frequency at which a weak external field can strongly perturb a nonlinear system will in general depend on the amplitudes of the various modes excited, so no matter how we set this frequency, the resonance will be detuned as the amplitudes change. This will prevent us from being able to drive the system slowly from one mode to another, say in a time T, unless the change in the resonant frequency during this transition is within a natural width of order $1/T$. This is a particularly promising way to look for departures from the linearity

---

*A talk on the topic of precision tests of quantum mechanics was presented by Professor S. Weinberg during the 1989 Oskar Klein Memorial Lectures. His article on the same topic, originally published in *Phys. Rev. Lett.* **62** (1989) 485, has therefore been included in this volume. Reproduced here with kind permission from the American Physical Society.

of quantum mechanics, because in the effort to set frequency standards, experimentalists[1] have already been able to drive hyperfine transitions in ions such as $^9\text{Be}^+$ with T as long as several seconds, and with good prospects of increasing this to several minutes. Thus such experiments (or related NMR or ESR experiments) can potentially set very stringent limits on any frequency shifts that might arise from nonlinearities in the equations of motion of the wave function.

In order to see what sort of nonlinearities might actually be expected to show up in such an experiment, and to make all this quantitative, it is essential to formulate some sort of nonlinear generalization of quantum mechanics. Recently there has been proposed[2,3] a generalized version of quantum mechanics that seems physically satisfactory, at least non-relativistically. For the purposes of the present paper, it is sufficient to consider a discrete system, like the hyperfine states mentioned above, for which the wave function can be taken to depend on only a discrete variable k. The time dependence of the wave function is assumed to be given by an equation of the form

$$i\frac{d\psi_k}{dt} = \frac{\partial h(\psi, \psi^*)}{\partial \psi_k^*}, \tag{1}$$

where h is a real function of ψ and ψ^*, satisfying the homogeneity requirement, that for any complex number λ,

$$h(\lambda\psi, \psi^*) = h(\psi, \lambda\psi^*) = \lambda h(\psi, \psi^*). \tag{2}$$

[This requirement ensures that if $\psi_k(t)$ is a solution of (1), then so is $\lambda\psi_k(t)$. It also plays an essential role in many aspects of the theory, including

---

[1] J. J. Bollinger, J. D. Prestage, S. M. Itano, and D. J. Wineland, *Phys. Rev. Lett.* **54** (1985) 1000. The familiar cesium atomic clock operates with a smaller value for T, and would give a less stringent absolute limit on nonlinearities.

[2] S. Weinberg, in *Proceedings of the International Symposium* on *Spacetime Symmetries*, Maryland, 24 May 1988 (to be published). University of Texas Report No. UTTG-15-88, 1988. There are problems with the construction in this reference of $h(\psi, \psi^*)$ for position-dependent wave functions, to be discussed in Ref. 3. However, these problems will not bother us in the present paper, where we only discuss discrete systems.

[3] S. Weinberg, *Ann. Phys. (N. Y.)* **194** (1989) 336.

the proper treatment of separated systems.[4]] These equations are of the Hamiltonian form, with $q_k \equiv \sqrt{2}\mathrm{Re}\psi_k$ and $p_k \equiv \sqrt{2}\mathrm{Im}\psi_k$. Equation (1) reduces to the usual linear time-dependence equation of quantum mechanics if we take $h(\psi, \psi^*)$ as bilinear, $h = \psi_k^* H_{kl}\psi_l$, but it is not necessary for h to be bilinear to be homogeneous. Small nonbilinear terms in h will produce small nonlinearities in the equation of motion (1).

Let us first consider a system like the $^9\mathrm{Be}^+$ ion, but in the absence of time-varying external fields. For the moment, we will also restrict ourselves to two-component systems, with $k = 1, 2$. The Hamiltonian function h will be taken to have the form

$$h = n\bar{h}(a), \tag{3}$$

where n is the norm,

$$n = |\psi_1|^2 + |\psi_2|^2, \tag{4}$$

and $\bar{h}$ is an arbitrary real function of the convenient action variable,

$$a = |\psi_2|^2/n. \tag{5}$$

[The form (3) will be automatic for the specific system to be studied here. More generally, it can be shown[3] that by a "canonical" transformation $\psi_k \rightarrow \psi_k'$, any two-component system satisfying (1) and (2) can be given a Hamiltonian of the form (3), while still preserving the equation of motion (1).] Nonlinear terms in $\bar{h}(a)$ yield nonbilinear (though homogeneous) terms in h.

In general, energy eigenstates are solutions of (1) for which the whole wave function oscillates with a common factor $\exp(-iEt)$. In our case, we easily find them to be (A)

$$\psi_2 = 0, \quad a = 0, \quad E = E_A = \bar{h}(0), \tag{6}$$

[4]See Ref. 3. The problem of dealing with separated systems has led other authors to limit possible nonlinear terms in the Schrödinger equation to a logarithmic form; see I. Bialynicki-Birula and J. Mycielski, *Ann. Phys. (N. Y.)* **100** (1976) 62. The homogeneity assumption (2) makes this unnecessary.

and (B)

$$\psi_1 = 0, \quad a = 1, \quad E = E_B = \bar{h}(1). \tag{7}$$

Now let us turn on a small time-varying perturbation Δh, of the sort that might drive a transition between states A and B. Since Δh is assumed very small, we ignore the possibility that it might include even smaller nonbilinear terms, and take it as the bilinear

$$\Delta h = g\psi_2^*\psi_1 e^{-i\Omega t} + \text{c.c.}, \tag{8}$$

where g is a small coupling parameter, and Ω is a frequency at our disposal. The time dependence of $\psi_k(t)$ is then given by using Eq. (1), with $h + \Delta h$ in place of h. These two complex equations can conveniently be boiled down to a single real equation for the action (5):

$$\frac{da}{dt} = \{4|g|^2 a(1-a) - [\Omega a - \bar{h}(a) + C]^2\}^{1/2}, \tag{9}$$

where C is an integration constant.

Now let us see if the perturbation can drive the transition $A \to B$ between the energy eigenstates (6) and (7). In order that the square root in (9) not have a negative argument at the starting point $a = 0$, it is necessary to take $C = \bar{h}(0)$. Then, in order that the square root not have a negative argument at the ending point $a = 1$, it is necessary also that

$$\Omega = \bar{h}(1) - \bar{h}(0) = E_B - E_A. \tag{10}$$

It is comforting to see that this result, related to energy conservation, holds here for arbitrary $\bar{h}(a)$ just as in ordinary quantum mechanics. With Ω tuned to this value, the transition will go all the way from $a = 0$ to $a = 1$ if and only if the argument of the square root is positive-definite for all intermediate a, i.e. if and only if

$$\{[\bar{h}(1) - \bar{h}(0)]a + \bar{h}(0) - \bar{h}(a)\}^2/a(1-a) < 4|g|^2 \tag{11}$$

for all a with $0 < a < 1$. In ordinary quantum mechanics $\bar{h}(a)$ is linear, so the left-hand side vanishes, and the transition does occur. More generally, seeing the transition occur provides an upper bound on the nonlinearities in $\bar{h}(a)$. Note that it is *not* necessary to verify with high precision that the transition goes *all* the way from $a = 0$ to $a = 1$; if the inequality (11) is violated, the transition will typically get no further than half-way. As mentioned earlier, in place of $|g|^2$ this bound can be written in terms of the time $T = \pi/2|g|$ that the transition would take in ordinary quantum mechanics. [Even when the inequality (11) is satisfied, nonlinearities in $\bar{h}(a)$ can show up as an asymmetry of the resonance line.]

As a variation on the detuning approach, one can try to make use of a technique due to Ramsey[5] that is used in setting frequency standards. As presently used, one first observes that for a given external rf field of frequency Ω, the transition $A \to B$ is driven in a certain time T. One then repeats the experiment, but now driving the transition only for a time $T/2$, then blocking the external rf field for a time $T' \gg T$, during which the system oscillates freely, and finally driving the transition again for a time $T/2$. In effect, this shifts the constant C in Eq. (9) by an amount $|g| \sin\{[\Omega - \bar{h}'(\frac{1}{2})]T'\}$, preventing the transition $A \to B$ unless $|\Omega - \bar{h}'(\frac{1}{2})T' \ll 1$. This in itself does not set any new limits on nonlinearities in $\bar{h}(a)$; all that is learned is that not only is Ω equal to $\bar{h}'(a)$ within an accuracy of order $1/T$ over the range $0 < a < 1$, but also $\Omega = \bar{h}'(\frac{1}{2})$ to within a greater accuracy, of order $1/T'$. However, we can try blocking the external rf field at several different times, thus verifying that $\Omega = \bar{h}'(a)$ at various *different* values of a.

What form do we expect for $\bar{h}(a)$ in the experiments of Ref. 1? The ^{9}Be nucleus has spin $j = \frac{3}{2}$, and is in a magnetic field, mostly due to the valence electron of the ^{9}Be$^+$ ion. Let us first consider the nuclear Hamiltonian function h_0 in the absence of the magnetic field. The only bilinear term in h_0 allowed by rotational invariance is just proportional to the norm n, and merely contributes a constant to $\bar{h}(a)$. The simplest nonbilinear term satisfying the homogeneity condition (2) would be proportional to a product of two ψ's and two ψ^*'s, divided by a single power of the norm. There are two rotationally invariant terms of this type (because the product of two ψ's can only have total spin $j = 3$ or 1, and likewise for two ψ^*'s) but one linear combination just gives a term proportional to n again, so we have essentially only one possible rotationally invariant term of this type. This

[5]N. F. Ramsey, *Molecular Beams* (Oxford Univ. Press, London, 1956), Chap. V, Sec. 4.

term is of the form

$$
\begin{aligned}
h_0 = \frac{\epsilon}{n} \sum_\sigma \Bigg\{ &2 \left| \sqrt{3}\psi\left(\frac{3}{2},\sigma\right)\psi\left(-\frac{1}{2},\sigma\right) - \psi\left(\frac{1}{2},\sigma\right)\psi\left(\frac{1}{2},\sigma\right) \right|^2 \\
&+ 3 \left| \psi\left(\frac{3}{2},\sigma\right)\psi\left(-\frac{3}{2},\sigma\right) - \psi\left(\frac{1}{2},\sigma\right)\psi\left(-\frac{1}{2},\sigma\right) \right|^2 \\
&+ 2 \left| \sqrt{3}\psi\left(\frac{1}{2},\sigma\right)\psi\left(-\frac{3}{2},\sigma\right) - \psi\left(-\frac{1}{2},\sigma\right)\psi\left(-\frac{1}{2},\sigma\right) \right|^2 \Bigg\}.
\end{aligned}
\tag{12}
$$

Here ϵ is a small coefficient with the dimensions of energy; n is the norm,

$$
n = \sum_{m,\sigma} |\psi(m,\sigma)|^2,
\tag{13}
$$

and $\psi(m,\sigma)$ is the component of the wave function with nuclear spin z component equal to m and with quantum numbers for everything else in the problem labeled σ. For the $^9\text{Be}^+$ ion, σ can be taken as the z component of the valence electron spin, so that $\sigma = \pm\frac{1}{2}$. If $\psi(m,\sigma)$ can be factored as $\psi_N(m)\psi_e(\sigma)$ then, because of its homogeneity, the energy eigenvalues and time dependence obtained from (12) do not depend on ψ_e. (This is an example of what was referred to earlier as a proper treatment of separated systems. As we shall see, in our case this factorization is a fair approximation.)

In the absence of any other terms, Eq. (12) would have the energy eigenvalues $\frac{9}{4}\epsilon, 2\epsilon, \frac{3}{4}\epsilon$, and zero, each corresponding to a rotationally invariant submanifold of eigenvectors. However, this pattern is completely changed by the external magnetic fields acting on the ^9Be nucleus.

Now let us include the hyperfine interaction of the ^9Be nucleus with the valence electron's spin, and the interaction of both nucleus and electron with an external magnetic field B. These are small relative to typical nuclear energies, so here we ignore possible nonbilinear terms, and take h as just the expectation value of the usual Hamiltonian of quantum mechanics:

$$
\begin{aligned}
h_{\text{QM}} = \sum_{m,\sigma}\sum_{m',\sigma'} \psi^*(m',\sigma)\psi(m,\sigma)[&\mu_e(\mathbf{J}_e)_{\sigma'\sigma} \cdot \mathbf{B}\delta_{m'm} \\
&+ \mu_N(\mathbf{J}_N)_{m'm} \cdot \mathbf{B}\delta_{\sigma'\sigma} + \kappa(\mathbf{J}_e)_{\sigma'\sigma} \cdot (\mathbf{J}_N)_{m'm}],
\end{aligned}
\tag{14}
$$

with κ representing the strength of the hyperfine interaction. By itself, this would give eight states, with energies as a function of $|B|$ represented by a typical Breit–Rabi diagram. For the relatively strong magnetic field used in Ref. 1, the energy eigenstates are nearly pure in m and σ, with admixtures limited to a few percent. The separation of these states is so much greater (presumably) than the small shifts due to nonlinearities that we can first solve for the energy eigenstates of (14) (by finding its stationary points on the surface of unit norm) and then evaluate the nonlinear term (12) for any mixture of these.

In the measurements of Ref. 1, one starts with the ^{9}Be$^+$ ion in an energy eigenstate with $m = -\frac{3}{2}, \sigma = \frac{1}{2}$, and drives a single-photon transition to the eigenstate with $m = -\frac{1}{2}, \sigma = \frac{1}{2}$ with an rf field tuned to this transition. No other components of the wave function are appreciably excited, so we can use our previous results for the two-component system. In our previous notation, taking $\psi_1 \equiv \psi(-\frac{3}{2}, \frac{1}{2})$ and $\psi_2 \equiv \psi(-\frac{1}{2}, \frac{1}{2})$, and dropping other components, Eq. (12) yields a nonlinear term in $\bar{h}(a)$,

$$\bar{h}(a) = 2\epsilon a^2. \tag{15}$$

[There are also much larger linear terms in $\bar{h}(a)$, arising from h_{QM}.] From Eq. (11), we see that if the transition $A \to B$ is observed to be driven in a time T, then $|\epsilon| < 2|g| \simeq \pi/T$. The measurements already performed, with T of order 1 s, set a bound on $|\epsilon|$ of order 10^{-15} eV, less than the binding energy per nucleon of the ^{9}Be nucleus[6] by a factor of order 10^{-21}. This may be improved by one or two orders of magnitude by the reduction of the

---

[6]The question naturally arises, whether to compare the limit on $|\epsilon|$ with the binding energy of the ^{9}Be nucleus, of the larger rest mass of the nucleon, or the much smaller hyperfine energy in the ^{9}Be$^+$ ion. The departures from quantum mechanics discussed here would arise from the internal energy of the *free* ^{9}Be nucleus, as given in Eq. (12), not from its interaction with external magnetic fields, so it would be pointless to compare $|\epsilon|$ with the hyperfine energy. Also, rotational invariance does not allow any nonbilinear homogeneous terms in the Hamiltonian function for a free particle of spin $\frac{1}{2}$, so it seems inappropriate to compare $|\epsilon|$ with the nucleon rest masses. On the other hand, because the ^{9}Be nucleus has spin $\frac{3}{2}$, rotational invariance does not prevent departures from quantum mechanics in the internal dynamics of the nucleus from producing nonbilinear terms in the Hamiltonian function of a free nucleus, such as that shown in Eq. (12). Thus it seems reasonable to compare $|\epsilon|$ with some energy characteristics of the internal dynamics of the nucleus, such as the binding energy per nucleon.

rf power to lengthen the transition time T, or by use of the Ramsey trick, with several free-precession times of several minutes at various stages in the transition. Line-splitting methods might allow one to do even better.

Acknowledgments

I am grateful to D. Wineland for helpful conversations regarding the experimental methods of Ref. 1. This research is supported in part by the Robert A. Welch Foundation and NSF Grant No. PHY-8605978.

QUANTUM THEORY AND FIVE-DIMENSIONAL
RELATIVITY THEORY*

Oskar Klein

in Copenhagen, Denmark
Received April 28, 1926

In the following pages I want to point out a simple connection between the proposed theory of Kaluza[1] regarding the connection between electromagnetism and gravitation on one hand and the suggested method of de Broglie[2] and Schrödinger[3] for the treatment of quantum problems on the other hand. Kaluza's theory attempts to connect the ten gravitational potentials g_{ik} of Einstein and the four electromagnetic potentials φ_i with the coefficients γ_{ik} of a line element of a Riemannian space, which besides the four usual dimensions also contains a fifth dimension. The equations of motion of charged particles then take the form of equations of geodesic lines also in electromagnetic fields. When these are interpreted as wave equations by considering the matter as a kind of propagating wave, then one is led almost automatically to a partial differential equation of second order which can be regarded as a generalization of the usual wave equation. If, now, such solutions to these equations are considered in which the fifth dimension appears purely harmonically with a definite period related to Planck's constant, one comes directly to the above-mentioned quantum theoretical methods.

1. Five-Dimensional Theory of Relativity

I begin by giving a short description of the five-dimensional relativity theory which connects closely to Kaluza's theory but differs in some points from it. Consider a five-dimensional Riemannian line element, for which we

*Original in Z. Phys. **37** (1926) 895, reproduced here with permission from Springer–Verlag. Translated from the German by Mrs Uta Schuch and Dr. Lars Bergström.

[1]Th. Kaluza, *Sitzungdber. Berl. Akad.* (1921), p. 966.
[2]L. de Broglie, *Ann. Phys.* (10) **3** (1925) 22. Thesis, Paris 1924.
[3]E. Schrödinger, *Ann. Phys.* **79** (1926) 361 and 489.

postulate a meaning independent of the system of coordinates. We write:

$$d\sigma = \sqrt{\sum \gamma_{ik} dx^i dx^k}, \tag{1}$$

where the symbol $\sum$, as everywhere in the following, describes a summation over the doubly appearing indices from 0 to 4. Here $x^0 \ldots x^4$ denote the five coordinates of the space. The 15 quantities γ_{ik} are the covariant components of a five-dimensional symmetric tensor. In order to transform these to the quantities g_{ik} and φ_i of the usual relativity theory we have to make certain special assumptions. First, four of the coordinates, let us say x^1, x^2, x^3, x^4, always have to characterize the usual space–time. Secondly, the quantities γ_{ik} must not depend on the fifth coordinate x^0. From this follows that the allowed coordinate transformations are restricted to the following group[4]:

$$\left. \begin{aligned} x^0 &= x^{0'} + \psi_0(x^{1'}, x^{2'}, x^{3'}, x^{4'}), \\ x^i &= \psi_i(x^{1'}, x^{2'}, x^{3'}, x^{4'}) \quad (i = 1, 2, 3, 4). \end{aligned} \right\} \tag{2}$$

In fact, we should have written a constant times $x^{0'}$ instead of $x^{0'}$. The restriction of the constant to the value unity is, however, quite inessential.

As one can easily show, γ_{00} is invariant under the transformations (2). The assumption $\gamma_{00} = \text{const.}$ is therefore allowed. It is tempting to suggest that only the ratios of γ_{ik} have physical significance. In this case this assumption is only a convention that is always possible. Leaving the unit of measure of x^0 indefinite for the time being, we set:

$$\gamma_{00} = \alpha. \tag{3}$$

One shows furthermore that the following differential quantities remain invariant under the transformations (2), namely[4]:

$$d\vartheta = dx^0 + \frac{\gamma_{0i}}{\gamma_{00}} dx^i, \tag{4}$$

$$ds^2 = \left(\gamma_{ik} - \frac{\gamma_{0i}\gamma_{0k}}{\gamma_{00}} \right) dx^i dx^k. \tag{5}$$

---

[4]Cf. H. A. Kramers, Proc. Amsterdam 23, Nr. 7, 1922, where a discussion of a simple proof for the invariance of $d\vartheta$ and ds^2 is given that is similar to the following considerations.

In these expressions the doubly appearing indices should be summed over from 1 to 4. For such sums we will, as usual, omit the summation sign. The quantities $d\vartheta$ and ds are connected to the line element in the following way:

$$d\sigma^2 = \alpha d\vartheta^2 + ds^2. \tag{6}$$

Because of the invariance of $d\vartheta$ and γ_{00}, it now follows that the four γ_{0i} ($i \neq 0$), if x^0 is held fixed, transform as the covariant components of an ordinary four-vector. If x^0 is also transformed, there appears additively the gradient of a scalar. This means that the quantities:

$$\frac{\partial \gamma_{0i}}{\partial x^k} - \frac{\partial \gamma_{0k}}{\partial x^i},$$

transform as the covariant components F_{ik} of the electromagnetic field tensor. The quantities γ_{0i} are thus from the point of view of invariance theory behaving as the electromagnetic potentials φ_i. Therefore we assume

$$d\vartheta = dx^0 + \beta \varphi_i dx^i, \tag{7}$$

that is

$$\gamma_{0i} = \alpha \beta \varphi_i \quad (i = 1, 2, 3, 4), \tag{8}$$

where β is a constant and where the φ_i are so defined that in orthogonal Galilean coordinates:

$$\left.\begin{aligned}(\varphi_x, \varphi_y, \varphi_z) &= A, \\ \varphi_t &= -cV,\end{aligned}\right\} \tag{9}$$

where A is the ordinary vector potential, V the ordinary scalar potential and c represents the speed of light.

We want to identify the differential ds with the line element of the usual standard relativity theory. We thus set

$$\gamma_{ik} = g_{ik} + \alpha \beta^2 \varphi_i \varphi_k, \tag{10}$$

where we want to choose g_{ik} so that in orthogonal Galilean coordinates:

$$ds^2 = dx^2 + dy^2 + dz^2 - c^2 dt^2. \tag{11}$$

Hereby the quantities γ_{ik} are brought back to known quantities. The problem is now to find such field equations for the quantities γ_{ik} that g_{ik} and φ_i for sufficient accuracy are given by the field equations of the standard relativity theory. We do not want to examine this difficult problem further here but we want to show that the ordinary field equations can be easily embraced from the viewpoint of the five-dimensional geometry. We form the invariant:

$$P = \sum \gamma^{ik} \left[\frac{\partial \left\{ \begin{matrix} i\mu \\ \mu \end{matrix} \right\}}{\partial x^k} - \frac{\partial \left\{ \begin{matrix} ik \\ \mu \end{matrix} \right\}}{\partial x^\mu} + \left\{ \begin{matrix} i\mu \\ \nu \end{matrix} \right\} \left\{ \begin{matrix} k\nu \\ \mu \end{matrix} \right\} - \left\{ \begin{matrix} ik \\ \mu \end{matrix} \right\} \left\{ \begin{matrix} \mu\nu \\ \nu \end{matrix} \right\} \right], \quad (12)$$

where γ^{ik} are the contravariant components of a five-dimensional funda-mental metric tensor and where $\left\{ \begin{matrix} rs \\ i \end{matrix} \right\}$ represents the Christoffel three-index symbol, that is

$$\left\{ \begin{matrix} rs \\ i \end{matrix} \right\} = \frac{1}{2} \sum \gamma^{i\mu} \left(\frac{\partial \gamma_{\mu r}}{\partial x^s} + \frac{\partial \gamma_{\mu s}}{\partial x^r} - \frac{\partial \gamma_{rs}}{\partial x^\mu} \right). \quad (13)$$

In the expression for P we have in mind that all the quantities are independent of x^0 and that $\gamma_{00} = \alpha$.

Let us now consider the integral, evaluated over a closed area of the five-dimensional space:

$$J = \int P \sqrt{-\gamma} dx^0 dx^1 dx^2 dx^3 dx^4, \quad (14)$$

where γ stands for the determinant of the γ_{ik}. We form δJ by varying the quantities γ_{ik} and $\frac{\partial \gamma_{ik}}{\partial x^l}$ where their boundary values are not to be changed. Here α should be considered to be a constant. The variational principle

$$\partial J = 0, \quad (15)$$

then leads to the following equations:

$$R^{ik} - \frac{1}{2} g^{ik} R + \frac{\alpha \beta^2}{2} S^{ik} = 0 \quad (i, k = 1, 2, 3, 4), \quad (16a)$$

and

$$\frac{\partial \sqrt{-g} F^{i\mu}}{\partial x^\mu} = 0 \quad (i = 1, 2, 3, 4), \quad (16b)$$

where R represents Einstein's curvature invariant, R^{ik} the contravariant components of Einstein's curvature tensor, g^{ik} the contravariant components of Einstein's fundamental tensor, S^{ik} the contravariant components of the electromagnetic energy-momentum tensor, g the determinant of the g_{ik} and finally $F^{i\mu}$ the contravariant components of the electromagnetic field tensor. If we set

$$\frac{\alpha\beta^2}{2} = \kappa, \tag{17}$$

where κ represents the gravitational constant used by Einstein, we see that the equations (16a) are in fact identical with the equations of relativity theory for the gravitational field, and (16b) are identical with the generalized Maxwell's equations of relativity theory for a matter-free field point.[5]

If we restrict ourselves to the usual schematic way of treating matter in electron theory and relativity theory, we can obtain the usual equations for the non-matter-free case in a similar way. We replace P in (14) by

$$P + \kappa \sum \gamma_{ik}\Theta^{ik}.$$

In order to define the Θ^{ik}, we first want to consider the tensor appropriate for an electron or a hydrogen nucleus:

$$\vartheta^{ik} = \frac{dx^i}{dl}\frac{dx^k}{dl}, \tag{18}$$

where the dx^i represent the changes of position of the particle, and dl is a certain invariant differential. The Θ^{ik} should be equal to the sums of the ϑ^{ik} for the different particles, per unit volume. We then get back to equations of the ordinary type which become identical to the ordinary field equations, if we set:

$$v_0\frac{d\tau}{dl} = \pm\frac{e}{\beta c}, \tag{19}$$

$$\frac{d\tau}{dl} = \begin{cases} \sqrt{M} \\ \sqrt{m} \end{cases}, \tag{20}$$

---

[5] See, e.g., B. W. Pauli, *Relativitätstheorie*, pp. 719 and 724.

where in general

$$v_i = \sum \gamma_{i\mu} \frac{dx^\mu}{dl} \tag{21}$$

are the covariant components of the five-dimensional velocity vector v^i, where

$$v^i = \frac{dx^i}{dl}. \tag{22}$$

Further, e represents the electric elementary quantum, M and m the masses of the hydrogen nucleus and electron, respectively. Here the upper symbol pertains to the nucleus, the lower to the electron. In addition,

$$d\tau = \frac{1}{c}\sqrt{-ds^2}$$

is the differential of proper time.

From the field equations, there follow naturally in an ordinary way the equations of motion for the matter particles and the continuity equation. The calculations which lead to this can be easily summarized from our point of view. As one can easily see, our field equations are, namely, equivalent to the following 14 equations:

$$P^{ik} - \frac{1}{2}\gamma^{ik}P + \kappa\Theta^{ik} = 0 \tag{23}$$

$(i, k = 0, 1, 2, 3, 4$, but not both of them zero), where the P^{ik} are the contravariant components of the reduced five-dimensional curvature tensor (corresponding to the R^{ik}). The equations in question now follow by forming the divergence of (23). From this follows that the charged particles move on five-dimensional geodesic lines which satisfy the conditions (19) and (20).[6] As one sees immediately, these conditions are therefore compatible with the equations of geodesic lines because x^0 does not appear in γ_{ik}.

Here must be recalled that there really do not exist sufficient reasons for the exact validity of Einstein's field equations. Nevertheless, it might

---

[6]The special values of $\frac{d\tau}{dl}$ are of course in this connection without significance. Important is here only $\frac{d\tau}{dl} = $ const.

not be without interest that all the 14 field equations can be summarized in such an easy way from the point of view of Kaluza's theory.

2. The Wave Equation of the Quantum Theory

We are now going to relate the theory of stationary states, and the corresponding characteristic deviations from the mechanics which appear in the modern quantum theory, to the five-dimensional theory of relativity. Let us consider for this purpose the following differential equation which should be related to our five-dimensional space and which can be considered as a simple generalization of the wave equation:

$$\sum a^{ik}\left(\frac{\partial^2 U}{\partial x^i \partial x^k} - \sum \begin{Bmatrix} ik \\ r \end{Bmatrix}\frac{\partial U}{\partial x^r}\right) = 0. \tag{24}$$

Here the a^{ik} signify the contravariant components of a five-dimensional symmetric tensor which should be certain functions of the coordinates. The equation (24) is valid independently of the coordinate system.

Let us first consider a wave propagation determined by (24) which corresponds to the limiting case of geometrical optics. We arrive at this if we set:

$$u = Ae^{i\omega\Phi}, \tag{25}$$

and assume ω so large that in (24) only those terms have to be taken into account that are proportional to ω^2. We then obtain:

$$\sum a^{ik}\frac{\partial\Phi}{\partial x^i}\frac{\partial\Phi}{\partial x^k} = 0, \tag{26}$$

an equation which corresponds to the Hamilton–Jacobi partial differential equation of mechanics. If we set:

$$p_i = \frac{\partial\Phi}{\partial x^i}, \tag{27}$$

the differential equations for the rays, as is well known, can be written in the following Hamiltonian form:

$$\frac{dp_i}{-\frac{\partial H}{\partial x^i}} = \frac{dx^i}{\frac{\partial H}{\partial p_i}} = d\lambda, \tag{28}$$

where

$$H = \frac{1}{2} \sum a^{ik} p_i p_k. \tag{29}$$

From (26) there also follows that

$$H = 0. \tag{30}$$

A different representation of this equation which corresponds to the Lagrangian form, follows through the circumstance that the rays can be regarded as geodesic zero lines of the differential form:

$$\sum a_{ik} dx^i dx^k,$$

where the a_{ik} represent reciprocal quantities to the a^{ik}, that is

$$\sum a_{i\mu} a^{k\mu} = \delta_i^k = \begin{cases} 1, & i = k \\ 0, & i \neq k \end{cases}. \tag{31}$$

If we now set

$$\sum a_{ik} dx^i dx^k = \mu (d\vartheta)^2 + ds^2, \tag{32}$$

we can achieve, by an appropriate choice of the constant μ, that our ray equations become identical to the equations of motion of charged particles. If we set, in order to see this:

$$L = \frac{1}{2} \mu \left(\frac{d\vartheta}{d\lambda} \right)^2 + \frac{1}{2} \left(\frac{ds}{d\lambda} \right)^2, \tag{33}$$

there follows

$$p_0 = \frac{\partial L}{\partial \frac{dx^0}{d\lambda}} = \mu \frac{d\vartheta}{d\lambda}, \tag{34}$$

and

$$p_i = \frac{\partial L}{\partial \frac{dx_i}{d\lambda}} = u_i \frac{d\tau}{d\lambda} + \beta p_0 \varphi_i \quad (i = 1, 2, 3, 4), \tag{35}$$

where $u_1 \ldots u_4$ represent the covariant components of the ordinary velocity vector. The ray equations now have the form:

$$\frac{dp_0}{d\lambda} = 0, \tag{36a}$$

$$\frac{dp_i}{d\lambda} = \frac{1}{2} \frac{\partial g_{\mu\nu}}{\partial x^i} \frac{dx^\mu}{d\lambda} \frac{dx^\nu}{d\lambda} + \beta p_0 \frac{\partial \varphi_\mu}{\partial x^i} \frac{dx^\mu}{d\lambda} \qquad (i = 1, 2, 3, 4). \qquad (36)$$

From

$$\mu d\vartheta^2 + ds^2 = \mu d\vartheta^2 - c^2 d\tau^2 = 0,$$

there follows

$$\mu \frac{d\vartheta}{d\tau} = c\sqrt{\mu}. \qquad (37)$$

Since, according to (34) and (36a), $\frac{d\vartheta}{d\lambda}$ and therefore also $\frac{d\tau}{d\lambda}$ are constant, we can choose λ such that

$$\frac{d\tau}{d\lambda} = \begin{cases} M \text{ for the hydrogen nucleus} \\ m \text{ for the electron.} \end{cases} \qquad (38)$$

Furthermore, in order to get to the ordinary equations of motion, we have to assume:

$$\beta p_0 = \begin{cases} +\dfrac{e}{c} \text{ for the hydrogen nucleus} \\[2ex] -\dfrac{e}{c} \text{ for the electron.} \end{cases} \qquad (39)$$

From (37) there follows then:

$$\mu = \begin{cases} \dfrac{e^2}{\beta^2 M^2 c^4} \text{ for the hydrogen nucleus} \\[2ex] \dfrac{e^2}{\beta^2 m^2 c^4} \text{ for the electron.} \end{cases} \qquad (40)$$

The equations (35) and (36) then completely agree with the ordinary equations of motion of charged particles in gravitational fields and electromagnetic fields. In particular, the quantities p_i, defined according to (35), are identical with the generalized momenta defined in the usual way, which is important for the following considerations. Since we can still choose β arbitrarily, we will set:

$$\beta = \frac{e}{c}. \tag{41}$$

It then follows simply that

$$p_0 = \begin{cases} +1 \text{ for the hydrogen nucleus} \\ -1 \text{ for the electron.} \end{cases} \tag{39a}$$

$$\mu = \begin{cases} \dfrac{1}{M^2 c^2} \text{ for the hydrogen nucleus} \\ \dfrac{1}{m^2 c^2} \text{ for the electron.} \end{cases} \tag{40a}$$

As one sees, for the square root in (37) we have to choose the positive sign in the case of the nucleus and the negative sign in the case of the electron. This is indeed rather unsatisfactory. But the fact that one obtains for a single value of μ two different classes of rays which in a certain way are related to each other like positively and negatively charged particles, could be understood as a hint that it might be possible to change the wave equation so that the equations of motion for both kinds of particles follow from a single set of values of the coefficients. We do not now want to enter into this question further, but are going to consider more closely the wave equation that follows from (32) in the case of the electron.

Since for the electron it was assumed that $p_0 = -1$, as a consequence of (27) we have to set:

$$\Phi = -x^0 + S(x^1, x^2, x^3, x^4). \tag{42}$$

De Broglie's theory now follows if we look for the standing waves compatible with the wave equation corresponding to a certain value of ω and thereby assume that the wave propagation proceeds according to the laws of geometrical optics. For that purpose we need the well-known law of the conservation of phase, which immediately follows from (28) and (30). Namely, it follows that

$$\frac{d\Phi}{d\lambda} = \sum \frac{\partial \Phi}{\partial x^i} \frac{dx^i}{d\lambda} = \sum p_i \frac{\partial H}{\partial p_i} = 2H = 0. \tag{43}$$

The phase is thus carried along by the wave. Let us now consider the simple case where Φ can be split into two parts, one of which depends only on a single coordinate, let us say x, which swings back and forth periodically with time. Then, a standing wave will be possible, which is characterized by the fact that a harmonic wave represented at a certain moment by (25) after one period in x meets in phase with that wave which results from the same solution (25) by inserting the new values of x^0, x^2, x^3, x^4. Because of the conservation of phase, the condition for this is simply:

$$\omega \oint p\,dx = n2\pi, \tag{44}$$

where n is an integer. If we set:

$$\omega = \frac{2\pi}{h}, \tag{45}$$

where h represents Planck's constant, the ordinary quantization condition for a separable coordinate then results. The analogous situation is of course true for an arbitrary periodic system. The ordinary quantum theory of periodic systems thus corresponds completely to the treatment of interference phenomena through the assumption that the waves propagate according to the laws of geometrical optics. It may also be emphasized that because of (42), the relations (44), (45) are invariant under the coordinate transformations (2).

Let us now also consider the equation (24) in the case where ω is not so large so that we only have to take into account terms to the second power in ω. We thereby restrict ourselves to the simple case of an electrostatic field. We then have in Cartesian coordinates:

$$\left.\begin{aligned} d\vartheta &= dx^0 - eV\,dt, \\ ds^2 &= dx^2 + dy^2 + dz^2 - c^2 dt^2. \end{aligned}\right\} \tag{46}$$

Therefore there follows

$$H = \frac{1}{2}(p_x^2 + p_y^2 + p_z^2) - \frac{1}{2c^2}(p_t + eVp_0)^2 + \frac{m^2c^2}{2}p_0^2. \tag{47}$$

In equation (24) we can neglect the quantities that are proportional to $\left\{\begin{smallmatrix} ik \\ \tau \end{smallmatrix}\right\}$, since according to (17) the three-index symbols are in this case small quantities proportional to the gravitational constant κ. We therefore obtain[7]

$$\Delta U - \frac{1}{c^2}\frac{\partial^2 U}{\partial t^2} - \frac{2eV}{c^2}\frac{\partial^2 U}{\partial t \partial x^0} + \left(m^2 c^2 - \frac{e^2 V^2}{c^2}\right)\frac{\partial^2 U}{\partial x^{02}} = 0. \qquad (48)$$

Since V depends only on x, y, z, we can set for U in agreement with (42) and (45):

$$U = e^{-2\pi i\left(\frac{x^0}{\hbar} - vt\right)}\psi(x, y, z). \qquad (49)$$

This inserted in (48), yields

$$\Delta\psi + \frac{4\pi^2}{c^2 h^2}[(hv - eV)^2 - m^2 c^4]\psi = 0. \qquad (50)$$

If we set additionally:

$$hv = mc^2 + E, \qquad (51)$$

we obtain the equation given by Schrödinger,[8] whose standing waves correspond, as known, to the values of E which are identical to the energy values calculated from Heisenberg's quantum theory. As one sees, E is in the limiting case of geometrical optics equal to the mechanical energy defined in the usual way. As Schrödinger emphasized, the frequency condition says, according to (51), that the light frequencies belonging to the system are equal to the differences that are formed from the different values of the frequency v.

3. Final Remarks

Like the papers of de Broglie, the above considerations have arisen from the endeavour to use the analogy between mechanics and optics, which

---

[7]Except for the appearance of x^0, which is negligible for this application, this equation differs from the Schrödinger equation by the way in which the time appears in (48). In support of this form of the quantum equation one can mention, in the case where V depends harmonically on time, that this equation has solutions which correspond to the dispersion theory of Kramers as do Schrödinger's solutions to the quantum theory of spectral lines, which can be shown by a simple perturbation calculation. I owe this remark to Dr. W. Heisenberg.

[8]Schrödinger, Ref. 3.

appears in the Hamiltonian method, for a deeper understanding of quantum phenomena. The similarity of conditions for the stationary states of an atomic system to the interference phenomena of optics indeed seems to indicate that this analogy has a real physical significance. Now, concepts such as point charge and material point are indeed strange in classical field physics. Of course, the hypothesis has often been maintained that the matter particles have to be interpreted as special solutions of field equations, which determine the gravitational field and the electromagnetic field. It is tempting to relate the mentioned analogy to this concept. Because according to this hypothesis, it is really not so strange that the motion of material particles has similarities to the propagation of waves. The analogy under discussion, however, is incomplete, as long as one considers the wave propagation in a space of only four dimensions. This already appears in the variable speed of material particles. But if one imagines, however, the observed motion as a kind of projection on the space–time of a wave propagation, which proceeds in a space of five dimensions we can, as we saw, make the analogy complete. In mathematical terms this means that the Hamilton–Jacobi equation cannot be interpreted as a characteristic equation of a four-dimensional but rather of a five-dimensional wave equation. In this way one is led to Kaluza's theory.

Although the introduction of a fifth dimension in our physical considerations might seem strange at the outset, a radical modification of the geometry based upon the field equations is suggested in a totally different way by the quantum physics. For, as is well known, it is less and less probable that the quantum phenomena allow a unified space–time description, whereas the possibility of describing these phenomena by a system of five-dimensional field equations probably cannot be excluded a priori.[9] Whether there is something real behind these indications of possibilities, the future of course will have to decide. In any case it must be emphasized that the way of treatment attempted in this note, concerning the field equations as well as the theory of stationary states, has to be regarded as provisional. This is in particular true for the schematic way of treating matter, mentioned on page 71, as well as the circumstance that the two kinds

---

[9]Remarks of this kind, which Prof. Bohr has made on several occasions, have had a decisive influence on the creation of the present note.

of charged particles are treated by different equations of Schrödinger's type. The question is also left quite open whether one can be content with the 14 potentials when describing physical processes, or whether the Schrödinger method means the introduction of a new state quantity.

I have been occupied with the considerations presented in this note at the Physical Institute of the University of Michigan, Ann Arbor, as well as at the here present institute for theoretical physics. At this point I also want to express my warmest thanks to Prof. H. M. Randall and Prof. N. Bohr.

THE ATOMICITY OF ELECTRICITY AS A QUANTUM THEORY LAW*

Oskar Klein

In the five-dimensional theory of the connexion between electromagnetism and gravitation first proposed by Th. Kaluza,[1] the equations of motion of an electrified particle may be shown to be the equations of geodetics belonging to the following line element:

$$d\sigma = \sqrt{(dx^0 + \beta\phi_i dx^i)^2 + g_{ik}dx^i dx^k},\qquad(1)$$

where x^1, x^2, x^3, x^4 are the co-ordinates of ordinary space time with the line element $g_{ik}dx^i dx^k$, while x^0 is a fifth co-ordinate, and the ϕ_i are the four co-variant components of the electromagnetic potential vector. If the constant β is given the value

$$\beta = \sqrt{2\kappa},\qquad(2)$$

κ being the Einstein gravitational constant, the 14 field equations of the Einstein theory may, moreover, be simply expressed by means of the curvature tensor belonging to this line element.

Let now $d\tau$ be the differential of proper time belonging to a particle of mass m and charge e; then the Lagrange function L for the geodetics

---

*Reproduced from *Nature* **118** (1926) 516, with permission from Macmillan Magazines Ltd.
[1] *Sitzungsberichte Berl. Akad.* (1921), p. 766; see also O. Klein, *Z. Phys.* **37** (1926) 875.

representing the motion of the particle may be given the form

$$L = \frac{1}{2}m \left(\frac{d\sigma}{d\tau}\right)^2. \tag{3}$$

Defining momenta in the ordinary way by putting

$$p_i = \frac{\partial L}{\partial(dx^i/d\tau)} \quad (i = 0, 1, 2, 3, 4), \tag{4}$$

p_0 is seen to be constant along a geodetic, since x^0 does not appear in L. In addition to the equation expressing this constancy, the system belonging to (3) contains four equations which indeed become identical with the equations of motion of the particle if we put

$$p_0 = \frac{e}{\beta c}. \tag{5}$$

With this choice of p_0 the momenta p_1, p_2, p_3, p_4 are further seen to agree with the ordinary definition of the momenta of an electrified particle.

   Now the charge e, so far as our knowledge goes, is always a whole multiple of the electronic charge, so that we may write

$$p_0 = \frac{N\epsilon}{\beta c}, \tag{6}$$

ϵ being the electronic charge and N a whole number, positive or negative. This formula suggests that the atomicity of electricity may be interpreted as a quantum theory law. In fact, if the five-dimensional space is assumed to be closed in the direction of x^0 with a period l, and if we apply the formalism of quantum mechanics to our geodetics, we shall expect p_0 to be governed by the following rule:

$$p_0 = N\frac{h}{l}, \tag{7}$$

N being now a quantum number, which may be positive or negative according to the sense of motion in the direction of the fifth dimension, and h the constant of Planck. Comparing (7) with (6), and making use of the value (2) of β, we get for the period l:

$$l = \frac{hc\sqrt{2\kappa}}{\epsilon} = 0.8 \times 10^{-30}\,\text{cm}. \tag{8}$$

The small value of this length together with the periodicity in the fifth dimension may perhaps be taken as a support of the theory of Kaluza in the sense that they may explain the non-appearance of the fifth dimension in ordinary experiments as the result of averaging over the fifth dimension.

In a former paper[2] the writer has shown that the differential equation underlying the new quantum mechanics of Schrödinger can be derived from a wave equation of a five-dimensional space, in which h does not appear originally, but is introduced in connexion with a periodicity in x^0. Although incomplete, this result, together with the considerations given here, suggests that the origin of Planck's quantum may be sought just in this periodicity in the fifth dimension.

---

[2] Z. Phys., l.c.

ON THE FIELD THEORY OF CHARGED PARTICLES

Oskar Klein*

1. Introduction

The discovery of the so-called heavy electron or mesoton and the role it is supposed to play in nuclear physics for explaining the origin of attractive forces at small distances — a role suggested by Yukawa[1] before the discovery — would seem to mean a considerable enlargement of the region of applicability of the field concept, which has hitherto been limited by the self-energy difficulties. In fact, while electrons could be treated unambiguously only down to distances which were large compared to the so-called radius of the electron, the new particle would seem to present no principal difficulties before it approaches its own "radius", which, due to its larger mass, is about two orders of magnitude smaller than that of the electron. Moreover, a characteristic length of just the order of magnitude of the electron radius is introduced as a "Compton" wave length corresponding to the mass of the new particle, which, as shown by Yukawa, gives the range of the forces due to the field of this charged particle.

The logical consistency of this enlargement of the field concept would seem to require the removal of the self-energy difficulty of the electron,

*Translated by Dr Lars Bergström from the original "Sur la Théorie des Champs Associés à des Particules Chargées." Original in *New Theories in Physics*, edited by International Institute of Intellectual Cooperation (Paris, 1939). Reprinted by permission of Kluwer Academic Publishers.

[1]Yukawa, *Proc. Phys. Math. Soc. Japan* **17** (1935) 48; Yukawa and Sakata, *Proc. Phys. Math. Soc. Japan* **19** (1937) 1084; Yukawa, Sakata and Taketani, *Proc. Phys. Math. Soc. Japan* **20** (1938) 319; Yukawa, Sakata, Kobayasi and Taketani, *Proc. Phys. Math. Soc. Japan* **20** (1938) 720.

at least down to distances approaching the radius of the new particle. Considering the order of magnitude and range of nuclear forces — the nuclear binding energies being comparable with the rest mass energy of the electron and the range of the forces of the order of the electronic radius — it would not seem unreasonable to assume that a theory explaining nuclear attractions would also account for the rest mass of the electrons, the attractive forces required as a compensation for the Coulomb repulsion being of a similar nature as the nuclear forces. A necessary condition for such an explanation — which would mean the removal of at least two orders of magnitude of the point from where the integrations over higher frequencies are "cut off" — is that the new forces belonging to the mesoton field are determined by means of the elementary electric charge, which determines the electromagnetic forces, so that no other independent constant than the mass of the new particle will appear in the theory. Here we shall not enter further into the interesting question of how to know if such a field theory, containing also the fields corresponding to the new charged particle, is unambiguously determined by the conditions imposed by the electronic self-energy and general invariance; but we will show how a theory of this kind may be built up as a logical generalization of the formalism of ordinary field theory. If this theory of the mesoton and its interaction with electromagnetic fields and ordinary particles makes sense, which is still doubtful, it ought to satisfy automatically the above demands as to the electronic self-energy.

For the invariant formulation of the field theory the five-dimensional representation has proved useful as a very direct and simple means of expressing the fundamental conservation theorems for energy, momentum and electric charge through their relation to space–time translation invariance and gauge invariance. In the original form of that representation, which leads exactly to the Einstein–Maxwell theory of gravitation and electromagnetism, the field quantities are supposed not to contain the new auxiliary coordinate x^0. The fact that this coordinate appears as the canonical conjugate (apart from a constant factor) to the electric charge suggests, however, as a natural generalization, the assumption that the field quantities also contain terms depending upon x^0 and represent fields associated with charged particles or "charged fields" which resemble those represented by the solutions of the quantum theoretical wave equations of electric particles.[2] The existence of an elementary quantum of electricity would hereby require — as a classical model later to be replaced by a suitable quantization — a periodicity in x^0 corresponding to the length $l_0 = \frac{he\sqrt{2\kappa}}{c}$ where h is Planck's quantum of action, c the vacuum velocity of light, e the elementary quantum of electricity and κ the Einstein gravitational constant.[3] The period being invariant under a gauge transformation of x^0, this model will satisfy automatically the conditions of gauge invariance.

The direct and general way it expresses the fundamental conservation and invariance theorems seems to make this representation a natural starting point for a general quantum field theory comprising also the charged fields which are supposed to correspond to the mesotons. The field quantities, which may be taken in a form adapted to the demands of a generally relativistic Dirac wave equation in five dimensions,[4] would moreover have to satisfy commutation relations corresponding to the Bose–Einstein statistics. While these quantities will represent particles with integer spin, the ordinary elementary particles would have to be represented by Dirac spinor wavefunctions. But just as the field quantities contain parts depending upon x^0 representing charged fields, and parts independent of x^0 representing the ordinary electromagnetic and gravitational fields, we shall assume that also the spinors consist of x^0-independent components representing neutral particles (neutron, neutrino) and of x^0-dependent components representing charged particles (proton, electron). Probably, the higher harmonics of the Fourier development with respect to x^0 which would correspond to multiply charged particles, have no physical significance. They may be avoided in a similar way as in the theory of electronic spin through the introduction of two-row matrices. This description of charged and neutral particles would be consistent with the way protons and neutrons are treated in Heisenberg's theory of nuclear constitution, the matrix representation just mentioned corresponding to the so-called isotopic spin. On the other hand, this matrix representation is closely connected with the gauge transformation.

---

[2] Klein, *Nature* **118** (1926) 516; *Z. Phys.* **46** (1927) 188.

[3] Klein, *Nature, l. c.*

[4] See Schrödinger, *Berl. Ber.* (1932), p. 105; and Pauli and Solomon, *J. Phys.* (7) **3** (1932) 452 and 582.

The theory that will be outlined and which may be derived from a variation principle — the Lagrangian containing the spinor as well as the tensor field quantities, will describe an interaction between the proton, the neutron, the electron and the neutrino through the intermediate of charged and uncharged fields, giving, as it would seem, a quantitative formulation of the considerations of Yukawa, Kemmer, Bhabha, Heitler and Fröhlich on nuclear fields of force.[5] In particular, the theory will contain no new physical constants other than the mass of the mesoton; and, in the first approximation, the nuclear interactions will, like the electromagnetic forces — neglecting all direct gravitational effects — depend upon e^2. Even if it is not yet possible to decide whether this theory may give a description of the mesotons or not, it can perhaps create some interest as a simple model for such a theory.

As to the rest mass of the new particle, which does not appear in the ordinary field equations, it might be introduced by the addition of a term to the Lagrangian without disturbing the invariance. But it is not impossible that a further development of the theory will make this somewhat arbitrary addition superfluous, the mass appearing as some sort of self-energy determined by the other lengths entering into the theory. As an example of the lack of simplicity of the present theory it should be mentioned that although the allowed transformations are limited to general four-dimensional co-ordinate transformations and gauge transformations, it makes use of invariants and tensors belonging to a general five-dimensional Riemann space. In the older form of the theory, where the physical quantities do not depend upon x^0, this inconvenience has been avoided by means of a projective treatment.[6] Even if it may seem doubtful whether this treatment could be extended to the generalized form of the theory developed below, this lack is probably more formal than real. As to the more or less arbitrary addition (in the field equations for the charged fields) and omission (in the Dirac equations) of mass terms connected with electric charges, this may have to do with a real limitation of the theory, these terms being perhaps — as indicated above — related to different stages of the self-energy

[5]Yukawa, *l. c.*; Kemmer, *Proc. R. Soc. A* **166** (1938) 127; Fröhlich, Heitler and Kemmer, *Proc. R. Soc. A* **166** (1937) 154; Bhabha, *Proc. R. Soc. A* **166** (1938) 501.
[6]See W. Pauli, *Ann. Phys.* **18** (1933) 305 and 337.

problem. It is doubtful whether the circumstance that the unitary treatment of gravitational, electromagnetic and charged fields yielded by the theory does not also comprise the spinor particles (the Lagrangian of which is simply added to the common Lagrangian of the fields mentioned) must be regarded as a deficiency. It should further be emphasized that although gravitational actions will be neglected in the following, the structure of the theory is essentially determined by the circumstance that it fulfils from the outset the claims of general relativity. This indirect effect of gravitation may perhaps be compared with the influence of spin on the constitution of atoms and molecules in the approximation where all direct effects of the spin momenta are neglected.

2. Mathematical Treatment of the Spinor Particles

Turning now to the mathematical formulation of the above remarks, we introduce a Riemann metric tensor $\gamma_{\mu\nu}$, μ and ν taking the values 0, 1, 2, 3, 4. Corresponding to the invariance towards general transformations of the space–time co-ordinates $x^1, x^2, x^3, x^4 = ict$ together with that towards the gauge transformation expressed by means of the transformation equation

$$x^{0'} = x^0 + \text{arbitrary function of space–time co-ordinates (A)},$$

we define this tensor through the following relations

$$\gamma_{00} = I, \quad \gamma_{0k} = \beta\chi_k, \quad \gamma^{kl} = g^{kl}, \quad k, l = 1, 2, 3, 4. \qquad (1)$$

Here the χ_k represent certain functions of x^0, x^1, x^2, x^3, x^4, to be more closely characterized below, β is a constant equal to $\sqrt{2\kappa}$ and g^{kl} is the ten contravariant components of the Einstein metric tensor. Further, we introduce a set of Dirac matrices γ_μ, satisfying the relations

$$\frac{1}{2}(\gamma_\mu\gamma_\nu + \gamma_\nu\gamma_\mu) = \gamma_{\mu\nu}, \qquad (2)$$

and the corresponding quantities $\gamma^\mu = \gamma^{\mu\nu}\gamma_\nu$, which fulfil the relations

$$\frac{1}{2}(\gamma^\mu\gamma_\nu + \gamma_\nu\gamma^\mu) = \delta_{\mu\nu}, \qquad (3)$$

and

$$\frac{1}{2}(\gamma^{\mu}\gamma^{\nu} + \gamma^{\nu}\gamma^{\mu}) = \gamma^{\mu\nu}.$$

The γ_{μ} and the γ^{μ} may be taken as linear combinations of five ordinary constant Dirac matrices ε_0, ε_1, ε_2, ε_3, ε_4, satisfying the relations

$$\frac{1}{2}(\varepsilon_{\mu}\varepsilon_{\nu} + \varepsilon_{\nu}\varepsilon_{\mu}) = \delta_{\mu\nu}, \tag{4}$$

and since $\gamma_{00} = I$, we may put

$$\gamma_0 = \varepsilon_0. \tag{5}$$

Further, it follows from the definition of the γ^{μ} that

$$\gamma_0 = \gamma_{0\nu}\gamma^{\nu} = \gamma^0 + \beta\chi_k\gamma^k,$$

or

$$\gamma^0 = \varepsilon_0 - \beta\chi_k\gamma^k, \tag{6}$$

where the summation with respect to k goes from 1 to 4.

We shall now consider the following expression

$$\varphi\gamma^{\rho}\psi_{i\rho} - \varphi_{i\rho}\gamma^{\rho}\psi, \tag{7}$$

well-known from the Lagrangian leading to the generally relativistic form of Dirac's wave equations, the summation with respect to ρ being here extended from 0 to 4, and

$$\psi_{i\rho} = \frac{\partial\psi}{\partial x^{\rho}} - \Gamma_{\rho}\psi, \qquad \varphi_{i\rho} = \frac{\partial\varphi}{\partial x^{\rho}} + \varphi\Gamma_{\rho} \tag{8}$$

are so-called covariant derivatives of ψ and φ respectively, the Γ_{ρ} being certain matrices, the necessity of which was first shown by Schrödinger. Introducing the expression (6) for γ^0 and omitting the term $\varphi\varepsilon_0\frac{\partial\psi}{\partial x^0} - \frac{\partial\varphi}{\partial x^0}\varepsilon_0\psi$ — the omission of which does not violate the invariance required, although from a formal point of view it is not altogether satisfactory — we get

$$\varphi\gamma^k(\nabla_k\psi) - (\nabla_k\varphi)\gamma^k\psi - \varphi(\gamma^{\rho}\Gamma_{\rho} + \Gamma_{\rho}\gamma^{\rho})\psi, \tag{7a}$$

where

$$\nabla_k = \frac{\partial}{\partial x^k} - \beta \chi_k \frac{\partial}{\partial x^0}. \tag{9}$$

From the assumption made above (and to be stated more precisely below) about the x^0 dependence of the quantities concerned, it follows that $\beta \frac{\partial}{\partial x^0}$ will not contain β and will have to be retained when — as we shall now do — all terms in the expression (7a) containing β or any power of it are omitted, which means that we neglect all direct gravitational effects. If we use a Cartesian co-ordinate system ($g^{kl} = \delta_{kl}$), only Γ_0 will be "finite" with this approximation and given by

$$\Gamma_0 = \frac{1}{2} \beta \frac{\partial \chi_k}{\partial x^0} \varepsilon_0 \varepsilon_k;$$

but the term $\gamma^0 \Gamma_0 + \Gamma_0 \gamma^0$ will be small of order β since ε_0 anticommutes with all four ε_k, and we are left with the expression

$$\varphi \gamma^k (\Delta_k \psi) - (\Delta_k \varphi) \gamma^k \psi, \tag{7b}$$

which, by a suitable choice of the γ^k, will also be correct in a non-Cartesian co-ordinate system, even in the non-Euclidean case.[7]

As Lagrangian L^0 for our spin particles we take

$$L^0 = -\hbar c [\varphi \gamma^k (\nabla_k \psi) - (\nabla_k \varphi) \gamma^k \psi] + 2Mc^2 \varphi \psi, \tag{10}$$

where the mass M of the particle is to be taken very nearly equal to the proton mass for the pair proton–neutron and probably equal to zero for the pair electron–neutrino.

In a Cartesian co-ordinate system we may put

$$\gamma^k = \varepsilon_k, \tag{11}$$

and further

$$\varphi = \psi^* \varepsilon_4, \tag{12}$$

ψ^* being the complex conjugate wave function to ψ.

---

[7] By a suitable choice of the γ^ρ we might have made the expression $\gamma^\rho \Gamma_\rho + \Gamma_\rho \gamma^\rho$ equal to 0 from the beginning, but such a choice would in general not correspond to $\gamma_0 = \varepsilon_0$.

The more general expression (10) may be used in a well-known way to derive the components T_{kl}^0 of the corresponding energy-momentum tensor which — since L^0 will have the numerical value 0 — may be defined by means of the relation

$$\delta L^0 = -T_{kl}^0 \delta g^{kl},$$

obtained by putting

$$\delta \gamma^k = \frac{1}{2}\gamma^l \delta g^{kl}. \tag{13}$$

This gives

$$T_{kl}^0 = \frac{\hbar c}{4}[\varphi\gamma_l(\nabla_k\psi) + \varphi\gamma_k(\nabla_l\psi) - (\nabla_k\varphi)\gamma_l\psi - (\nabla_l\varphi)\gamma_k\psi], \tag{14}$$

where, in a Cartesian co-ordinate system,

$$\gamma_k = \varepsilon_k. \tag{11a}$$

In conformity with what has been said above about the x^0 dependence of the quantities concerned we now put

$$Z_k = \begin{pmatrix} A_k, & \tilde{B}_k \\ B_k, & A_k \end{pmatrix}, \quad \beta\frac{\partial\chi_k}{\partial x^0} = \frac{ie}{\hbar c}\begin{pmatrix} 0, & -\tilde{B}_k \\ B_k, & 0 \end{pmatrix}, \tag{15}$$

$$\psi = \begin{pmatrix} \psi_N \\ \psi_p \end{pmatrix}, \quad \beta\frac{\partial\psi}{\partial x^0} = \frac{ie}{\hbar c}\begin{pmatrix} 0 \\ \psi_p \end{pmatrix},$$

$$\varphi = (\varphi_N, \varphi_p), \quad \beta\frac{\partial\varphi}{\partial x^0} = -\frac{ie}{\hbar c}(0, \varphi_p), \tag{15a}$$

χ_1, χ_2, χ_3 being Hermitian matrices and χ_4 a Hermitian matrix multiplied by i. The A_k will be seen to be the components of the electromagnetic potential four-vector, while the B_k and $\tilde{B}_k$ describe the charged field. After a simple calculation we get from (10) and (13)

$$L^0 = -\hbar c\left[\varphi_N\gamma^k\frac{\partial\psi_N}{\partial x^k} - \frac{\partial\varphi_N}{\partial x^k}\gamma^k\psi_N + \varphi_p\gamma^k\frac{\partial\psi_p}{\partial x^k}\right.$$

$$\left. - \frac{\partial\varphi_p}{\partial x^k}\gamma^k\psi_p - 2\frac{ie}{\hbar c}A_k\varphi_p\gamma^k\psi_p\right.$$

$$- \frac{ie}{\hbar c}(\tilde{B}_k \varphi_N \gamma^k \psi_p + B_k \varphi_p \gamma^k \psi_N)$$

$$+ 2Mc^2 (\varphi_N \psi_N + \varphi_p \psi_p) \bigg], \tag{16}$$

$$T_{kl}^0 = \frac{\hbar c}{4} \bigg[\varphi_N \gamma_k \frac{\partial \psi_N}{\partial x^l} + \varphi_N \gamma^l \frac{\partial \psi_N}{\partial x^k} - \frac{\partial \varphi_N}{\partial x^l} \gamma_k \psi_N - \frac{\partial \varphi_N}{\partial x^k} \gamma_l \psi_N$$

$$+ \varphi_p \gamma_k \frac{\partial \psi_p}{\partial x^l} + \varphi_p \gamma_l \frac{\partial \psi_p}{\partial x^k} - \frac{\partial \varphi_p}{\partial x^l} \gamma_k \psi_p - \frac{\partial \varphi_p}{\partial x^k} \gamma_l \psi_p$$

$$- 2\frac{ie}{\hbar c}(A_k \varphi_p \gamma_l \psi_p + A_l \varphi_p \gamma_k \psi_p)$$

$$- \frac{ie}{\hbar c}(\tilde{B}_k \varphi_N \gamma_l \psi_p + \tilde{B}_l \varphi_N \gamma_k \psi_p$$

$$+ B_k \varphi_p \gamma_l \psi_N + B_l \varphi_p \gamma_k \psi_N) \bigg]. \tag{17}$$

To determine the current-density four-vector S^{0k}, we make use of the relation

$$S^{0k} = \frac{1}{2} \frac{\partial L^0}{\partial A_k}. \tag{18}$$

In this way we get from (16)

$$S^{0k} = ie\varphi_p \gamma^k \psi_p. \tag{19}$$

In a Cartesian co-ordinate system this gives, according to (12),

$$S^{0k} = ie\psi_p^* \varepsilon_4 \varepsilon_k \psi_p, \tag{20}$$

and especially

$$S^{04} = ie\psi_p^* \psi_p,$$

in conformity with Dirac's original assumption about $\psi^*\psi$. We see also that only ψ_p and not ψ_n contributes to the electric current and density.

From the expression (16) for L^0 and by means of the variation principle

$$\delta \int L^0 \sqrt{g} dx = 0, \quad g = |g_{kl}|, \quad dx = dx^1 dx^2 dx^3 dx^4, \tag{21}$$

the variations of φ and ψ vanishing at the border of the space–time region under consideration, we obtain the following wave equations for ψ_N, φ_N, ψ_p, φ_p, where for the sake of simplicity we have used a Cartesian co-ordinate system

$$\varepsilon_k \left(\frac{\partial \psi_N}{\partial x^k} - \frac{ie}{2\hbar c} \tilde{B}_k \psi_p \right) = \frac{Mc}{\hbar} \psi_N,$$

$$\left(\frac{\partial \varphi_N}{\partial x^k} + \frac{ie}{2\hbar c} B_k \varphi_p \right) \varepsilon_k = -\frac{Mc}{\hbar} \varphi_N,$$

$$\varepsilon_k \left(\frac{\partial \psi_p}{\partial x^k} - \frac{ie}{\hbar c} A_k \psi_p - \frac{ie}{2\hbar c} B_k \psi_N \right) = \frac{Mc}{\hbar} \psi_p, \tag{22}$$

$$\left(\frac{\partial \varphi_p}{\partial x^k} + \frac{ie}{\hbar c} A_k \varphi_p + \frac{ie}{2\hbar c} \tilde{B}_k \varphi_N \right) \varepsilon_k = -\frac{Mc}{\hbar} \varphi_p.$$

These equations are seen to be of the Dirac type, the terms containing the B field, which does not appear in the ordinary equations, describing the action on the spinor particles due to the charged fields.

3. Treatment of Charged and Neutral Fields

In order to obtain field equations for the A and B fields we shall now regard the quantity Γ, which is the five-dimensional analogue to the Lagrangian of the Einstein gravitational field equations, namely

$$\Gamma = \gamma^{\mu\nu} \left[\left\{ \begin{matrix} \mu\tau \\ \sigma \end{matrix} \right\} \left\{ \begin{matrix} \nu\sigma \\ \tau \end{matrix} \right\} - \left\{ \begin{matrix} \mu\nu \\ \rho \end{matrix} \right\} \left\{ \begin{matrix} \rho\tau \\ \tau \end{matrix} \right\} \right], \tag{23}$$

$\left\{ \begin{smallmatrix} \mu\nu \\ \rho \end{smallmatrix} \right\}$ being the well-known bracket expressions

$$\left\{ \begin{matrix} \mu\nu \\ \rho \end{matrix} \right\} = \frac{1}{2} \gamma^{\rho\tau} \left(\frac{\partial \gamma_{\tau\mu}}{\partial x^\nu} + \frac{\partial \gamma_{\tau\nu}}{\partial x^\mu} - \frac{\partial \gamma_{\mu\nu}}{\partial x^\tau} \right). \tag{24}$$

To begin with we consider only the case where gravitation may be neglected; we use a Cartesian co-ordinate system, where the Einstein metric tensor is given by

$$g_{kl} = \delta_{kl}, \tag{25}$$

and further we neglect all quantities containing higher powers of β than the second. Using the expressions (1) we get, after a simple calculation,

$$\Gamma = -\frac{\beta^2}{4} \chi_{rs} \chi_{rs}, \tag{26}$$

where

$$\chi_{rs} = \nabla_r \chi_s - \nabla_s \chi_r, \tag{27}$$

∇_r being the operator defined in (9).

In a non-Cartesian co-ordinate system the corresponding formula will be (where we have still neglected those terms in g_{kl} which depend upon x^0)

$$\Gamma = -\frac{\beta^2}{4} g^{kr} g^{ls} \chi_{kl} \chi_{rs} + G, \tag{28}$$

where G is the Einstein Lagrangian formed by means of the g_{kl} in a similar way as (23) is formed by means of the $\gamma_{\mu\nu}$. Apart from the terms containing x^0, where we have neglected all terms of gravitational order of magnitude, the formula (28) is exact and contains the well-known representation in five dimensions of the Maxwell–Einstein field theory, the χ_{rs} being, apart from the x^0 terms, the components of the antisymmetric tensor of the electromagnetic field. Although (28) is invariant with regard to general relativity and therefore may be expected to give a correct description of the influence of ordinary gravitational fields, it will probably not give a consistent treatment of the finer gravitational effects, since the neglected terms — among others those corresponding to charged g_{kl} fields — are of the same gravitational order of magnitude. For our present purpose, however, which is to determine the energy-momentum tensor of the charged field in the non-gravitational case, it will suffice.

Substituting now for the χ_r the expression (15) we get

$$\chi_{rs} = \begin{pmatrix} A_{rs}, & \tilde{B}_{rs} \\ B_{rs}, & A_{rs} \end{pmatrix}, \tag{29}$$

where

$$A_{rs} = F_{rs} + \frac{ie}{\hbar c}(B_r\tilde{B}_s - B_s\tilde{B}_r) \quad \text{and} \quad F_{rs} = \frac{\partial A_s}{\partial x^r} - \frac{\partial A_r}{\partial x^s}, \quad (30)$$

F_{rs} being the components of the ordinary electromagnetic field tensor and

$$B_{rs} = \delta_r B_s - \delta_s B_r, \quad \delta_r = \frac{\partial}{\partial x^r} - \frac{ie}{\hbar c}A_r,$$

$$\tilde{B}_{rs} = \tilde{\delta}_r \tilde{B}_s - \tilde{\delta}_s \tilde{B}_r, \quad \tilde{\delta}_r = \frac{\partial}{\partial x^r} + \frac{ie}{\hbar c}A_r. \quad (31)$$

We must now introduce the diagonal term of $\chi_{rs}\chi_{rs}$, namely $A_{rs}A_{rs} + B_{rs}\tilde{B}_{rs}$, into our Lagrangian (26), giving, if we go over to the generally relativistic form,

$$\Gamma = -\frac{\kappa}{2}g^{kr}g^{ls}(A_{kl}A_{rs} + B_{kl}\tilde{B}_{rs}) + G, \quad (32)$$

where we have replaced $\frac{\beta^2}{2}$ by κ. To obtain the total Lagrangian we have further to add a term corresponding to the rest mass μ of the mesoton, namely $-\kappa\frac{\mu^2 c^2}{\hbar^2}g^{kl}B_k\tilde{B}_l$ and also the quantity κL^0 which gives

$$L = \Gamma + \kappa L^0 - \kappa\frac{\mu^2 c^2}{\hbar^2}g^{kl}B_k\tilde{B}_l. \quad (33)$$

In the variation principle

$$\delta\int L\sqrt{g}dx = 0, \quad (34)$$

the quantities to be varied are now $\psi_N, \varphi_N, \psi_p, \varphi_p, A_k, B_k, \tilde{B}_k$, and $g^{kl}, k, l = 1, 2, 3, 4$, all variations vanishing at the border of the region. The variation with respect to ψ and φ give, as we have already seen, the equations (22). The variation with respect to g^{kl} gives us the Einstein field equations with right sides containing the energy-momentum tensor, namely

$$G_{kl} = R_{kl} - \frac{1}{2}g_{kl}R = -\kappa T_{kl}, \quad (35)$$

where

$$T_{kl} = T_{kl}^0 + T_{kl}^e, \tag{36}$$

T_{kl}^0 being the quantities defined by (17) while

$$T_{kl}^e = A_{kr}A_{lr} + \frac{1}{2}(B_{kr}\tilde{B}_{lr} + B_{lr}\tilde{B}_{kr}) - \frac{1}{4}\delta_{kl}(A_{rs}A_{rs} + B_{rs}\tilde{B}_{rs}). \tag{37}$$

R_{kl} are the components of the well-known Einstein curvature tensor and R the corresponding invariant. In the formula for T_{kl}^e (the components of the energy-momentum tensor due to the A and B fields) we have used, for the sake of simplicity, a Cartesian co-ordinate system. If the terms containing the B's are omitted, we obtain the usual formula for the electromagnetic energy-momentum tensor.

A variation of the A_k gives after a simple calculation, where again we use a Cartesian co-ordinate system,

$$\frac{\partial F_{kl}}{\partial x^i} = S_k = S_k^0 + S_k^e, \tag{38}$$

with

$$S_k^e = \frac{ie}{\hbar c}\left[\frac{\partial}{\partial x^i}(\tilde{B}_k B_l - \tilde{B}_l B_k) + \frac{1}{2}(\tilde{B}_{kl}B_l - B_{kl}\tilde{B}_l)\right], \tag{39}$$

while

$$S_k^0 = ie\varphi_p\varepsilon_k\psi_p \tag{19a}$$

is the spinor particle current given above. We see that the equations (39) are the ordinary Maxwell equations for the electromagnetic field, the right side containing the current-density four-vector, to which the charged B field contributes by the quantities S_k^e.

Finally we get wave equations for B_k and $\tilde{B}_k$ by varying B_k and $\tilde{B}_k$ respectively:

$$\delta_l B_{kl} - 2\frac{ie}{\hbar c}A_{kl}B_l + \frac{\mu^2 c^2}{\hbar^2}B_k = ie\varphi_N\varepsilon_k\psi_p,$$

$$\tilde{\delta}_l \tilde{B}_{kl} + 2\frac{ie}{\hbar c}A_{kl}\tilde{B}_l + \frac{\mu^2 c^2}{\hbar^2}\tilde{B}_k = ie\varphi_p\varepsilon_k\psi_N. \tag{40}$$

We may easily verify the conservation theorem for the total electric charge. Indeed, a simple calculation gives

$$\frac{\partial S_k^e}{\partial x^k} = \frac{ie}{2\hbar c} [B_l(\tilde{\delta}_k \tilde{B}_{kl}) - \tilde{B}_l(\delta_k B_{kl})],$$

or, by means of (40),

$$\frac{\delta S_k^e}{\delta x^k} = \frac{e^2}{2\hbar c} (B_k \varphi_p \varepsilon_k \psi_N - \tilde{B}_k \varphi_N \varepsilon_k \psi_p). \tag{41}$$

From (22) we get similarly

$$\frac{\partial S_k^0}{\partial x^k} = -\frac{e^2}{2\hbar c} (B_k \varphi_p \varepsilon_k \psi_N - \tilde{B}_k \varphi_N \varepsilon_k \psi_p), \tag{41a}$$

showing that

$$\frac{\partial S_k}{\partial x^k} = \frac{\partial S_k^0}{\partial x^k} + \frac{\partial S_k^e}{\partial x^k} = 0.$$

Thus the total charge is conserved, but not the charge of the spinor particles or that of the charged fields separately. At the same time the number of spinor particles (charged or uncharged) is conserved. In fact, the sum of the probability current-density vectors belonging to the neutral and the charged spinor particles respectively, $i\varphi_N \varepsilon_N \psi_N + i\varphi_p \varepsilon_k \psi_p$, fulfils the continuity equation in that

$$\frac{\partial}{\partial x^k} (i\varphi_N \varepsilon_k \psi_N) = \frac{e}{2\hbar c} (B_k \varphi_p \varepsilon_k \psi_N - \tilde{B}_k \varphi_N \varepsilon_k \psi_p).$$

The equations (40) for the B field are similar to, though not identical with the equations given by Proca[8] and applied by several authors to the problem of the mesoton. The most important difference is the appearance of non-linear terms in (40) due to the quantities $\frac{ie}{\hbar c}(B_r \tilde{B}_s - B_s \tilde{B}_r)$ in the A_{rs}.

---

[8]Proca, *J. Phys. Radium* **7** (1936) 347.

These same quantities are also seen to appear in the expression for the current-density vector, where they give rise to a magnetic spin moment. In fact, the component of the magnetic moment due to that part of the current in the direction of x^1 is given by the following expression

$$\frac{ie}{\hbar c}\int(\tilde{B}_2 B_3 - \tilde{B}_3\tilde{B}_2)dr, \quad dr = dx^1 dx^2 dx^3,$$

where the integral has to be taken over all space; the expressions $\frac{ie}{\hbar c}(\tilde{B}_k B_l - \tilde{B}_l B_k)$ thus defining the density of magnetic spin moment.[9]

One has to remember that the Lagrangian L^0 may belong either to the pair neutron–proton or to the pair neutrino–electron. In (33) L^0 should therefore, strictly speaking, be the sum of two different L^0's, one referring to the heavy and the other to the light spinor particles, each with its own set of ψ's. This we have omitted for the sake of shortness and because the corresponding completion is quite obvious. But it is worthwhile to notice that the complete Lagrangian will imply an interaction of heavy and light spinor particles not only through the intermediate of the electromagnetic field but also through the B field, an interaction which will entail the occurrence of β-processes, the probability of which may be calculated on the basis of the theory developed in this report.

This theory, however, will not be completely founded before we have given rules for its quantization, a question we shall briefly touch upon here. First of all, the ψ corresponding to the different kinds of spinor particles have to satisfy well-known quantization rules, being the application of the formalism of Jordan–Wigner to the wave equation of Dirac and corresponding to Fermi statistics. Moreover, the electromagnetic quantities have to be quantized in the usual way, so that the remaining problem is the formulation of quantization rules for the B field. Here we meet with the same formal difficulty as in the case of the electromagnetic field, that is that the time derivatives of B_4 and $\tilde{B}_4$ are not present in the Lagrangian. We may, however, get over this difficulty by means of a general method to be described elsewhere. As a result one obtains the following

---

[9] See Proca, *J. Phys. Radium, l.c.*

commutation relations[10]

$$[\tilde{B}_{4k}(r), B_l(r')] = 2\hbar c \delta_{kl} \delta(r - r'),$$
$$[B_{4k}(r), \tilde{B}_l(r')] = 2\hbar c \delta_{kl} \delta(r - r'),$$

(42)

where r and r' indicate two space points to be taken at the same time t, $\delta(r-r')$ being the well-known singular function. Further, the B_k commute with all the B_k, $\tilde{B}_k$ and all the B_{kl}, the $\tilde{B}_k$ similarly with all the B_k, $\tilde{B}_k$ and all the $\tilde{B}_{kl}$, while B_4 and $\tilde{B}_4$ commute with all the quantities mentioned.

As an interesting application of the commutation rules we may study the commutation expression for the total charge Q^e of the B field with one of the quantities B_k, $\tilde{B}_k$. We have

$$Q^e = -i \int S_4^e(r') dr',$$

(43)

S_4^e being given by (39). A simple calculation gives now

$$[Q^e, B_k(r)] = \frac{e}{2\hbar c} \int [\tilde{B}_{4k}(r'), B_k(r)] B_k(r') dr' = e B_k(r),$$

(44)

and similarly

$$[Q^e, \tilde{B}_k(r)] = -\frac{e}{2\hbar c} \int [B_{4k}(r'), \tilde{B}_k(r)] \tilde{B}_k(r) dr' = e \tilde{B}_k(r).$$

(44a)

Since

$$\frac{\partial B_k}{\partial x^0} = \frac{ie}{\beta \hbar c} B_k, \quad \frac{\partial \tilde{B}_k}{\partial x^0} = -\frac{ie}{\beta \hbar c} \tilde{B}_k,$$

we see that $\frac{Q^e}{\beta c}$, as far as the B field is concerned, plays the role of the canonical conjugate to χ^0, which is a special case of a general rule according to which x^0 is canonically conjugated to the total charge divided by βc. At the same time the relations (44) are seen to be closely related to the fact that the charge is always a whole positive or negative multiple of the elementary electric charge. They express, moreover, the gauge transformation properties of the B_k and $\tilde{B}_k$.[11]

---

[10]Similar relations for the Proca equations are given by Kemmer and by Bhabha, *Proc. R. Soc., l.c.*

[11]See Heisenberg and Pauli, *Z. Phys.* **59** (1930) 168.

*

*   *

PROFESSOR MOELLER recalled that discussions of the experimental results relative to the interactions between the particles constituting nuclei appeared to show the existence of forces between two protons and between two neutrons of the same order of magnitude as the force between one proton and one neutron. If one wishes to explain these forces by the intermediary of a field of particles capable of being emitted or absorbed by protons and neutrons, we must then admit, in addition to Yukawa's charged particles, neutral particles of the same mass. It appears hard to see how such neutral particles can naturally find a place in a general scheme of the type proposed by Professor Klein.

PROFESSOR KLEIN answered that if, in the Lagrangian (28), one replaces the χ_{rs} given by (29), by the following expression

$$\chi_{rs} = \begin{pmatrix} A_{rs} - C_{rs}, \ \tilde{B}_{rs} \\ B_{rs}, \ A_{rs} + C_{rs} \end{pmatrix}, \tag{29a}$$

where

$$C_{rs} = \frac{\partial C_s}{\partial x^r} - \frac{\partial C_r}{\partial x^s},$$

and, in the Lagrangian (10), the χ_r given by (15), by the corresponding expression,

$$\chi_r = \begin{pmatrix} A_r - C_r, \ \tilde{B}_r \\ B_r, \ A_r + C_r \end{pmatrix}, \tag{15a}$$

the C field corresponds to neutral particles to which it is possible to give any mass, for instance that of the mesoton. It seems, however, that such particles would give rise, according to the Lagrangian concerned, to repulsive forces and not to attractive forces between protons.

FROM MY LIFE OF PHYSICS*

Oskar Klein

Professor V. Fock, Physical Institute of the University of Leningrad, USSR:

Professor Klein's activity in the field of physics began about the same time as mine — in the twenties — or even earlier. From about 1920 he worked for a long time with Niels Bohr. Professor Klein's name met with mine in 1926, when we published independently and nearly at the same time (Klein somewhat earlier) an attempt to find a relativistic wave equation for a charged particle which would generalize de Broglie's and Schrödinger's wave equations. The spin was then not taken into account and the true equation for the electron was found later by Dirac, but the equation that bears (as it undoubtedly should) Klein's name can be interpreted as the equation for particles with zero spin (for bosons) and is still of importance.

Klein's name is also associated with the relativistic formula for the scattering of electromagnetic radiation by free electrons (the Klein–Nishina formula).

Much discussed in the early thirties was the Klein paradox connected with states of negative energy. The solution of this paradox came with Dirac's theory of positrons.

Klein's transformation of the commutation relations and his ideas on the isotopic spin were many years later developed by Yang–Mills. Professor Klein's scientific interests in Einstein's relativity and gravitation theory never ceased. In his early work he investigated questions connected with the five-dimensional formulation. His general attitude towards unlimited

---

*Lecture held at the International Centre for Theoretical Physics; in *From a Life of Physics* (World Scientific, 1989).

cosmological applications of Einstein's gravitation theory is, so far as I know, critical.

Last I must mention another important part of Professor Klein's work, namely his activity on the Nobel Prize Committee.

From Childish Curiosity

Being received with so much kindness and generosity at this coast of the Adriatic sea brings to mind the arrival of Odysseus at a land — situated in this same sea — after much toil and suffering. It lies very near to my heart to think of this, having loved and often reread the Odyssey ever since my childhood — in Swedish translation. And some years ago I was given a Greek edition with an English translation, from which I can guess at the meaning of some of the Greek words. As you know, Homer stresses over and over again the tribulations through which his hero had to pass. Now, none of us physicists has conquered Troy, but some have solved problems requiring a similar kind of ingenuity as that used by the man of many devices. A main problem for us theoreticians rather resembles that represented by Charybdis and Scylla, between which Odysseus was forced to steer. Thus, speculation is certainly a necessary part of theoretical work, just as much as building on experimental facts. Still, it drags many of us into a mental whirlpool not unlike the hydrodynamical one of Charybdis, from which the escape feels like a miracle. On the other hand, sticking too closely to facts — Scylla had six hard ones — may be equally deadly, when using them as building stones for a theory. There are many examples of this even from the days of Aristotle.

Returning to the reception at this friendly coast, you will remember that in Odysseus' time a stranger would be met with a number of questions like: from where do you come, what is your name, who are your parents, what sort of life do you lead, a pirate's or a peaceful one? Now, is this not almost the evening speakers' situation? I must try to make the best of it, my answer having one advantage over that of Odysseus, which lasted many hours, filling about a quarter of the Odyssey, namely comparative shortness.

My name and whence I come are already answered. To the question about parents I would have liked to give a much fuller answer than time will allow, because of their deep influence on my whole outlook on life not to be separated from that on physics. Some eleven years before I was born my parents moved from Germany to Stockholm, where my father became chief rabbi at the Jewish congregation. He was born in the little town of Humennéh in the Carpathian mountains, his parents having a small shop on its outskirts. He left home at an early age, studied the Talmud in Eisenstadt and got his doctor's degree in Heidelberg, finishing his rabbinic studies under the guidance of Abraham Geiger, the famous founder of a deeply liberal movement in Judaism. There was no dividing line between my father's way of regarding his profession and his deep engagement in research, mostly concerning the origin of Christendom in relation to contemporary trends in Judaism. Being very far from a narrowly literal belief in the Bible text, he shared the dream of the Hebrew prophets of an era of universal peace for mankind. He died shortly before the First World War, which spared him the deep disappointment it caused to others of a similar mind.

In contrast to many theoretical physicists, my life in physics had not a philosophical background but came as a result of childish curiosity. On the one hand this made me for several years a bit of a young naturalist, collecting all sorts of things from shells and butterflies to stars — the latter with the help of my mother's opera glasses. On the other hand it made me ask my seven-year older brother — a great authority for me — all sorts of impossible questions about connections and origins, somewhat like Kipling's elephant child.

When All Was New

By and by I began to read scientific books, at first mostly on biology, then on chemistry, my father helping me to find such books, among them Darwin's *Origin of Species* and *Descent of Man*. Just before I was 16 (summer 1910) he introduced me to the great physicist Svante Arrhenius, whom he knew "als ein humaner Mensch" a high praise in his mouth. Arrhenius kindly let me do some experimental work in his laboratory, lent me books and advised me as to reading. These were wonderful times when all was new and my eagerness almost unlimited.

In my last year at Arrhenius' laboratory (1917–18), after I had finished my university studies and also my rather long military service, I spent much time reading papers on the new quantum theory and also on statistical mechanics. Thereby I was, of course, amazed by Bohr's explanation of the spectroscopic Rydberg constant by means of a simple atomic model. But I was far from understanding the deep background of this result, being more impressed by the explicitly mathematical papers of Sommerfeld, Einstein and Debye. So, when I was granted a fellowship for studies abroad, I chose primarily Einstein and Debye, the latter more for dipoles, on which I had been working, than for quantum theory. But, since Bohr was so near, I wrote to him first, asking whether I might come for a shorter stay in Copenhagen, which he kindly accepted. This was in the spring of 1918, and there I went and stayed, with some interruptions, most of them in Stockholm, the longest at the University of Michigan, until I started as a professor of theoretical physics at the University of Stockholm in the winter of 1931. And I never came to Einstein or Debye.

When I came to Copenhagen, Bohr had only one, but a very efficient, theoretical collaborator, namely Kramers, and he had no institute but just a room at the Engineering College, and we were allowed to sit in the library there. The first part of his paper on spectral lines had just appeared, which, with Kramer's help, I studied thoroughly. And Bohr himself commented further on his ideas, especially during a long walking trip north of Copenhagen, when he also talked about many other things more or less connected with physics: his general philosophical attitude and his father's view on the relation between life and physics, both themes containing the germs of his later complementarity viewpoint.

I was then toiling with the forces between ions in strong electrolytes on the lines of Bjerrum and dipoles on the line of Debye, trying to apply Gibbsian statistical mechanics to these problems; and Bohr showed me his deeper view of this subject, telling me how Gibbs' general canonical distribution gave the very definition of temperature. All this meant a new epoch for me, and an essentially happy one, although I had more troubles than results from my own work, which, however, led to my thesis on generalized Brownian motion, meant as a foundation for a theory of solutions of interacting particles.

During these early years the essential work of Bohr and Kramers turned around quantization, the way of sorting out, among the mechanically possible orbits of electrons in atoms, those suitable to represent the quantum states — although Bohr already knew that this could be but a provisional procedure, a preparation for a rigorous quantum mechanics. But astonishingly much was reached in this way — for which Bohr introduced the word correspondence — through his exceptional gift of obtaining largely correct results from unfinished theory and insufficient experimental facts, verily a strangely successful passage between Charybdis and Scylla.

Deeper Background

In these surroundings it was natural to dream about the deeper background of these strange quantum rules with their whole quantum numbers. I shall dwell a little on this because it was the primary source of several attempts of mine, which slowly, after many mistakes, led me to my present views on relativity theory and its relations to quantum theory and cosmology.

Some time in the early twenties, from studying Fresnel's work on the wave theory of light, I was struck by the fact that whole numbers in physics appear in one out of two ways, either through atomism, the usual way of regarding quanta, or through wave interference. Slowly this trend of thought began to take a somewhat more definite form due to my reading the review in Whittaker's *Analytical Dynamics* of Hamilton's original way to what was later called the Hamilton–Jacobi equation — known to us as quantizers from Jacobi's *Vorlesungen über Mechanik*. According to Hamilton, who started from optics, this equation determines the propagation of a wave front in the manner of Huygens, forming a link between geometrical optics and mechanics. Thus a quantum state belonging to a particle moving in a closed orbit came to look as if a wave (in the geometrical approximation) interfered with itself, the number of wavelengths on the orbit being equal to the quantum number. I had known this for some time, when in the summer of 1923, in connection with my writing a little book on optics, I noticed that this could be interpreted as the condition for a standing wave, i.e. a proper vibration.

Towards the end of that summer I got married and in the beginning of September my wife and I went to Ann Arbor in the U.S.A., where through the mediation of Walter Colby — now for many years a dear friend of ours — I had been appointed an instructor in theoretical physics at the University of Michigan. There we stayed for almost two years during which time, apart from lectures and other teaching, I worked hard on different problems, more or less connected with the study of molecular spectra going on at the Department of Physics under the most competent and unselfish guidance of H.M. Randall, the head of the department.

Whirlpool of Speculation

The next autumn, however, I gave a lecture course on electromagnetism, towards the end of which I derived the general relativistic Hamilton–Jacobi equation for an electric particle moving in a combined gravitational and electromagnetic field. Thereby, the similarity struck me between the ways the electromagnetic potentials and the Einstein gravitational potentials enter into this equation, the electric charge in appropriate units — appearing as the analogue to a fourth momentum component, the whole looking like a wave front equation in a space of four dimensions. This led me into a whirlpool of speculation, from which I did not detach myself for several years and which still has a certain attraction for me.

The strong impresssion this made on me came from the attempt I mentioned to find a wave background to the quantization rules. Thus, for some time I had played with the idea that waves representing the motion of a free particle had to be propagated with a constant velocity, in analogy to light waves — but in a space of four dimensions — so that the motion we observe is a projection on our ordinary three-dimensional space of what is really taking place in four-dimensional space. This idea seemed to me to get certain support by some occasional utterings of Bohr. Thus, while most strongly stressing the impossibility of describing quantum phenomena by a space–time theory of the ordinary type — regarding them as "a knot on existence which might be shifted but not removed" — he sometimes said that such a theory would perhaps be possible in a space of four dimensions. This had nothing to do with so-called parapsychological phenomena, of which, since my early youth, I have remained most sceptical — like Bohr.

Apart from being entangled in these speculations there was another reason which made me hesitate to start an investigation, by means of a wave equation, of the standing waves corresponding to the quantum states of a particle in a field of force. Thus, from the book of Hadamard on wave propagation I knew that the relation between a wave-front equation and a second-order wave equation is not unique, that on top of the linear second-order terms there might be non-linear terms. And I believed that such terms might have some bearing on the particle nature of electrons. This is certainly foolish, but was not obviously so at that time. But Dirac may well say that my main trouble came from trying to solve too many problems at a time!

I did not try until more than a year later, when again in Copenhagen, to determine the states of standing waves of the linear wave equation corresponding to the Hamilton–Jacobi equation of a harmonic oscillator. But due to lack of time and poor mathematics I had not succeeded in this when Schrödinger's paper containing the similar solution for the electron in the hydrogen atom appeared.

In Five Dimensions

Returning to my Ann Arbor attempts, I became immediately very eager to see how far the mentioned analogy reached, first trying to find out whether the Maxwell equations for the electromagnetic field together with Einstein's gravitational equations would fit into a formalism of five-dimensional Riemann geometry (corresponding to four dimensions plus time) like the four-dimensional formalism of Einstein, which, by the way, I knew very superficially at the time and now tried to learn by means of Pauli's excellent book. It did not take me a long time to prove this in the linear approximation, assuming a five-equation, according to which an electric particle describes a five-dimensional geodesic.

However, I was not satisfied with this result but did much work, implying lengthy calculations, on establishing the rigorous equations, which took me most of the summer of 1925, after our return to Denmark. To my surprise they came out to agree exactly with Einstein's gravitational equations and the Maxwell equations in general relativistic form, when the proportionality factor connecting the charge of the particle with the fourth momentum component was appropriately chosen, the sign being such as to make the extra dimension space-like. Taking at that time the geometrical picture of a four-dimensional space literally, this relation, together with the experimental fact that any electric charge is a whole number multiple (positive or negative) of an elementary unit of charge, led me to believe that this space is closed in the direction of the fourth dimension, the circumference being 0.8×10^{-30} cm, far beyond the smallest distances observed. According to this picture, physical quantities ought to be periodic functions of the extra coordinate, measurable quantities being averages taken over this small circumference, the higher overtones corresponding to states of high electric charge. This, I thought, would be the reason why ordinary physical space is only three-dimensional. Moreover, I believed at that time that the periodicity in question was the root of the quantal aspect of nature.

At the end of the summer we went to Sweden to visit my mother before going back to Denmark, where Bohr had obtained a fellowship permitting me to stay in Copenhagen during a leave of absence from Michigan. However, due to my getting seriously ill — an infectious hepatitis — we had to stay in Sweden for a year, arriving in Copenhagen only at the beginning of March 1926. During that half year, when so much happened in physics (Heisenberg's breakthrough in quantum mechanics, Goudsmit's and Uhlenbeck's paper on electron spin, Pauli's matrix theory of the hydrogen atom, to mention the most important ones), I had hardly done any work on reading. But a few weeks before we went to Copenhagen, during a recreation trip, I had written out the summer's work on five-dimensional theory, leaving my quantal speculations for work in Copenhagen. As already mentioned, I started there with the simplest kind of wave equation and tried to work out the stationary states of the harmonic oscillator, when Schrödinger's first wave mechanical paper appeared.

When Pauli came to Copenhagen some weeks later, I showed him my manuscript on five-dimensional theory and after reading it he told me that Kaluza some years before had published a similar idea in a paper I had missed. So I looked it up but — as with de Broglie's thesis, which Bohr had shown me in the summer of 1925 — I read it rather carelessly but quoted both, of course, in the paper I then wrote in a spirit of resignation. In the summer of 1925 I had found Kramers in a state of despondency after the failure of the Bohr–Kramers–Slater statistical interpretation of quantum theory — although, at the same time he had given an important and beautiful contribution to correspondence theory by his dispersion formula. I tried to cheer him up by pointing out in a letter that certainly I considered science a subject of great importance, just as important as the play of children. We had both small children at the time. I thought that now I had to apply this to myself.

In the paper I tried, however, to rescue what I could from the shipwreck, and at the same time to learn as much as possible from Schrödinger and also from de Broglie, whose beautiful group velocity consideration impressed me very much even if by and by I saw that it did not essentially differ from my own way by means of the Hamilton–Jacobi equation. From Schrödinger I learnt in the first place his definition of the non-relativistic expressions for the current-density vector, which it was then easy to generalize to that belonging to the general relativistic wave equation. In this, after Schrödinger's success with the hydrogen atom, I definitely made up my mind to drop the possible non-linear terms, although I was still far from certain that this was more than a linear approximation. Also I derived the energy-momentum components, which in the five-dimensional formalism belonged to the current-density vector. These I published much later, due to the appearance in the meantime of a paper by Schrödinger containing the corresponding non-relativistic expressions.

Kind Scepticism

Not long after having sent the paper to *Zeitschrift für Physik* I got to my great surprise and pleasure a letter from H.A. Lorentz — for whom I had a special veneration, having read his excellent book on differential and integral calculus for science students when I was 16 and later several of his writings on different parts of physics. He extended a most kind invitation to spend some weeks in Leiden in June that year and tell about my attempts regarding five-dimensional theory. As I learnt later, the background for this was that L.H. Thomas, whom I had given a copy of my paper, passed through Leiden on his way from Copenhagen to Cambridge, showing it to Ehrenfest, who with characteristic impulsiveness asked Lorentz to invite me. Before leaving Copenhagen I had the privilege to give a talk about my ideas to Bohr and Heisenberg, to which they listened with kind scepticism. The stay in Leiden was most animated and refreshing after a rather hard year. Lorentz came to my lectures, taking part in the discussions in his quiet and lucid way. Among the physicists I met, in the first place Ehrenfest and his wife, I shall mention Uhlenbeck and Goudsmit, which led to much pleasant discussion.

One day Ehrenfest brought us a new paper by Dirac containing a quantum dynamical theory of the Compton effect which we tried to read, but without success. It occurred to me that it could be treated the Schrödinger way in analogy with the transitions from one atomic stationary state to another, the states being here those of an electron in a box under the influence of a radiation field. Uhlenbeck and I began to calculate, but, although it looked all right qualitatively, we never reached the quantitative result in the short time then at our disposal. I had planned to continue this when again at home, but a lot of problems were coming up, among others the derivation of Kramers' dispersion formula in the Schrödinger way — the possibility of which Heisenberg had pointed out to me when I showed him the relativistic (time-dependent) wave equation soon after Schrödinger's first paper had appeared. Before my paper on the correspondence treatment of wave mechanics appeared, where the procedure was sketched but still not carried through, Gordon had published a paper, where with a similar background he had performed the calculation with the same result as that given in the mentioned paper by Dirac.

Argument on Radiation

In September the same year there was the remarkable visit by Schrödinger to Copenhagen already mentioned by Heisenberg,[1] when he told us about his general views on wave mechanics, which he regarded as a very close analogy to the situation in optics, when corpuscular theory was replaced by wave theory. This led to very animated objections from Bohr and Heisenberg, a most important argument being that this way of describing the radiation emitted by atoms would not agree with Planck's formula for the temperature radiation, the very foundation stone of quantum theory. Finally Schrödinger declared himself convinced, but in his later years he came more or less back to his original ideas, which was hard to understand. On the other hand, we all in Copenhagen admired Schrödinger's many applications of wave mechanics to a number of problems. But for a time

---

[1] In *From a Life of Physics*, eds. H. A . Bethe *et al.* (World Scientific, 1989).

there was a certain competition between wave mechanics on the one hand and matrix mechanics based on Heisenberg's original ideas on the other. But by and by it became clear, especially through the work of Dirac, Born and Jordan, that they were just different aspects of one and the same theory. And in the spring of 1927 time was ripe for Bohr's complementarity view, which got a strong impetus from Heisenberg's paper on the uncertainty relations. About this I shall only say that Bohr's attitude to physics reminded me decidedly of my father's to religion.

Returning to my own work, I agreed so far with Bohr's and Heisenberg's objections to Schrödinger that a kind of, so to say practical, quantization is needed in order to get wave mechanics to agree with experimental facts, as is seen from my paper mentioned above (Elektrodynamik und Wellenmechanik vom Standpunkt des Korrespondenzprinzips), the writing of which proceeded very slowly under numerous discussions with Bohr. Still, I hoped, as is seen from its last section, that a further development of five-dimensional theory would give a foundation for this approximate treatment. Hence, after the paper was finally off to the *Zeitschrift für Physik*, I began to look more closely into this matter. First, I convinced myself that Schrödinger's emission theory leads to the Rayleigh formula and not to that of Planck, in agreement with what Bohr and Heisenberg had maintained. Then I tried to see whether anything like the quantal treatment of the many-particle problem might come out of five-dimensional theory. But the result was discouraging.

Then there came a new paper by Dirac containing an application of the quantization procedure on a radiation field considered in the Rayleigh manner as a system of uncoupled oscillators corresponding to its resolution into standing waves. This made me try to use the same procedure on a Schrödinger field coupled electrostatically to itself by means of a scalar potential derived in the usual way from the density. I preferred this approach, starting from a space–time formulation, to that based on a quantum theory translation of general Hamiltonian mechanics of a system of particles. Apart from a disturbing term, where the non-commuting factors had the wrong order, the result of my calculations agreed with Schrödinger's wave equation in configuration space. This was in March 1927, just a year after I became acquainted with the new quantum mechanics. From then on I was convinced that a generalized field theory, whether four- or five-dimensional, ought to be a quantum field theory and not a theory of the classical type.

In the autumn Jordan, who independently had started the same approach, came to Copenhagen. He showed me that my trouble had to do with the electrostatic self-energy, which in this particular case was cancelled by an interchange of the non-commuting factors, and we wrote our joint paper. About the same time, I wrote a new paper on five-dimensional theory, where I took the standpoint just mentioned — to the satisfaction of Pauli who had been very angry when reading my correspondence theory paper.

Jordan's and my paper treated only the case of symmetric quantization but, while we wrote it, Jordan told me of his approach to antisymmetric quantization, later treated in detail by him and Wigner. This impressed me immensely. During the past year I had tried to consider the Pauli principle in the wave theory way but had not been able to find a mathematical expression for this all-or-nothing situation, which was now so beautifully realized by Jordan's approach.

Not long after this, Dirac sent Bohr the first draft of his paper on the electron, which came as an amazing surprise. And early in the new year Bohr sent me to Cambridge to learn more about it. This impressed me so much that for a time I abandoned completely my general relativity speculations. And when Pauli, around Easter time, came to Copenhagen we drank a bottle of wine on the death of the fifth dimension — which for a time he and Heisenberg had used in their joint, not yet published work on quantum electrodynamics. But we both had our come-backs. That spring Nishina returned to Copenhagen after a stay in Hamburg and we worked until late in the summer on the Compton effect according to the electron theory of Dirac.

After a short attack of "five-dimensional" in the summer of 1935, I had a more violent one in 1937 — due to the Yukawa theory — on which I gave a paper at a conference in Warsaw in 1938, published in its proceedings. My trouble was, as in my earlier papers on five-dimensional theory, that I tried prematurely to connect the theoretical attempt with the very insufficient knowledge of what is now called elementary particle physics. This time I replaced the periodicity hypothesis — according to which the fields would be represented by Fourier series in the extra coordinate — by means of a two-row matrix representation, corresponding to zero and unit charge only, indicated in a note to the 1927 paper. Moreover, for the first time I used the general relativity form of the Dirac equation as the starting point for a generalized theory. Due to my abandonment of general relativistic speculation after the appearance of Dirac's electron theory, I had paid very little attention to this form, developed in detail by a number of physicists. But more and more I have come to regard this general relativistic equation as the natural starting point for a generalized quantum field theory, also containing non-gravitational interactions, in the first place the electromagnetic ones.

Leaving these rather speculative ideas, I shall say a few words on my position regarding the cosmological attempts of Einstein and his followers. An analysis of Einstein's way to general relativity theory shows that its essential foundation is the principle of equivalence, according to which the ordinary laws of physics, i.e. those of special relativity, must hold in a frame of reference freely falling in a gravitational field which is practically homogeneous within the region and during the time of using this frame. Now, in the first place, such a frame has no relation to the universe at large, being practically realized by a satellite at any place above the earth. Secondly, the principle of equivalence implies that physical quantities get their meaning through measurements made in a frame where gravitation may be neglected. This applies to the mass of any particle or body, having therefore no relation to the totality of mass in the universe, as assumed by Mach.

The Liberty of Doubt

Now, the main basis for Einstein's cosmological attempts was this very appealing idea of Mach, and he seemed not to notice its incompatibility with the principle of equivalence. As I have tried to show in some earlier papers and in a paper soon to be published, this takes away the *a priori* arguments for relativistic cosmology, on which Einstein put so much weight. Hence, the interesting *a posteriori* arguments — the so-called fireball radiation, predicted from cosmology by Dicke and the helium content of certain stars, predicted from a cosmological model due to Gamow — means a challenge to those, who like me, try to replace cosmology by a theory along the lines of ordinary physics, regarding the system of galaxies as the first known specimen of a higher-order type of stellar systems.

I should like to finish this talk on unfinished attempts by stressing the great admiration I feel for Einstein's contributions to physics, in the first place to relativity theory and quantum theory, which is certainly not incompatible with the observation that he also shared the universal condition of mankind of making mistakes when trying something new. A study of the history of science — not the history of philosophy — shows that the natural attitude of a scientist is to be inspired by the great predecessors, just as they themselves were by their predecessors, but always taking the liberty of doubting when there are reasons for doubt.

SCIENTIFIC BIBLIOGRAPHY OF OSKAR KLEIN

*(This bibliography is based on one made
by Professor Klein himself)*

"Über die Löslichkeit von Zinkhydroxyd in Alkalien," *Reports from the Nobel Institute of the Royal Swedish Academy of Sciences*, Stockholm, **2**, Nr. 18 (1912) 1; and *Z. Anorg. Chem.* (1912) 157.

"Beitrag zur Kenntnis der Dielektrizität unter besonderer Berücksichtigung der Theorie der molekularen Dipole," *Reports from the Nobel Institute of the Royal Swedish Academy of Sciences*, Stockholm, **3**, Nr. 24 (1917) 1.

"Gefrierpunkte binärer wässeriger Lösungen von Elektrolyten" (with O. Svanberg), *loc. cit.* **4**, Nr. 1 (1918) 1.

"Om det osmotiska trycket hos en elektrolyt," *loc. cit.* **5**, Nr. 6 (1919) 1.

"Calculation of scattered radiation from a plate exposed to a beam of X-rays," *Philos. Mag.* **37** (1919) 207.

"Zur statistischen Theorie der Suspensionen und Lösungen" (Inaugural dissertation for the doctor's degree, Stockholm), *Ark. Mat. Astr. Fys.* **16**, Nr. 5 (1921) 1.

"Über Zusammenstösse zwischen Atomen und freien Elektronen" (with S. Rosseland), *Z. Phys.* **4** (1921) 46.

"Über die gleichzeitige Wirkung von gekreuzten, homogen elektrischen und magnetischen Feldern auf das Wasserstoffatom," *Z. Phys.* **22** (1924) 109.

"Vad vi veta om ljuset I, II," *Natur och Kultur*, Stockholm (1925).

"Quantentheorie und fünfdimensionale Relativitätstheorie," *Z. Phys.* **37** (1926) 895.

"The atomicity of electricity as a quantum theory law," *Nature* **118** (1926) 516.

"Elektrodynamik und Wellenmechanik vom Standpunkt des Korrespondensprinzips," *Z. Phys.* **41** (1927) 407.

"Zum Mehrkörperproblem der Quantentheorie" (with P. Jordan), *Z. Phys.* **45** (1927) 751.

120   *O. Klein*

"Zur fünfdimensionalen Darstellung der Relativitätstheorie," *Z. Phys.* **46** (1927) 188.

"The scattering of light by free electrons according to Dirac's new relativistic dynamics" (with Y. Nishina), *Nature* **122** (1928) 398.

"Über die Streuung von Strahlen durch freie Elektronen nach der neuen relativistischen Quantendynamik von Dirac" (with Y. Nishina), *Z. Phys.* **52** (1929) 853.

"Die Reflexion von Elektronen an einem Potentialsprung nach der relativistischen Dynamik von Dirac ," *Z. Phys.* **53** (1929) 157.

"Zur Frage der Quantelung des asymmetrischen Kreisels," *Z. Phys.* **58** (1929) 730.

"Zur quantenmechanischen Begründung des zweiten Hauptsatzes der Wärmelehre," *Z. Phys.* **72** (1931) 767.

"Zur Berechnung von Potentialkurven für zweiatomige Moleküle mit Hilfe von Spektraltermen," *Z. Phys.* **76** (1932) 226.

"Einsteins relativitetsteori i allmäntillgänglig form," *Natur och Kultur*, Stockholm (1933).

"Zur Frage der quasimechanischen Lösung de quantenmechanischen Wellengleichung," *Z. Phys* **80** (1933) 792.

"Orsak och verkan i den nya atomteorins belysning," *Natur och Kultur*, Stockholm (1935). "Eine Verallgemeinerung der Diracschen relativistischen Wellengleichung," *Ark. Mat. Astr. Fys.* **25A**, Nr. 15 (1936) 1.

"Entretiens sur les idées Fondamentales de la physique moderne," Herman et CIE, Paris (1938).

"Sur la théorie des champs associés a des particules chargées," *Les Nouvelles Théories de la Physique*, Collection Scientific, Institute International de Coopération Intellectuel, Paris (1939), p. 81. Proceedings of Symposium in Warsaw, 30 May–3 June, 1938.

"Quelques remarques sur le traitement approximatif du problème des électrons dans un réseau cristallin par la mécanique quantique," *J. Phys.* (Paris) **9** (1938) 1.

Philosophy and Physics, Theoria, Stockholm (1938), p. 59.

"Considerations on the kinetics of respiration with special reference to the inhibition caused by carbon monoxide" (with J. Runnström) *Ark. Kemi, Mineral. Geol.* **14A** (1940) 1.

"On the meson pair theory of nuclear interaction," *Ark. Mat. Astr. Fys.* **30A**, Nr. 3 (1943) 1.

"On the magnetic behaviour of electrons in crystals," *loc. cit.* **31A**, Nr. 12 (1944) 1.

"On the origin of cosmic radiation," *loc. cit.* **31A**, Nr. 14 (1944) 1.

"On the origin of the abundance distribution of chemical elements" (with G. Beskow and L. Treffenberg), *loc. cit.* **33B**, Nr. 1 (1945) 1.

"On the specific heat of the superconductive state," *loc. cit.* **33B**, Nr. 2 (1945) 1.

"Some remarks on the quantum theory of the superconductive state" (with J. Lindhard), *Rev. Mod. Phys.* **17** (1945) 305.

"Meson fields and nuclear interaction," *Ark. Mat. Astr. Fys.* **34A**, Nr. 1 (1946) 1.

"On a case of radiation equilibrium in general relativity theory and its bearing on the early stage of stellar evolution," *loc. cit.* **34A**, Nr. 19 (1947) 1.

"Mesons and nucleons," *Nature* **161** (1948) 897.

"Sur la théorie termodynamique de la fréquence des eléments chimiques," *Rayons X et Structure Atomique*, Conf. Collège de France (1949).

"On the thermodynamical equilibrium of fluids in gravitational fields," *Rev. Mod. Phys.* **21** (1949) 531.

"On the statistical derivation of the laws of chemical equilibrium," *Nuovo Cimento*, serie IX, **6** (1949) 171.

"Theory of superconductivity," *Nature* **169** (1952) 578.

"On the emission of sound waves from an electron in a metal and the theory of superconductivity," *Ark. Fys.* **5** (1952) 459.

"Från klassisk fysik till kvantteori," Verdandi Småskrifter, Bonniers, Stockholm (1952).

"On a class of spherically symmetric solutions of Einstein's gravitational equations," *Ark. Fys.* **7** (1954) 487.

"Some cosmological considerations in connection with the problem of the origin of the elements," *Mém. Soc. Roy. Sci. Liège* **14** (1954) 42, and discussion p. 70.

"Generalisations of Einstein's theory of gravitation considered from the point of view of quantum field theory," Proceedings Fünfzig Jahre Relativitätstheorie, Bern (1955), *Helv. Phys. Acta Suppl.* **4** (1956) 58.

"Quantum theory and relativity," in *Niels Bohr and the Development of Physics*, London (1955), p. 96.

"On the Eddington relations and their possible bearing on an early state of the system of galaxies," *loc. cit.* (1956) 147.

"Some remarks on general relativity and the divergence problem of quantum field theory," *Nuovo Cimento*, serie X, **6** (1957) 344.

"Some remarks on the inverse on theorems of quantum field theory," *Nucl. Phys.* **4** (1957) 677.

"The Dirac theory of the electron in general relativity theory," *K. Norske Vidensk. Forhandl.* **31** (1958) 47.

"Some considerations regarding the earlier development of the system of galaxies," Inst. Internat. Phys. Solvay, Cons. Physique, Bruxelles (1958) 1.

"Zur Theorie der Elementarteilchen," Deutscher Physikertag, Berlin, Physik Verlag, Mosbach, Baden (1959), p. 1.

"On the systematics of elementary particles," *Ark. Fys.* **16** (1959) 191.

"On the treatment of the gravitational field in connection with the generally relativistic Dirac equation," *Ark. Fys.* **17** (1960) 517.

"Einige Probleme der allgemeinen Relativitätstheorie," in *Werner Heisenberg und die Physik unserer Zeit*, Viewig und Sohn, Braunschweig (1961), p. 58.

"Mach's principle and cosmology in their relation to general relativity," in *Recent Developments in General Relativity*, Pergamon Press (1962), p. 293.

"Matter and antimatter annihilation and cosmology" (with H. Alfvén), *Ark. Fys.* **23** (1962) 187.

"Some general aspects of Einstein's theory of relativity," *Astrophys. Norvegica* **9** (1964) 161.

"Who was Jordanus Nemorarius?" *Nucl. Phys.* **57** (1964) 345.

"Remark concerning the basic SU(3) triplets," *Phys. Rev. Lett.* **16** (1966) 63.

"Boundary conditions and general relativity," in *Preludes in Theoretical Physics in Honour of V. Weisskopf*, North Holland (1966), p. 23.

"Instead of cosmology," *Nature* **211** (1966) 1337.

"Glimpses of Niels Bohr as scientist and thinker," in *Niels Bohr*, North Holland (1967), p. 74.

"On the foundations of general relativity theory and the cosmological problem," *Ark. Fys.* **39** (1969) 157.

"Antisymmetrische Feldquantisierung und functionale Integration," *Z. Naturforsch.* **22a** (1967) 1291.

"A tentative program for the development of quantum field theory as an extension of the equivalence principle of general relativity theory," *Nucl. Phys.* **21B** (1970) 253.

"Arguments concerning relativity and cosmology," *Science* **171** (1971) 339.

The complete list contains another 33 contributions, most of them intended for educated layman readership and in Swedish, with the exception of 4 in Danish.

PART II

PART II.

PREFACE

The Oskar Klein Memorial Lecture and Seminar in 1990 were given by Hans A. Bethe, professor at Cornell University, Ithaca, USA, and Nobel Prize winner in Physics in 1967. The 1991 speaker was Alan H. Guth, professor at Massachusetts Institute of Technology, Cambridge, USA.

Bethe's lecture in the series of Oskar Klein Memorial Lectures at Stockholm University was given on October 9, 1990, and Guth's lecture on June 4, 1991. These two lectures, intended for both students and staff, are presented in this volume. The invited speakers also gave scientific seminars of a somewhat more advanced nature. These are also included here. Two of Oskar Klein's papers are presented in English translation for the first time.

Neutrinos are at the heart of Hans A. Bethe's contributions. In the lecture, he deals with the problems to theoretically account for the observations of neutrinos from the sun. He concludes that the solar model cannot be blamed for the discrepancies between the observed flux of neutrinos and the predictions of the standard solar model. His preferred explanation is one based on neutrino oscillations between neutrinos of different flavours.

The theme of Bethe's scientific seminar was the theory of supernovae, and for anyone interested in gaining an understanding of the processes going on in the collapse and explosion of a large star this lecture is to be recommended. The main problem for the theory has been to reproduce the outgoing shock wave and the ejection of matter. According to Bethe we now qualitatively understand that "the supernova is powered by the absorption of neutrinos".

The invited speaker for 1991 was Alan H. Guth, originator of the inflation scenario for the big bang. His lecture and scientific seminar dealt with various aspects of the origin of the universe. In the written version of his lecture, "The Big Bang and Cosmic Inflation", he has included the latest observations from the COBE satellite and their theoretical interpretation. The lecture is a goldmine for anyone willing to grasp the essentials of these ideas despite not being an expert. "Inflation allows the entire universe to develop from almost nothing" would be hard to understand without Guth's expert guidance. Towards the end he elegantly points out that the driving force behind the big bang expansion — something Oskar Klein devoted most of his late efforts to finding — is coming directly from the Klein–Gordon equation. The seminar dealt with the eventual possibility of creating a new universe in the laboratory. A quote from the seminar reads: ". . . the false vacuum bubble appears from the outside to be a black hole. From the inside, however, it appears to be an inflation region of false vacuum with new space being created as the region expands. The region completely disconnects from the original spacetime, forming a new, isolated closed universe."

In this volume, two papers by Klein are included in the English translation from the original German. Both papers have to do with early applications of Dirac's theory for the electron. The modern reader will be surpised to see how Klein and Nishina in 1928 derived their famous formula for Compton scattering. They treated the electromagnetic field as a classical wave and the electron as a Dirac field. No particles, only waves, appeared in this so-called semiclassical treatment. They got the kinematics right and they got the angular distribution right for the scattered radiation. The theory with light as particles, first advocated in 1992 by Compton himself, up to 1928 never succeeded in getting the differential cross section right; only the kinematical relations came out in agreement with measurements.

In the historically oriented paper "The Klein–Nishina Formula", an account is given of the collaboration between a Swedish theoretician and a Japanese experimentalist. New studies of the Nobel archives shed light upon the prevailing view on the struggle between a corpuscular theory for light and the wave theory. In those days, physicists were predisposed either to accept light as waves, rejecting light as particles, or to accept the reverse. One new unexpected finding is that Compton was not awarded the 1927 Nobel Prize in Physics for his discovery of the particle nature of light.

The second paper by Klein deals with the surprising results obtained when he calculated the reflection of Dirac electrons at a steep potential wall. The paper contains the "Klein paradox", namely that under certain conditions more electrons are reflected than are impinging upon the wall. The negative energy states, which Klein avoided by his method to solve the Compton scattering problem, but which were important for the reflection problem, were in those days considered unphysical. Klein's paper on wall reflection of electrons was considered a severe blow to the Dirac theory. Today there is no more any paradox. The reason for the phenomenon is the creation of electron–positron pairs in the strong electric field at the wall surface.

Some autobiographic material, which has been translated from the Swedish and included here, goes beyond what was contained in *The Oskar Klein Memorial Lectures*, Vol. 1 (ed. Gösta Ekspong, World Scientific, Singapore, 1991). One may here get acquainted with Klein's interest in cosmology and his attempts to find the driving force behind the expanding system of galaxies — the meta-galaxy in his terminology. The idea that there may be several isolated expanding meta-galaxies was not foreign to him. The reader of Guth's contributions will find similarities.

Stockholm, Sweden; March 1993
Gösta Ekspong

THEORY OF NEUTRINOS FROM THE SUN

Hans A. Bethe

Newman Laboratory of Nuclear Studies
Cornell University
Ithaca, New York 14853, USA

1. Introduction

I knew Oskar Klein. He introduced me in 1967 when I received the Nobel Prize and also I had a very nice summer excursion with him a couple of years later. I am happy to remember him.

My subject today is neutrinos from the sun. To start with I will show you the sum of the reactions that take place in the sun which I suppose are familiar to many of you. The main reaction which keeps the sun shining, which it did yesterday, is the combination of two protons to form a deuteron with the emission of a positive electron and a neutrino.

$$H + H \rightarrow D + e^+ + \nu, \tag{1}$$

$$D + H \rightarrow He^3 + \gamma, \tag{2}$$

$$He^3 + He^3 \rightarrow He^4 + 2H, \tag{3a}$$

$$\text{or } He^3 + He^4 \rightarrow Be^7 + \gamma. \tag{3b}$$

The deuterons then undergo lots of further reactions, only some of which are written down here, and some of the deuterons end up as beryllium-7.

$$Be^7 + e^- \rightarrow Li^7 + \nu \tag{4a}$$

$$\text{or } Be^7 + H \rightarrow B^8 + \gamma. \tag{4b}$$

Beryllium-7 normally decays by capturing an electron which makes the familiar nucleus lithium-7 and of course a neutrino. Very rarely the beryllium-7 will capture a proton and that makes a nucleus boron-8, which of course is a very unusual nucleus having five protons and three neutrons. As quickly as it can, it gets rid of a bit of charge and transforms in the following manner:

$$B^8 \rightarrow Be^{8*} + e^+ + \nu. \tag{5}$$

Alternative to (1):

$$H + H + e^- \rightarrow D + \nu. \tag{1b}$$

So we have at least three reactions which give rise to neutrinos. The neutrino energy from the first reaction is about 0.4 MeV; from beryllium it is 0.8 MeV and from boron you have a continuous spectrum which goes all the way up to 14 MeV.

2. Detection of Solar Neutrinos

So you have very energetic neutrinos and less energetic ones. Well, it clearly would be very nice to observe these neutrinos. It is unfortunately very difficult to observe neutrinos because they have so little interaction with ordinary matter. However, it can be done. It was pointed out by J. Bahcall many years ago that it would be good to use chlorine-37 for this detection. Chlorine-37 when it captures a neutrino can give an analogue state in argon-37 which makes this transition relatively easy, but only for neutrinos of more than 6 MeV. Anyway, this is the standard detector for these neutrinos. Argon-37 has a 35-day half life. It then decays back into chlorine-37 and it does that by capturing a K-electron and the hole in the K-shell is then filled with electrons from the L-shell. You can see the L-electron which is ejected in this manner. This experiment has been done by Raymond Davis for about 20 years. He uses a tank car full of carbon chloride. Then that makes argon-37 — very little of it, a few atoms — so he uses a carrier. He puts in a mixture of helium and argon to flush out the argon which has been formed. He puts that gas into a counter in order to count the emitted Auger electrons.

3. What Is Wrong?

Well, he has done that very patiently.[1] Indeed he found counts, he found neutrinos. However, he found only about one quarter of the predicted number. So, in one respect this was very satisfactory; the neutrinos were actually observed. In another respect it presented a puzzle because he saw only a small fraction of the predicted number. What is wrong? Well, the first possibility is that the experimenter might have been wrong. But Davis is a very careful experimenter and he checked it over and over again and it remained true that he got much fewer neutrinos than expected. Second, it is possible that the theory of the sun might be wrong. It might be that the temperature at the centre of the sun could be lower than had been calculated. Bahcall and his collaborators studied that and he went through with them very carefully and found that there was no mistake, probably. Then finally the nuclear physics experiments might have been wrong. Maybe the cross section for making boron-8 was much smaller than had been assumed. Fowler's group at Caltech went through that with great care and even I looked at it a little and we all came to the conclusion that it was all right and that there were no mistakes there.

This investigation was summarized by Bahcall and others in a very big paper in *Reviews of Modern Physics* in 1982 and you get the feeling that everything is all right about the theory.[2,3] So still we have this discrepancy. They invented something called the solar neutrino unit (SNU) — doesn't matter what that is — and the calculation says there should be $7.6(1 \pm 0.4)$ SNU and the experiment gave 2.1 ± 0.3 SNU. Some of these 7.6 come from boron-8 and so it was desirable to find a way to measure the other neutrinos. For this purpose, for a long time, it was suggested one should use gallium-71. This transforms into germanium and after many, many years of argument two experiments are now going on using gallium.

Gallium is unfortunately very, very expensive. There are two experiments. One is in Russia in the Caucasus. Baksan is the name of the laboratory and this experiment actually has been going on since the beginning of this year and even a little longer. A preliminary report was given in the spring of this year. The other experiment is a joint experiment of Germany, Italy and France in the Gran Sasso in Italy. This is also going on but so far there is no report. Obviously the Russians are richer than the West Europeans. The West Europeans have 30 tons of gallium. The

Table 1.   Predicted rates in SNU.

	pp	pep	^{7}Be	^{8}B	Total
^{37}Cl	0	0.2	1.0	6.0	7.6
^{71}Ga	71	3	34	14	132

Russians now have 30 tons but expect to get another 30 tons. We in America are much poorer and couldn't afford any of these experiments, but we are collaborating in the SAGE experiment in Russia.

In Table 1 the predicted rates in SNU are shown.

The total predicted for chlorine is 7.6 SNU, of which 6 come from boron-8, 1 from beryllium, 0.2 from proton–proton and the remainder come from the carbon–nitrogen reactions which I have not mentioned.

For gallium the total expected is 132 SNU. Relatively little comes from boron, a lot from the proton–proton reaction and a lot from beryllium. This is something you should keep in mind. As I mentioned already, if there were no boron-8 the rest would be 1.6 SNU, which is very close to the observed 2.1 SNU. The first and most natural hypothesis is that the boron-8 for some reason isn't there. Well, how could that be? It could be, if the sun had a lower temperature near its centre. The standard solar model says that the central temperature is a little over 15 million degrees. If you lower that to something over 13 million degrees then the boron-8 production would go down tremendously, since it goes with the 18th power of the temperature and so everything would be fine.

The trouble is that if you lower the central temperature then you also lower the rate of the proton–proton reaction on which everything depends. We know that rate very well, because after all we get so much energy, so much radiation from the sun, and that is essentially the rate of the proton–proton reaction. This is just saying in one word what I said previously, namely looking at all the possibilities Bahcall and others found it just can't be done. There is just no way how the astrophysical calculations about the sun could have been so wrong as to give this result.

4. Neutrino Conversion

Maybe the trouble is not in the sun. If it isn't the sun, it could be the neutrinos. The sun makes enough neutrinos, but something happens to the neutrinos. For quite a long time people have said, well, there are three types of neutrinos just as there are three types of leptons. Leptons, as you know, are charged particles which do not have strong interaction. We know three leptons, namely electron, μ-meson and τ-meson. The electron has a mass of about half a million eV, a μ-meson about 100 MeV and a τ-meson 1700 MeV. To each type of lepton there should belong a neutrino. In fact, we have known that since the 1960's and the 1988 Nobel Prize was given for that discovery.

There should be three types of neutrinos — electron neutrino, μ-neutrino and τ-neutrino — and their masses should go with the lepton mass, that is the mass of the τ-neutrino should be bigger than that of the μ-neutrino, which in turn is bigger than the mass of the electron neutrino. There is no experimental confirmation of that; it is just reasonable to assume that. First of all it is reasonable to assume that neutrinos, like any other particle, have some mass and by mass, of course, I mean rest mass. It is reasonable, in analogy with other particles, to assume that neutrinos in vacuum are not the same as the neutrinos which you get from nuclear reactions. These two assumptions are reasonable to make.

We know that from the K-meson for instance. If the neutrinos in vacuum are called ν_1, ν_2, ν_3, then ν_1 is almost equal to the electron neutrino but not equal. If that is so we should be able to observe in the laboratory oscillations of neutrinos, where you start from one flavour and then end up after some distance in the other flavour. This has not been observed in spite of many attempts to do so. I think at the end of this lecture you will understand why it has not been observed.

The decisive step was taken in 1985 by two Russian theorists, Mikheyev and Smirnov.[4] It is enough here to consider just two neutrino types, the e- and the μ-neutrino. Well, they say that the e-neutrino is some factor times ν_1 + some other factor times ν_2, and the factors are chosen so that the sum of the squares is one so as to normalize and the μ-neutrino is orthogonal to the electron neutrino.

$$\begin{aligned} |\nu_e\rangle &= \cos\theta|\nu_1\rangle + \sin\theta|\nu_2\rangle, \\ |\nu_\mu\rangle &= -\sin\theta|\nu_1\rangle + \cos\theta|\nu_2\rangle. \end{aligned} \tag{6}$$

Theta is known as the mixing angle. The ν_1 and ν_2, the free space neutrinos, have certain masses m_1 and m_2. If I then transform this to the representation of electron- and μ-neutrino I get a mass matrix whose important part is the second part where the difference of the mass squares comes in. In ν_1, ν_2 representation:

$$M = m^2 = \frac{1}{2}(m_1^2 + m_2^2)\begin{pmatrix} 1 & 0 \\ 0 & 1 \end{pmatrix} + \frac{1}{2}(m_2^2 - m_1^2)\begin{pmatrix} -1 & 0 \\ 0 & 1 \end{pmatrix}.$$

Transforming this to the ν_e, ν_μ representation gives

$$M = m^2 = \frac{1}{2}(m_1^2 + m_2^2)\begin{pmatrix} 1 & 0 \\ 0 & 1 \end{pmatrix} + \frac{1}{2}(m_2^2 - m_1^2)\begin{pmatrix} -\cos 2\theta & \sin 2\theta \\ \sin 2\theta & \cos 2\theta \end{pmatrix}$$

(the first line is ν_e, the second is ν_μ).

That's all well and good. Now, if you put the neutrinos in matter then the neutrinos will interact with the electrons and this interaction will come about both by the neutral weak interaction and by the charged weak interaction. The neutral weak interaction is the same for all types of neutrinos and therefore it doesn't change the relative masses, but the charged weak interaction gives a peculiar interaction for the electron neutrino:

$$\frac{G}{\sqrt{2}}\nu_e\gamma_\lambda(1+\gamma_5)\bar{e}e\gamma_\lambda(1+\gamma_5)\bar{\nu}_e = \frac{G}{\sqrt{2}}\nu_e\gamma_\lambda(1+\gamma_5)\bar{\nu}_e(e\gamma_\lambda\bar{e}+\cdots),$$

$$(7)$$

using the Fierz transformation. For electrons at rest only $\gamma_4 \neq 0$; for neutrinos

$$\gamma_4 = 1, \quad (1+\gamma_5) = 2; \quad e\gamma_4\bar{e} = N_e, \text{ so } V = G\sqrt{2}N_e. \quad (8)$$

The electron neutrino can exchange with an electron and that was discovered earlier by Wolfenstein,[5] which Smirnov made use of. This interaction is equivalent to giving the electron neutrino an extra mass due to the possibility of its interaction with the electron. This extra interaction is equivalent to a repulsive potential for electron neutrinos which is written down as the last equation. Now what does that mean? We know that generally in special relativity theory

$$k^2 + m^2 = (E - V)^2 \approx E^2 - 2EV.$$

If you have a repulsive potential then instead of energy squared you have energy minus potential squared (k is the momentum). If I neglect V squared, which is extremely small, this equation is equivalent to changing the mass of the electron neutrino, namely adding to it twice the neutrino energy times its repulsive potential,

$$m_i^2 = 2EV = 2\sqrt{2}GN_eE = 2\sqrt{2}G\frac{Y_e}{m}\rho E \equiv A. \tag{9}$$

The repulsive potential we know from Wolfenstein; G is the universal Fermi interaction, next is the density of electrons which is the ordinary density times Y_e, the ratio of electrons to nucleons. I call that product A, which is an extra mass squared for the electron neutrino. Thereby the mass matrix, which I wrote down before, gets an extra term:

$$M = m^2 = \frac{1}{2}(m_1^2 + m_2^2)\begin{pmatrix} 1 & 0 \\ 0 & 1 \end{pmatrix} + \frac{1}{2}\begin{pmatrix} A - \Delta\cos 2\theta & \Delta\sin 2\theta \\ \Delta\sin 2\theta & -A + \Delta\cos 2\theta \end{pmatrix},$$

where $\Delta = m_2^2 - m_1^2$. I have written it in such a way that its trace is zero, with $(1/2)A$ for the electron neutrino and $-(1/2)A$ for the μ-neutrino.

The quantity A is proportional to the density of matter. Therefore the electron neutrino mass changes with density — with the matter density. It is useful just to plot as in Fig. 1 the masses against the density of matter. At low density the μ-neutrino has a higher mass than the electron neutrino. At high density the reverse is the case. I get a curve, that's the upper mass level, which at high density represents an electron neutrino, at low density a μ-neutrino. What is going to happen? A neutrino — an electron neutrino — is emitted at high density in the sun and as the neutrino goes out of the sun it slides down this curve and when it comes out of the sun it isn't an electron neutrino, it's a μ-neutrino. You can't observe μ-neutrinos of low energy, because the only way you could observe them is if they interacted with the nucleus and got converted back into a μ-meson. In order to do that it would need an energy of 100 MeV. None of the neutrinos from the sun have that much energy, so with any detector like chlorine or gallium it is impossible to observe μ-neutrinos. It will be possible for neutrinos to be formed as observable electron neutrinos at high density in the sun, but when they come out they are μ-neutrinos and cannot be observed. In principle, it would be possible that we might not observe any neutrinos from the sun at all. In fact, things are not as bad as that; we do observe some neutrinos.

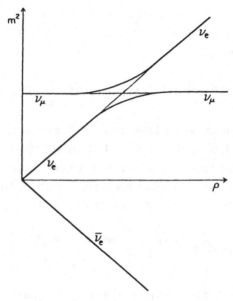

Fig. 1.   The masses of two flavours of neutrinos as a function of matter density. The curves
nearly cross for a particular value of the density. The mass of the electron antineutrino is
also shown.

Antineutrinos have an attractive potential and therefore their effective
mass will go down with increasing density and so there will not be any
crossing, any resonance, between μ- and electron antineutrinos. The change
is of course a resonance and this resonance exists for neutrinos and not for
antineutrinos. That is the theory by Mikheyev and Smirnov of 1985 and
the simplest hypothesis now is again to say let us eliminate the boron-8
neutrinos. Suppose all the boron-8 neutrinos are converted into μ-neutrinos
as they go through the sun and through the resonance. This is the conclusion
I drew in 1986 and taking the energy of the boron neutrinos to be 10 MeV and
using the formula which I had before — well, it's the resonance condition —
I find that the difference of the squares of the masses of the two neutrinos
has to be less than 10^{-4} (eV)^2. So assuming that the μ-neutrino mass is
much bigger than the electron neutrino mass this means that the mass of the
μ-neutrino is less than 10^{-2} eV compared with 500,000 eV for the electron
itself. A tiny, tiny mass but not zero.

All right, well, that would be a reasonable hypothesis. You then have to
assume further that the beryllium neutrinos are not converted, so the Δm^2

must be greater than 10^{-5} (eV)^2 for that purpose. In other words beryllium neutrinos would be formed below the resonance energy (Fig. 1), below the resonance density, and they would remain electron neutrinos all the way.

About the same time as I did that analytic work, Rosen and Gelb confirmed this solution by numerical calculation but they also found a second solution which also would explain the chlorine observation.[6] The second solution would have a lower difference of mass between the two neutrino types and a larger mixing angle (θ). In the second solution the neutrinos from boron-8 are only partially converted and so are those from beryllium-7. Figure 2 shows the result by Rosen and Gelb. Plotted horizontally is the energy in MeV and vertically is the probability for a neutrino of that energy to arrive on earth. And you see the stars are the solution which I found. In that solution Δm^2 is 10^{-4} (eV)^2 and boron-8 neutrinos which are high energy have very little chance to arrive on earth, but low energy neutrinos have a very good chance.

In the other solution, the Rosen–Gelb solution, the reverse is the case. How does this other solution work? This has been investigated by a number of people. The first one was Haxton at Seattle[7] and then Parke at Fermi National Laboratory.[8] The best solution was given by Pizzochero,[9] a very young Italian physicist who has been working at Stony Brook for the last

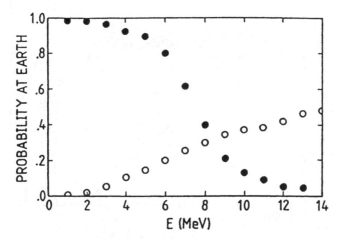

Fig. 2. Probability that electron neutrinos from the sun arrive at Earth vs. neutrino energy, E. The symbol ○ is for $\Delta m^2 = 1.1 \cdot 10^{-6}$ eV^2; ● for $\Delta m^2 = 1.0 \cdot 10^{-4}$ eV^2 (sin^2 $2\Theta = 10^{-1.5}$ has been assumed).

two years. He says, let us assume that in the sun the density is an exponential function of the distance from the centre. And that is very nearly true. The distance, R_s, over which the density decreases by a factor e is accurately known for the sun to be $R_s = 6.6 \cdot 10^9$ cm (66,000 km). With this density distribution a neutrino of a given flavour will survive the resonance, that is, remain of the same flavour — an electron neutrino remains an electron neutrino with a probability $p = e^{-C/E}$, where E is the energy of the neutrino and the constant is very simply given by $C = \pi R_s \Delta m^2 \sin^2 \theta$. That is the general theory. Unfortunately, nobody knows and quotes Pizzochero's paper but they quote the other papers, which are more complicated, give a more complicated result and are less accurate.

Most people have done numerical calculations and Fig. 3 shows one of the early and very good numerical calculations. Plotted here are the contours which would give you a certain number of SNU's in the Davis experiment. Remember Davis observed 2.1 SNU and the central curve here is precisely that. And you can have your choice. Horizontally plotted is the mixing angle and vertically the Δm^2. You can for 2.1 SNU have three different alternatives. The first solution is constant Δm^2. The second is a sloping solution where $\Delta m^2 \sin^2 \theta$ is constant, the Rosen–Gelb solution.

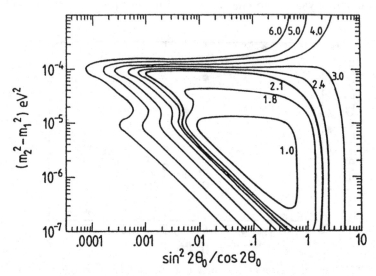

Fig. 3.   Expected results for the chlorine experiment. Iso-curves in SNU in dependence of neutrino mass and mixing angle.

And finally a solution, particularly found by the French group, with a large mixing angle and fairly arbitrary mass difference. I think it is unlikely to be true because everything we know about other particles. It is true that the free particle and the particle which comes out of interactions are different, but the mixing angles generally are small. I remind you of the Cabbibo angle which is the mixing angle for weak interactions. The Cabbibo angle has a $\sin^2 \theta = 0.05$, so I consider it thereotically very unlikely that for neutrinos it should be different and that the mixing angles should be very large as this solution indicates.

From the theoretical point of view I say the alternatives are the horizontal line and the sloping line in Fig. 3. From the Davis experiments you couldn't tell whether you are on the horizontal line or on the sloping line. But then came the new experiment — real information always comes from experiments and we theorists only have to read the experimental results and try to make the best of them.

5. A New Experiment

The new experiment was done at Kamioka in Japan by a big team, most of them Japanese, but some of them American. In this laboratory one had observed the neutrinos from the supernova explosion SN1987A. They also can observe neutrinos from the sun. How do they do it? They have a big water tank about the size of this room and in this water tank — very pure water — the neutrinos collide with electrons. They project electrons and then you can measure both the energy and the direction of these electrons. And very reasonably the Kamiokande people say, let's just take those electrons which come almost from the sun, not quite because of course the electrons do not go exactly in the same direction as the neutrinos, which come almost from the sun. They take an angular distribution and subtract the background which is averaged over all directions and then they get some excess electrons which come more or less from the sun and that is what they report.[10] And there are not 50 authors but only 38. They have a minimum electron energy of 7.5 MeV and therefore they can only observe boron-8 neutrinos and they observe lots and lots of counts from which it follows that boron-8 neutrinos exist, are made, and come to the earth. So the first explanation, namely that the boron-8 is absent, is just wrong. This one experiment proved the hypothesis to be wrong. Therefore no matter what you do to the distribution of temperature in the sun you cannot explain the solar neutrino experiment. It cannot be explained by any strange assumptions about the sun. Also it cannot be explained by taking the horizontal line in Fig. 3, because in that case the boron-8 neutrinos would be converted into μ-neutrinos and be unobservable. That leaves only the slanting line.

The two experiments together, Kamiokande and Davis, show you that you have to be on the slanting line and show you also that you have to attribute the result to a conversion of neutrinos into unobservable ones. You cannot explain it by assuming that the sun doesn't behave right. The Kamiokande people put their result in terms of the ratio of what they observed to the prediction from the standard solar model. And they say that the ratio is 0.46 ± 0.06 (systematic error) ± 0.05 (statistical error). They observed for a little over 1000 days. The chlorine experiment on the other hand gave a ratio to the prediction of not 0.46 but 0.27 ± 0.04. These two numbers are different and therefore you have to explain that.

Thus the conclusions:

1. Boron-8 neutrinos do reach the earth.
2. The MSW theory (Mikheyev–Smirnov–Wolfenstein) with relatively high Δm^2 is excluded. Therefore that leaves the MSW with low Δm^2 — less than 10^{-5} $(eV)^2$. It is significant that the chlorine experiment gives a lower ratio to prediction than Kamiokande, which precisely is what the slanting line gives you. I showed you the prediction by Rosen and Gelb. The low energy neutrinos have less probability to remain electron neutrinos than the high energy neutrinos. That is the striking and important result.

6. A Proposed Solution

Well, seeing this, Bahcall and myself wrote a paper[11] which should be published in *Physical Review Letters*. We use the probability of escaping the resonance. According to Pizzochero, this probability is $p = e^{-C/E}$. We determined the constant C to give the observed rate in the chlorine experiments and that comes out to be $C = 10.5 \pm 5.5 \, \text{MeV}$. The reason why it is so uncertain is that the standard solar model doesn't give the amount of boron-8 very accurately and so we need that large uncertainty.

So this being so and taking now the beryllium neutrinos which have an energy of about 1 MeV, the survival probability ($e^{-C/E}$ with $E = 1$ MeV) is negligible or just exceedingly small. So we do not get any beryllium neutrinos to the earth. We do not get any proton–proton neutrinos to the earth. They all become μ-neutrinos. I shall modify that a little bit in a few minutes. But in first approximation that's it. Now if we take the constant C from the Davis chlorine experiment we can then calculate what the Kamiokande result should be and the answer is 0.46. Exactly equal to the measurement and it has no reason to be that exact, since I showed you that there was an experimental error of 0.06 or so. We don't think that this can be an accident and we think that this is a very good way to verify the sloping solution, the Rosen–Gelb solution.

I said already that with the probability being $e^{-C/E}$, the probability of getting any beryllium neutrinos is very, very small, $p = 5 \cdot 10^{-6}$. What about the gallium experiment? Bahcall, in particular, made propaganda for years to have the gallium experiment done. But now we say, if this is the solution then the gallium experiment won't give any answer. In fact, even in gallium, most of the counts will be due to boron-8 neutrinos. At this stage we would say that in SAGE and GALLEX we would get only 4 SNU instead of 130, about 3% of the theoretical prediction. I'll come back to that in a minute. Now using the constant C we can calculate the product

$$\Delta m^2 \sin^2 \theta = (1.0 \pm 0.5)10^{-8} \ (\text{eV})^2. \tag{11}$$

That neutrino mass is not very large and therefore that's the reason why it is essentially impossible in the laboratory to observe the oscillation of neutrinos between different flavours. Remember, that Δm^2 presumably is the mass of the μ-neutrino. The mass of the electron neutrino will be much, much less still.

If I take for θ the Cabibbo angle, then $\Delta m^2 = 2 \cdot 10^{-7}$ (eV)^2, which means that the μ-neutrino will have a mass of about half a milli-electron volt. I said that this prediction was not quite correct because we must remember that from the sun to the earth the neutrinos have to go through vacuum. At the resonance in the sun the electron neutrinos, let us say coming from beryllium, are converted into μ-neutrinos, but when they come out into vacuum, I must again say that they are partly ν_1 and partly ν_2. And then the ν_1 and ν_2 will propagate in space and when they arrive at the earth they can again be observed, and when they are observed I am interested in what's the likelihood that they go back to electron neutrinos. Well, it's easy to calculate that the probability is

$$\Phi = 2 \sin^2 \theta \cos^2 \theta. \tag{12}$$

The total flux of converted neutrinos, namely those which have become μ-neutrinos, is $132 - 5 = 127$, which were preserved. Therefore the predicted count of gallium is really $4.5 + 2 \cdot 127 \sin^2 \theta \cos^2 \theta$. We don't know the value of the mixing angle θ. That's very nice, because if they can measure accurately enough in the gallium experiment then they can determine this number and if they can do so then we shall know the mixing angle θ. We know already the product $\Delta m^1 \sin^2 \theta$ and then we also know the Δm, the difference in mass.

It will be quite difficult to do the gallium experiment accurately enough, but I have talked to the American collaborators in the SAGE experiment, which is the Russian experiment, and they say, well, it's difficult but we'll try. If $\sin^2 \theta$ for instance is 0.05, then instead of 4 SNU they should observe 16 SNU. That still is difficult. At the present time the SAGE experiment has given a preliminary answer. Namely they answer, we see nothing which is very nice and in agreement with our opinion and the maximum is maybe 50 SNU. It's a long way from 50 to 16 in the presence of background, but I think there is hope.

There is also an upper limit to Δm^2 which says it is certainly less than 10^{-4} (eV)^2. We are very hopeful that the gallium people actually can do as well as I would like them to do and can tell us the mixing angle. If they can then we shall know for the neutrino, the μ-neutrino probably, both the mass and the mixing angle and that would be, of course, a very important result.

Note Added in Proof

By now (May 1993) results have been published by both GALLEX[16] and by SAGE. GALLEX finds

$$83 \pm 19 \text{ (stat)} \pm 8 \text{ (syst) SNU.} \tag{13}$$

This is well below the prediction from the solar model, 131 ± 6 SNU. Their result is fit well by the MSW theory with

$$\Delta m = (2 \pm 0.6) \times 10^{-3} \, \text{eV}. \tag{14}$$

The SAGE result is somewhat lower than (13), but within the given limits of error.

7. Final Comments

Two small additions. Davis has claimed that there is a correlation between his neutrino measurements and sun spots.[12] The Japanese group has, for a thousand days, when the solar activity increased 15-fold, found no such correlation. They have measured first at a very quiet sun and now at a very active sun and found no difference. Since I like there to be no correlation I believe Kamiokande. At present there are just two experiments contradicting each other. A number of Russian theorists have tried to explain the Davis result of correlation to sun spots by giving the neutrino a complicated kind of magnetic moment.[13] One important experiment in the planning stage and approved is the so-called SNO experiment in Sudbury, Ontario. Using a big tank of heavy water rather than light water, you can disintegrate the deuteron by the neutral weak current giving proton plus neutron or by the charged current giving proton plus proton and that would be very interesting.

They can also see electron recoils which give different spectra from μ-neutrinos and from electron neutrinos, so they will be able to confirm or dispute our idea that many of the neutrinos will be converted. Once you get the total number of neutrinos by these experiments you can determine how much boron-8 is actually produced in the sun and thereby you can confirm the standard solar model. And that essentially was the aim of the neutrino experiment. People wanted to see to the centre of the sun and neutrinos permit you to do that. Already now Bahcall has calculated that the standard solar model is very nicely confirmed. But only to $\pm 15\%$.

Finally some theory. There is a theory which is known as the see-saw mechanism, which was proposed independently by two groups.[14,15] That says that the mass of a neutrino of flavour i is proportional to the square of the mass of the corresponding charged particle, chosen to be a lepton or a quark:

$$m(\nu_i) = \frac{m^2(q_i)}{M_{GUT}},$$

where M_{GUT} is the mass of the Grand Unified Theory.

If we believe the mass of the μ-neutrino to be $5 \cdot 10^{-4}$ eV and then choose for the charged particle not the muon but the charmed quark with mass $m(q_{ch}) = 1.5$ GeV, then the mass of Grand Unified Theory would be of the order of 10^{13} GeV, which is the sort of thing that particle theorists like. If that theory is right, then the mass of the electron neutrino would be 10^{-8} eV, so it is not very likely for you to find that by looking at the upper limit of the spectrum of electrons from the decay of hydrogen-3. The mass of the τ-neutron may be 10 eV.

I have tried to show you that the problem of the solar neutrinos can be understood. That it's done in terms of the MSW theory. That we know then accurately the product of the mass of the μ-neutrino times the sine of the mixing angle and that we can hope with some luck that the gallium experiment will tell us the mixing angle itself and thereby will tell us everything at least about the μ-neutrino. That still is not the whole story but it is a nice part of the story.

References

1. R. Davis Jr., B. T. Cleveland and J. K. Rowley, *AIP Conf. Proc. Nr* **72** (1981).
2. J. N. Bahcall, W. F. Huebner, S. W. Lubow, P. D. Parker and R. G. Ulrich, *Rev. Mod. Phys.* **54**, 767 (1982).
3. J. N. Bahcall, B. T. Cleveland, R. Davis and J. K. Rowley, *Ap. J.* **292**, 79 (1985).
4. S. P. Mikheyev and A. Y. Smirnov, *Nuovo Cimento* **C9**, 17 (1986).
5. L. Wolfenstein, *Phys. Rev.* **D17**, 2369 (1978).
6. S. P. Rosen and J. M. Gelb, *Phys. Rev.* **D34**, 969 (1986).
7. W. C. Haxton, *Phys. Rev. Lett.* **57**, 1271 (1986).
8. S. J. Parke, *Phys. Rev. Lett.* **57**, 1275 (1986).
9. P. Pizzichero, *Phys. Rev.* **D36**, 2293 (1987).
10. K. S. Hirata *et al.*, *Phys. Rev. Lett.* **63**, 16 (1989); **65**, 1297 (1990).
11. J. N. Bahcall and H. A. Bethe, *Phys. Rev. Lett.* **65**, 2233 (1990).

12. B. Cleveland *et al.*, *Proc. 25th Int. Conf. on High Energy Physics* (Singapore, 1990) (World Scientific, 1991), p. 667.

13. M. B. Voloshin, M. I. Vysotsky and L. B. Okun, *Sov. Phys. JETP* **64**, 446 (1986).

14. T. Yanagita, *Prog. Theor. Phys.* **B315**, 66 (1978).

15. M. Gell-Mann, P. Ramond and Slansky, in *Supergravity* (North-Holland, Amsterdam, 1979).

16. GALLEX collaboration, *Phys. Lett.* **B285**, 376 and 390.

SUPERNOVA THEORY

Hans A. Bethe

Newman Laboratory of Nuclear Studies
Cornell University
Ithaca, New York 14853, USA

Supernovae are very spectacular phenomena. In a short time, a supernova produces as much energy as the sun in its entire lifetime of about 10^{10} years, namely about 10^{51} ergs (1 foe for fifty-one ergs). It is generally agreed that all elements heavier than helium came into the world from supernova explosions. There are two types of supernovae: type I originates from white dwarfs, type II from massive stars ($\geq 8M_\odot$, where $M_\odot$ is the mass of the sun) at the end of their life. Only type II will concern us here.

Nuclear reactions provide the energy of stars, converting H to He, then to C and O, then to Si, and finally to Fe or some element near it. The reactions proceed fastest near the centre of the star. Once Fe has been reached, no further energy can be extracted from the nuclei. Pressure is then provided in the core by the Fermi pressure of electrons. But there is a limit to this, the Chandrasekhar mass,

$$M_{Ch} = 5.8Y_e^2 M_\odot, \tag{1}$$

where Y_e is the number of electrons per nucleon. (There are some corrections to this, for various reasons.) Typically, in the core of a star late in its evolution, $Y_e = 0.45$, and after corrections, $M_{Ch} = 1.3M\odot$. Once the mass of the iron core exceeds this limit, the core collapses due to gravitation.

The collapse is nearly adiabatic and can therefore be calculated easily. However, one change takes place: electrons are captured by nuclei because their Fermi energy exceeds the energy which has to be added to a nucleus to convert one of its protons into a neutron. In this capture, a neutrino is emitted. If this process went on indefinitely, Y_e would become very small, and so would the Chandrasekhar mass, (1). Fortunately, however, once the density exceeds about 10^{12} g cm^{-3}, neutrinos get trapped because they are scattered by the nuclei present, a process due to the weak neutral current. The trapped neutrinos cause the reverse reaction: they are captured by nuclei, and an electron is emitted. An equilibrium is reached, and the final lepton fraction (number of electrons plus neutrinos per nucleon) is about $Y_L = 0.38$, giving (after suitable corrections, including general relativity)

$$M_{Ch} \simeq 0.7 \, M_\odot. \tag{2a}$$

Neutrinos are also scattered by electrons, which decreases their energy. They can then escape trapping more easily, as shown by Bruenn and others; Y_L is decreased to about 0.36 and

$$M_{Ch} \simeq 0.6 \, M_\odot. \tag{2b}$$

The collapse is stopped when the density at the centre becomes appreciably bigger than nuclear density,

$$\rho_0 \simeq 2.5 \cdot 10^{14} \, \text{g cm}^{-3}. \tag{3}$$

Then nuclei merge and form continuous nuclear matter which is quite incompressible. Infall is stopped and a pressure wave moves out. This turns into a shock wave when it reaches the surface of the Chandrasekhar sphere; the inside of that sphere remains unshocked. The material outside the sphere bounces back vigorously. This is called the "prompt shock".

The prompt shock starts moving with a velocity of $2-3 \times 10^9$ cm/s, then gradually slows down. It dissociates the existing heavy nuclei into nucleons. This process costs a lot of energy, about 9 MeV per nucleon $= 8.5 \cdot 10^{18}$ ergs per g. In addition, once that dissociation takes place, the protons eagerly capture electrons, emitting neutrinos and thus losing further energy. The shock has only limited energy, perhaps 10^{52} ergs, and so it can penetrate only a limited mass, about 10^{33} g (0.5 solar masses); but this number is quite uncertain. It is therefore important that the Chandrasekhar mass of the unshocked core be large. We found above that this mass is about $0.6 M_\odot$ so that the shock can penetrate out to about $1.1\ M_\odot$.

Computations have shown that the shock works better if general relativity is used. This is because, in general relativity, strong gravitational effects are enhanced, and one gets stronger collapse followed by stronger rebound. In fact, Baron, Cooperstein and Kahana (1985) obtained shocks which penetrated through the iron core. It was believed that if the shock penetrated the Fe, it would go through the entire star and expel its mass in a supernova display. The reason for this belief was that in the Si + S region surrounding the core, and even more in the O region, nuclear reactions were likely to release energy rather than absorb it. This assumption was never put to the test because the shock never progressed this far in the computation, but the question has now become irrelevant.

Unfortunately, the calculations of Baron *et al.* were invalidated by Bruenn's results on the collisions between neutrinos and electrons during the infall. As I mentioned earlier, these collisions interfere with the trapping of neutrinos, and substantially reduce the mass of the unshocked core. With this reduced mass, the shock gets stuck much earlier. Baron and Cooperstein (1990) found that even with a soft equation of state for nuclear matter, and with general relativity, the prompt shock will fail for Fe core masses above $1.1\ M_\odot$, and even for this low mass it will work only if some other conditions are also favourable. On the other hand, the calculations of pre-supernova evolution now appear to be rather firm. In spite of some corrections in the physics, the mass of the Fe core remains the same as it has been since 1985, viz. about $1.3\ M_\odot$. Thus it seems rather definite that the prompt shock will not work.

James Wilson in 1982 discovered an alternative mechanism. He left his computer running overnight, and found that the shock, previously stalled, revived. His analysis showed that this was due to the absorption of neutrinos at intermediate distances, around 100 km. These neutrinos are emitted by the core, which will ultimately become a neutron star. At this stage, the core contains the energy set free by the gravitational collapse, typically $3 \cdot 10^{53}$ ergs or 300 foes. This energy is now mostly in the form of heat, with temperatures ranging up to 50 MeV or about $5 \cdot 10^{11}$ K. The core has to get rid. of this heat in order to become a neutron star. The only way to do so is by emitting neutrinos. Photons have much too strong an interaction with ordinary matter: they can get out of the star only when the matter itself is ejected.

The core settles down into a sphere with a radius of some 20 km in which neutrinos diffuse in much the same way as photons do in ordinary stars. From the surface of this sphere neutrinos are emitted. It is called the neutrino sphere, in analogy to the photosphere of an ordinary star. At the neutrino sphere, the density is of the order of 10^{11} g cm^{-3}, and the temperature of the order of 5 MeV. Beyond this sphere, the neutrinos essentially move freely. However, they still have a certain chance to be absorbed by protons and neutrons, and give their energy to the matter in the process. Thereby this matter gets heated, develops pressure and drives the shock out.

This mechanism of Wilson gave successful supernova shocks. A characteristic feature was the formation of a bubble, a large region between neutrino sphere and shock front which contained very little matter, but had high pressure caused by a high intensity of electromagnetic radiation and electron pairs. This bubble propels the shock.

While the Wilson theory gave a successful shock, it had three drawbacks:

1. The shock was successful in some computer runs, but not in others. In 1989, Wilson and Mayle concluded that the flux of neutrinos from the core was not quite sufficient; so they artificially increased it by about 50% by postulating "salt finger" convection inside the core. As later developments have shown, this postulate is no longer necessary.
2. The explosion energy was found to be about 0.4 foe, whereas the observations on Supernova 1987A indicate 1.5 foe; so the mechanism gives insufficient energy.
3. The bubble has very low density, much lower than the material outside which is to be propelled in the shock; density ratios of 10 or more are common. Such a configuration is subject to Rayleigh-Taylor instability.

The third drawback has led to resolution of the problem. Rayleigh-Taylor instability itself does not work: it would lead to mixing of the material on the two sides of the interface between the bubble and the outside, but these materials are of the same kind — nucleons, electrons and photons. Instead, convection would take place. The condition for convection is that entropy decreases with increasing radius. This is well fulfilled: the entropy in the bubble equals several hundred while near the shock it is only of order 10.

As we know from observations on the sun as well as from laboratory experiments, convection takes place in cells; the hot material goes up on one side of the cell, the cold material down on the other. After moving a certain distance, the mixing length, the hot and the cold stream get mixed, and in the next cell above the same procedure occurs. In standard stellar evolution calculations, a very small entropy gradient suffices to drive convection. But in the supernova, where everything happens in seconds, much higher entropy gradients are needed. In an example, I found that the entropy (in units of k_B per nucleon) has to change by about 50 units from the shock to the neighbourhood of the neutrino sphere.

The first task for the neutrino heating plus convection is to turn the shock wave around. After the prompt shock stalls, material accretes to the core because the matter originally at large distance continues to fall in. This accreting material moves inward behind the shock, albeit with much smaller velocity than it had outside the shock. The rate of accretion determines the product

$$\frac{\lambda}{4\pi} = \rho r^2 v. \tag{4}$$

Knowing the mass distribution before the supernova collapse, and assuming more-or-less free fall, we find that

$$\rho r^2 v = 2 \cdot 10^{31} t^{-1} \, \mathrm{g\,s^{-1}}. \tag{5}$$

In computations, one finds typically that

$$v = 5 \cdot 10^8 \, \mathrm{cm\,s^{-1}}, \tag{6}$$
$$\rho r^2 = 4 \cdot 10^{22} \, \mathrm{g\,cm^{-1}}. \tag{7}$$

When the accreting material gets heated by neutrinos, the first task the heat has to accomplish is to turn the inward motion v into an outward motion. To examine the requirement for this, we use the Hugoniot relations,

$$(u_2 - u_1)^2 = (p_2 - p_1)(V_1 - V_2), \tag{8}$$
$$2(E_2 - E_1) = (p_2 + p_1)(V_1 - V_2). \tag{9}$$

Here quantities with subscript 1 refer to the outside of the shock, with subscript 2 to the inside. E is the internal energy per unit mass, p the pressure, $V = \frac{1}{\rho}$ the specific volume, and u the outward velocity, so $u_2 = -v$. We want to increase u_2 (algebraically), and since u_1 is large and negative, this means increasing $(u_2 - u_1)^2$. It is clear from (8) and (9) that this requires a sufficiently large $E_2 - E_1$. In order to make $u_2 > 0$, i.e. turn around the motion behind the shock, we must have

$$E_2 \geq E_2\,_{\min} = E_1 + \frac{1}{2}u_1^2 + p_1(V_1 - V_2). \tag{10}$$

For a quantitative estimate, I used a computation by Wilson and Mayle in which the shock was at $R = 217\,\mathrm{km}$. It gave

$$E_2\,_{\min} = 10.6 \cdot 10^{18}\ \mathrm{ergs\ g^{-1}}, \tag{11}$$

while the actual internal energy was

$$E_2 = 9.2 \cdot 10^{18}. \tag{12}$$

Since $E_2 < E_2\,_{\min}$, $u_2 < 0$, i.e. the material behind the shock moves inward. The energies E_2 and $E_2\,_{\min}$ may be compared with the gravitational potential,

$$GM/R = 9.9 \cdot 10^{18}. \tag{13}$$

It is not surprising that E_2 is about equal to GM/R because the material outside the shock has acquired its internal and kinetic energy by falling down in the gravitational field. In order to turn the shock around, we need to raise the internal energy behind the shock by about 15%. This we propose to do by convecting energy from inside the shock.

According to the mixing length theory of convection, the velocity of the energy stream relative to the material is

$$v_c = \left(\frac{p\, d\ln S}{\rho\, d\ln p}\right)^{\frac{1}{2}}. \tag{14}$$

The first factor in this is essentially the sound velocity. The entropy gets built up by neutrino absorption. The first requirement is that v_c becomes greater than v, the inward velocity of the material. Once this happens, the entropy generated by neutrino absorption moves outward rather than inward. I found that this requires raising S by 5 units, which can be accomplished in about 0.05 seconds.

The second requirement is to bring enough energy to the inside of the shock to turn it around. According to our discussion above, 0.2 GM/R per unit mass should be enough. Multiplied by the accretion rate λ, this is about 0.5 foe/s. It requires an entropy maximum about 10 units above the entropy behind the shock.

There is one process which interferes with the buildup of energy and entropy: the capture of electrons by nuclei. The rate of energy loss due to this process is

$$-\dot{w}_e = 2.0 \cdot 10^{18} T^6 \text{ ergs g}^{-1} \text{ s}^{-1}. \tag{15}$$

Here a factor T^4 arises from the energy density of electrons and positrons, and a factor T^2 from the cross section. The energy absorbed from neutrinos is

$$\dot{w}_\nu = 5.3 \cdot 10^{19} F_{52} r_7^{-2} \text{ ergs g}^{-1} \text{s}^{-1}, \tag{16}$$

where F_{52} is the flux of neutrinos in units of 10^{52} ergs s^{-1}. Energy gain from neutrinos equals energy loss by electron capture when

$$T^3 r_7 = 5.1 \, F_{52}^{\frac{1}{2}}. \tag{17}$$

We call this the gain point, R_g. At larger r the temperature T decreases rapidly, so the energy loss by electron capture is less than the energy gain. The material in the star may then be divided into three regions: (1) from the neutrino sphere to the gain point, (2) from the gain point to the entropy maximum, and (3) from that maximum to the shock. In region I all the energy gained from the neutrinos is lost by electron capture, in region II the energy gain dominates and entropy is built up, and in region III there is convection. All the energy generated in region III is convected to the shock front. So the *energy* available to the *supernova* shock is just the *total energy supplied by neutrinos to the region beyond the entropy maximum* (corrected for electron capture).

The energy gain, per unit neutrino flux, is

$$\alpha = F^{-1} \int \dot{w}_\nu d\tau = 5.3 \cdot 10^{-33} 4\pi \int \rho r^2 dr \, r_7^{-2} = 6.7 \cdot 10^{-18} \int \rho dr. \tag{18}$$

The radial integral over density can be calculated separately for small and for large shock radius R. For a small shock radius, $R = 300\,\mathrm{km}$, I assume that the density behaves as r^{-2}, as in Eq. (7); then

$$\int \rho dr \simeq \rho_g \tau_g. \tag{19}$$

This product can be calculated by combining (17) with the equation of state of a gas of electromagnetic radiation, which gives

$$\rho_g r_g = 2.6 \cdot 10^{16} \frac{F_{52}^{\frac{1}{2}}}{S_r}. \tag{20}$$

Using $F_{52} = 7$ and $S_\tau = 10.5$, and making a small correction for electron capture, we find that

$$\alpha = 2.4\%. \tag{21}$$

For a large shock radius, $R = 4000\,\mathrm{km}$, I have solved the equations of equilibrium explicitly. One of them is the hydrostatic equation

$$\frac{dp}{dr} = -\frac{GM}{r^2}\rho, \tag{22}$$

and the other gives the energy current due to convection,

$$J = 16\pi p r^2 \left(\frac{p}{\rho}\right)^{\frac{1}{2}} \left(\frac{d\ln S}{d\ln p}\right)^{\frac{3}{2}}. \tag{23}$$

I assumed that the pressure is dominated by radiation; then

$$\rho = 1.40 \cdot 10^{-11} \frac{p^{\frac{3}{4}}}{S}. \tag{24}$$

The result of the integration is given in Table 1. The table shows that ρr is nearly constant, so

$$\int \rho dr \simeq 0.35 \cdot 10^{15} \ln\left(\frac{R}{r_g}\right). \tag{25}$$

Thus neutrino absorption occurs about equally, all over the region from the gain point to the shock. Putting in all corrections, we find that

$$\alpha = 0.7\%. \tag{26}$$

Table 1.   Distribution of interesting quantities for shock
radius of 4000 km.

r(km)	$p_{24}r_8^2$	$\rho_8 r_7$	S	T (MeV)	$w_{24}r_8^2$
4000	20	0.67	10	0.32	45
2000	10.2	0.50	11.3	0.38	19.6
1000	6.4	0.40	13.9	0.48	10.2
500	4.6	0.35	18.1	0.62	6.1
250	4.0	0.32	24.7	0.84	5.0
125	3.6	0.32	33	1.17	3.8
62.5	3.7	0.35	43	1.65	3.4
31.2	4.0	0.42	53	2.4	2.8
15.6	5.6	0.63	64	3.7	3.0

Averaging the two results, at small and large shock radius, we get

$$\alpha_{av} = 1.6\%. \tag{27}$$

The total gravitational energy liberated when a mass of 1.6 $M_\odot$
collapses into a neutron star is about $3 \cdot 10^{53}$ ergs. This energy is emitted
about equally in each of the three neutrino flavours, ν_e, ν_μ and ν_τ; thus
$1 \cdot 10^{53}$ erg is in $\nu_e + \bar{\nu}_e$. This is confirmed by the measurement of the $\bar{\nu}_e$
from Supernova 1987A in Japan and Ohio. Then the predicted supernova
energy is

$$E_{th} = 1.6 \cdot 10^{51} \text{ ergs}. \tag{28}$$

The best number from the observation of the Doppler effect of the light
from 1987A is

$$E_{obs} = 1.5 \cdot 10^{51} \text{ ergs}. \tag{29}$$

The agreement is very good.

The computation in Table 1 also gives the energy per unit volume,
internal plus gravitational:

$$w = 3p - \left(\frac{GM}{r}\right)\rho. \tag{30}$$

This is seen to be positive for all r, indicating that all the material involved
can escape from the star. This is all the material which is involved in the

convection. Thus, the separation of ejecta from the neutron star should take place at the entropy maximum, or slightly inside it.

The energy density w can now be integrated over the volume, and the result is

$$E = 4\pi \int wr^2 dr = 4\pi R^3 p_s, \tag{31}$$

where p_s is the pressure at the shock. This can actually be proved, for $\gamma = 4/3$, by the virial theorem. Since $p_s = w_i/3$, with w_i the energy density,

$$E = \frac{4\pi}{3} R^3 w_i. \tag{32}$$

Thus the internal energy behind the shock is equal to the total energy in the ejecta, divided by the volume inside the shock. This relation has been used for many years in studies of nucleosynthesis.

Nucleosynthesis takes place in material which, before the supernova event, was ^{16}O, ^{28}Si or ^{32}S. Therefore, synthesis leads to the most strongly bound nucleus with equal numbers of neutrons and protons. This is ^{56}Ni. As the supernova expands, this nucleus decays into ^{56}Co and this in turn into ^{56}Fe. Ni is relatively short-lived (6 days), but Co has a half-life of 77 days. Its radioactive decay powered the light emission from SN 1987A after the first month or two, and indeed for a long time the light emission followed accurately a 77-day decay time. The two principal γ-rays from ^{56}Co, 0.84 and 1.25 MeV, were observed.

As we have seen, the supernova is powered by the absorption of neutrinos. These are released from the proto-neutron star rather slowly, in several seconds, as we know both from theoretical studies of their diffusion and from the observations in the neutrino detectors at Kamiokande and 1 MB. Therefore the energy of the shock wave also builds up rather slowly, over several seconds. But its velocity is quite high from the beginning, more than 10^9 cm/s (10^4 km/s), because it runs into very dilute material.

I believe we now understand qualitatively how a supernova of type II works. The way is now open to investigate many important questions. For example, How does the mechanism work as a function of the mass of the original star? What are the details of nucleosynthesis (which, after all, provides all the elements beyond He, including all those we live on)?

References

1. E. A. Baron, J. Cooperstein and S. Kahana, *Nucl. Phys.* **A440**, 744 (1985).
2. E. A. Baron and J. Cooperstein, *Ap. J.* **353** (1990).
3. S. W. Bruenn, *Ap. J.* **340**, 955; **341**, 955 (1989).
4. J. R. Wilson and R. L. Bowers, *Ap. J.* **263**, 366 (1982).

THE BIG BANG AND COSMIC INFLATION

Alan H. Guth*

Center for Theoretical Physics
Laboratory for Nuclear Science and Department of Physics
Massachusetts Institute of Technology
Cambridge, Massachusetts 02139, USA

A summary is given of the key developments of cosmology in the 20th century, from the work of Albert Einstein to the emergence of the generally accepted hot big bang model. The successes of this model are reviewed, but emphasis is placed on the questions that the model leaves unanswered. The remainder of the paper describes the inflationary universe model, which provides plausible answers to a number of these questions. It also offers a possible explanation for the origin of essentially all the matter and energy in the observed universe.

1. Introduction

It is a pleasure to visit Stockholm and to speak in honor of Oskar Klein, certainly one of the great theoretical physicists of the 20th century. All physicists are familiar with the many important developments that are identified with Klein's name, such as the *Klein*–Gordon equation, the *Klein* paradox, the *Klein*–Nishina formula, the Kaluza–*Klein* theory, and *Klein*–Jordan second quantization.

However, although these contributions are extraordinarily impressive, I have to admit that Oskar Klein is not nearly as well known as he should be, especially among the younger generation of physicists. I believe that many of these physicists, at least in my own country, could not tell you

*This work was supported in part by funds provided by the US Department of Energy (DOE) under contract #DE-AC02-76ER03069.

what Klein's first name was. And many are not even aware that all of the items on the list above were in fact done by the same person.

One cannot be certain how to explain this failure in public relations, but I have a theory, and a proposal to solve the problem. My theory is that the name "Klein" is just too simple, too short, too generic and too prosaic to capture anyone's attention. To investigate this theory scientifically, I did some sociological research — in the Boston telephone directory.[1] (Boston is clearly one of the most unusually average cities of the world, and in any case I had a copy of the Boston telephone directory in my closet.) The directory lists no less than 179 Klein's, ranging alphabetically from Adam Klein to Yasue Klein. Testing against a carefully selected control group, my research showed that there are 0 Gell-Mann's, 0 Feynman's, 0 Dirac's, 0 Einstein's, 0 Bohr's, 0 De Broglie's, 0 Schrödinger's, and only 1 Guth (i.e. me). Clearly, Prof. Klein's hard-working public relations agents have a serious problem to overcome.

In the face of this relentless uphill battle, I suggest that the situation can be significantly ameliorated by asking a simple question: if Oskar Klein were alive today, what would he use for a bitnet address? I contend that he would very likely do what so many of my friends with common last names have done — that is, he would append his first initial to his last name, giving

Oklein@sesuf51

If we all agree to follow this usage consistently, we can add the following *distinctive* terms to the physicists' vocabulary: the *Oklein*–Gordon equation, the *Oklein* paradox, the *Oklein*–Nishina formula, Kaluza–*Oklein* theory, and *Oklein*–Jordan second quantization. I have checked out this proposal with my professional friends on Madison Avenue, and they confirm that the strategy is likely to improve Prof. Klein's name recognition by at least 86%.

So, having settled that issue, I would now like to move on to the main topic of this talk, a subject which was of great interest to Oskar Klein: the very early history of the universe. First, I should explain that while I will be talking about cosmology, the perspective will be that of a particle theorist. In the late 1970's, I joined a small drove of particle theorists who began to dabble in studies of the early universe. We were motivated partly by the intrinsic fascination of cosmology, but also by new developments in elementary particle physics itself.

The particle physics motivation arose primarily from the advent of a new class of theories, known as "grand unified theories." These theories were invented in 1974, but it was not until about 1978 that they became a topic of widespread interest in the particle physics community. Spectacularly bold, these theories attempt to extend our understanding of particle physics to energies of about 10^{14} GeV (1 GeV = 1 billion electron volts $\approx$ rest energy of a proton). This amount of energy, by the standards of your local power company, may not seem so impressive — it is about what it takes to light a 100-watt bulb for one minute. The grand unified theories, however, attempt to describe what happens when that much energy is deposited on a single elementary particle. Such an extraordinary concentration of energy exceeds the capabilities of the largest existing particle accelerators by 11 orders of magnitude.

To get some feeling for how high this energy really is, imagine trying to build an accelerator that might reach these energies. One can do it in principle by building a very long linear accelerator. The largest existing linear accelerator is at Stanford, with a length of about 2 miles and a maximum energy of about 50 GeV. The output energy is proportional to the length, so a simple calculation shows how long an accelerator would have to be to reach an energy of 10^{14} GeV. The answer is almost exactly one light-year!

Unfortunately, both the US Department of Energy and NASA seem to be very unreceptive to proposals for funding a one-light-year-long accelerator. Consequently, if we want to see the most dramatic new implications of the grand unified theories, we are forced to turn to the only laboratory, to which we have any access, that has ever reached these energies. That "laboratory" appears to be the universe itself, in its very infancy. According to the standard hot big bang theory of cosmology, the universe had a temperature corresponding to a mean thermal energy of 10^{14} GeV at a time of about 10^{-35} second after the big bang. No wonder particle theorists suddenly became interested in the very early universe.

The first half of this lecture will review the standard hot big bang model of the early universe, while the second half will discuss developments that have taken place since 1978 — developments which have been motivated mainly by ideas from particle physics.[2]

2. The Big Bang Theory

Cosmology in the 20th century began with the work of Albert Einstein.[3] In March 1916, Einstein completed a landmark paper titled "The Foundation of the General Theory of Relativity."[4] The theory of general relativity is, in fact, nothing more or less than a new theory of gravity. Complex but very elegant, the theory describes gravity as a distortion of the geometry of space and time. Unlike Newton's theory of gravity, general relativity is consistent with the ideas of special relativity, which Einstein had introduced in 1905. While the rest of the world waited to be persuaded, Einstein was immediately convinced that he had found the correct description of gravity.

In less than a year after the publication of general relativity, Einstein applied it to the universe as a whole.[6] However, in carrying out these studies Einstein discovered something that surprised him a great deal: it was impossible to build a static model of the universe consistent with general relativity. Einstein was perplexed by this fact. Like his predecessors, he had looked into the sky, saw that the stars appeared motionless, and erroneously concluded that the universe is static.

In fact, the same problem that Einstein discovered in the context of general relativity also existed in Newtonian mechanics, although it had not been fully understood until the work of Einstein. The problem is nonetheless fairly simple to understand: if masses were distributed uniformly and statically throughout space, then everything would attract everything else and the entire configuration would collapse. (Netwon himself had wrestled with this problem, but had incorrectly convinced himself that, if the universe is infinite, the forces in opposite directions would cancel and the collapse would be avoided.[8]) Einstein nonetheless remained convinced that the universe was static. He therefore modified his equations of general relativity, adding what he called a "cosmological term" — a kind of universal repulsion that prevents the uniform distribution of matter from collapsing under the normal force of gravity. Of course, if one is constructing a theory of gravity, one does not want to introduce a repulsive force that cancels the attractive force under all circumstances. The cosmological force, however, has a different distance dependence than the attractive force. In the Newtonian approximation, the attractive force is proportional to $1/r^2$, where r is the distance, while the cosmological force grows with distance, proportional to r. In this way the cosmological force can stabilize the universe against collapse, while its effects are negligible on the scale of the solar system or smaller. The cosmological term, Einstein found, fits neatly into the equations of general relativity — it is completely consistent with all the fundamental ideas on which the theory was constructed.

Einstein's picture of cosmology remained viable for about a decade, until astronomers began to measure the velocities of distant galaxies. They then discovered that the universe is not at all static. To the contrary, the distant galaxies are receding from us at high velocities.

The pattern of the cosmic motion was codified at the end of the 1920's by Edwin Hubble, in what we now know as Hubble's law.[9] Hubble's law states that each distant galaxy is receding from us with a velocity which is, to a high degree of accuracy, proportional to its distance. Thus one can write

$$v = Hl,$$

where v is the recession velocity, l is the distance to the galaxy, and the quantity H is known as the Hubble "constant." (I put the word "constant" in quotation marks to call attention to its inaccuracy. The quantity was called a "constant" by astronomers, presumably because it remains approximately constant over the lifetime of an astronomer. The value of H changes, however, as the universe evolves; so, from the point of view of a cosmologist, it is not a constant at all.) Hubble's law is not a precise law of nature, but the observed velocities of galaxies are generally found to be very close to the values predicted by Hubble's equation.

The value of the Hubble constant is not very well known. The recession velocities of the distant galaxies are no problem — they can be determined very accurately from the Doppler shift of the spectral lines in the light coming from the galaxies. The distances to the galaxies, on the other hand, are very difficult to determine. These distances are estimated by a variety of indirect methods, and the resulting value of the Hubble constant is thought to be uncertain by a factor of about two. It is believed to lie somewhere in the range

$$H \approx \frac{0.5 \text{ to } 1}{10^{10} \text{ years}}.$$

Notice that the Hubble constant has the units of inverse time; when the Hubble constant is multiplied by a distance, the result has the units of distance per time, or velocity. In particular, if the expression for H above is multiplied by a distance in light-years, the result is a velocity measured as a fraction of the velocity of light. Alternatively, one can use

$$H \approx (15 \text{ to } 30) \text{ km/sec per million light-years}$$

to obtain a velocity in kilometers per second.[a]

The development of cosmology in the 20th century was somewhat confused by the fact that Hubble badly overestimated the value of the Hubble constant, reporting a value of 150 km/sec per million light-years. This mismeasurement had important consequences. In the context of the big bang model, an erroneously high value for the expansion rate implies an erroneously low value for the age of the universe. Hubble's value for the Hubble constant implied an age of about two billion years, a number which conflicts with geological evidence that the Earth is significantly older. This contradiction no doubt contributed to the popularity of the steady-state model during the middle of the 20th century. Not until 1958 did the measured value of the Hubble constant come within the currently accepted range, due primarily to the work of Walter Baade[10] and Allan Sandage.[11] The age of the universe is now estimated to be between 10 and 20 billion years.

Once it is noticed that the other galaxies are receding from us, there are two conceivable explanations. The first is that we might be in the center of the universe, with everything moving radially outward from us like the spokes on a wheel. In the early 16th century such an explanation would have been considered perfectly acceptable. Since the time of Copernicus, however, astronomers and physicists have become instinctively skeptical of this kind of inference, so alternative explanations are sought. In this case, an attractive alternative can be found.

The alternative explanation can be called *homogeneous expansion*, and it is illustrated in Fig. 1. The three pictures are intended to show

---

[a]Astronomers generally quote the Hubble constant in km/sec per Megaparsec, where 1 Megaparsec = 3.26 million light-years. In these units the Hubble constant lies between 50 and 100.

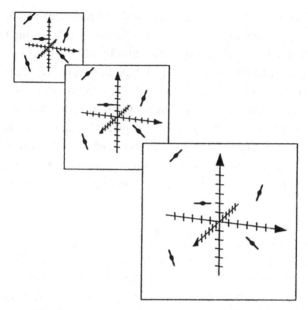

Fig. 1. The expanding homogeneous universe. The three pictures show snapshots of the same region of the universe, taken at three successive times. Each picture is essentially a photographic blowup of the previous picture, with all the distances enlarged by the same percentage. An observer on any galaxy would conclude that all the other galaxies are receding from him.

successive snapshots of a region of the universe. Each picture is essentially a photographic blowup of the previous picture, with all distances enlarged by the same percentage. According to this explanation, there is nothing special about our galaxy — or any other galaxy. All galaxies are approximately equivalent, and are spread more or less uniformly throughout all of space. (In this description, there is no center and no edge to the distribution of galaxies.) As the system evolves; from diagram (a) to (b) to (c), *all* intergalactic distances are enlarged. Thus, regardless of which galaxy we are living on, we would see all the other galaxies receding from us. Furthermore, this picture leads immediately to the conclusion that the recession velocities obey Hubble's law. Since all distances increase by the same *percentage* as the system evolves, larger distances increase by a larger amount. The apparent velocity of a galaxy is proportional to the amount by which the distance from us increases, and hence it is proportional to the distance.

The homogeneous expansion can be described mathematically in a simple way by introducing what is called a "comoving" coordinate system. This is a coordinate system which is marked off with grid marks that expand with the universe, as shown in Fig. 1. A typical galaxy will then maintain constant values of its x, y, and z coordinates as the universe expands. The comoving coordinate system thus provides a map of the universe that remains accurate even as the universe expands. If the universe doubles in size, one simply finds the words "1 grid mark = 25 light-years" in the lower right corner and replaces them with "1 grid mark = 50 light-years." The expansion is thereby incorporated into an overall scale factor, R, which depends on time:

$$l_{\text{phys}}(t) = R(t)l_{\text{coord}},$$

where l_{coord} is the (time-independent) distance between any two galaxies measured in grid marks, $R(t)$ is the scale factor, and $l_{\text{phys}}(t)$ is the physical distance between the two galaxies. When the value of $R(t)$ doubles, it means that all of the distances in the universe have doubled. (Since the Hubble expansion pattern is not rigorously exact, the comoving coordinates of real galaxies are not exactly time-independent, but instead vary slowly with time.)

Figure 1 shows that while the universe is expanding, the size of the individual galaxies does not increase with time. It is believed that shortly after the big bang, the matter that makes up the galaxies was in the form of a nearly uniform gas undergoing nearly uniform expansion. The galaxies then formed from the gravitational clumping of matter, and their size is now stabilized by the dynamics of internal motions and gravitational binding.

When one extrapolates this picture backward in time, one finds an instant in the past when all the galaxies must have been on top of each other and when the density of the universe would have been infinite. Such an event is called a "singularity," and in this case the singularity is the instant of the "big bang" itself, some 10 to 20 billion years ago. The time of the big bang is very uncertain for two reasons: we do not know the Hubble constant very accurately, and we are also uncertain about the mass density of the universe. (The mass density is important in calculating the history of the universe, because it determines how fast the cosmic expansion is slowing down under the influence of gravity.)

The reader should be warned, however, that the calculation implying an infinite density at the instant of the big bang is not to be trusted. As one looks backward in time with the density going up and up, one is led further and further from the conditions under which the laws of physics as we know them were developed. Thus, it is quite likely that at some point these laws become totally invalid, and then it is a matter of guesswork to discuss what happened at earlier times. Nonetheless, as the history of the universe is extrapolated backward, the density increases without limit for as long as the known laws of physics apply. Indeed, most cosmologists today are reasonably confident in our understanding of the history of the universe back to one microsecond ($t \approx 10^{-6}$ second). The goal of the cosmological research involving grand unified theories is to extend our understanding back to $t \approx 10^{-35}$ second.

The time scale that I am using is defined so that the extrapolated infinite density occurs at $t \equiv 0$. Even though the extrapolation to infinite density is not reliable, it still seems to be the most convenient way to define what we mean by $t = 0$. Despite this uncertainty about the zero point of time, we still believe that the time scale used in cosmology gives a meaningful measure of the time intervals between those events that are described. (The inflationary theory, as I will discuss later, gives a much more complicated picture of what happened at $t = 0$, *and even earlier.*)

Having discussed the key features of the big bang theory, we can now ask what evidence can be found to support it. In addition to Hubble's law, there are two significant pieces of observational evidence in favor of the big bang theory.

The first is the observation of the cosmic background radiation. To understand the origin of this radiation, begin by recalling that the temperature of a gas rises when the gas is compressed. For example, a bicycle tire is warmed when it is inflated by a hand pump. Similarly, a gas cools when it is allowed to expand. In the big bang model, the universe has been expanding throughout its history, which means the early universe must have been much hotter. (In fact, the standard calculation implies that the temperature was infinite at the instant of the big bang. This calculation, however, like the calculation of the infinite density, should not be taken as definitive.)

All hot matter emits a glow, just like the glow of hot coals in a fire. Thus, the early universe would have been permeated by the glow of light emitted by the hot matter. As the universe expanded, this light would have cooled, and hence red-shifted (i.e. decreased in frequency). Today, the universe would still be bathed by the radiation, a remnant of the intense heat of the big bang — now red-shifted into the microwave part of the spectrum. This prediction was confirmed in 1964 when Arno A. Penzias and Robert W. Wilson[12] of the Bell Telephone Laboratories discovered a background of microwave radiation with an effective temperature of about 3°K.

At the time of their discovery, Penzias and Wilson were not looking for cosmic background radiation. Instead, they were searching for astronomical sources capable of producing low-level radio interference. They discovered a hiss in their detector, and found that the hiss was present no matter what direction in the sky the antenna was pointed. The hiss seemed likely to be merely electrical noise from the detector, but Penzias and Wilson carefully examined all sources of noise, concluding that the signal was really coming from the sky.

Penzias and Wilson measured the intensity of the radiation at only one frequency, but other experimenters set out to determine whether the spectrum of the radiation agrees with the thermal spectrum predicted by the big bang theory. The data on the spectrum available in the early 1980's is summarized in Fig. 2, a graph adapted from *The Big Bang*, by Joseph Silk.[13] The diagram shows the original point found by Penzias and Wilson, and also a number of other more recent measurements. The solid line is the spectral curve expected for thermal radiation, in this case at a temperature of 2.9°K. It is clear that the data agrees well with the theoretical prediction, but note that there are at this point no measurements on the steep part of the curve on the high frequency side (at the right).

The earth's atmosphere poses a serious problem for measuring the high frequency side of the curve, so the best measurements must be done from balloons, rockets, or satellites. In 1988 a rocket probe was launched by a collaboration between the University of California at Berkeley, and Nagoya University in Japan.[14] Their results are shown in Fig. 3. Note that the points labeled 2 and 3 are much higher than the thermal spectrum predicts. Fitting these points individually to a temperature, the authors find:

$$\text{Point 2:} \quad T = 2.955 \pm 0.017°K,$$
$$\text{Point 3:} \quad T = 3.715 \pm 0.027°K.$$

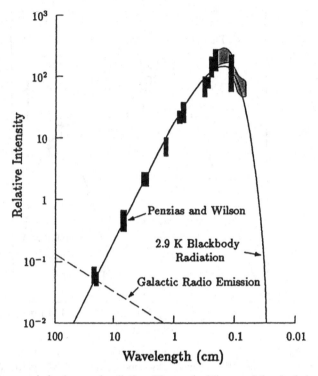

Fig. 2. The cosmic background radiation. The vertical lines and the shaded region show the results of intensity measurements at a variety of wavelengths. The data shown here was all gathered before 1981. The uncertainties are large, but the data is consistent with the theoretical curve, corresponding to thermal radiation at a temperature of 2.9°K. Note that there are no measurements on the right of the curve, where the theoretical prediction shows a very sharp drop in intensity.

These numbers correspond to discrepancies of 12 and 16 times the estimated uncertainty, respectively, from the temperature of $T = 2.74°K$ which fits the lower frequency points. In terms of energy, the excess intensity seen at high frequencies in this experiment amounts to about 10–15% of the total energy in the cosmic background radiation. Cosmologists were stunned by the extremely significant disagreement with predictions. Some tried to develop theories to explain the radiation, without much success, while others banked on the theory that it would go away. The experiment looked like a very careful one, however, so it was difficult to dismiss. The most likely source of error in an experiment of this type is the possibility that the

170 A. H. Guth

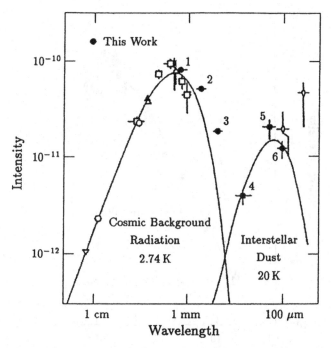

Fig. 3. The submillimeter excess background radiation. The Berkeley–Nagoya rocket experiment of 1988 measured the cosmic background radiation at the high frequency end of the spectrum, and found significantly more intensity than was expected. The discrepancy is seen in the points labeled 2 and 3, which lie far above the theoretically expected curve. (The vertical axis shows frequency times intensity per unit frequency, in units of watts per cm^2 per steradian.)

detectors were influenced by heat from the exhaust of the launch vehicle — but the experimenters very carefully tracked how the observed radiation varied with time as the detector moved away from the launch rocket, and it seemed clear that the rocket was not a factor.

The same group tried to check their results with a second flight a year later, but the rocket failed and no useful data was obtained.

In the fall of 1989 NASA launched the Cosmic Background Explorer, known as COBE. This marked the first time that a satellite was used to probe the background radiation. Within months, the COBE group announced their first results at a meeting of the American Astronomical Society in Washington, DC, January 1990. The detailed preprint, with a cover sheet showing a sketch of the satellite, was released the same day.

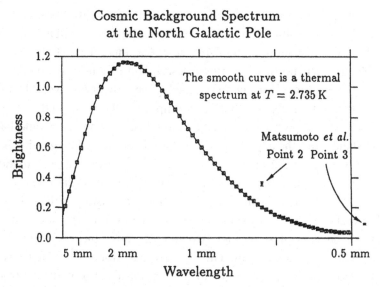

Fig. 4.   The cosmic background radiation as seen by COBE. By far the most precise determination of the spectrum to date indicates that it agrees beautifully with the thermal spectrum predicted by the big bang. The earlier points measured by Matsumoto *et al.* are shown on the same scale for comparison. (The vertical axis shows the intensity per unit frequency times the speed of light, in units of 10^{-11} watts per cm^2 per steradian per hertz, times cm/sec.)

The data showed a perfect fit to the thermal spectrum,[15] with a temperature of $2.735 \pm 0.06°K$, and no evidence whatever for the "submillimeter excess" that had been seen by Matsumoto *et al.* The data was shown with estimated error bars of 1% of the peak intensity, which the group regarded as very conservative. The graph is shown here as Fig. 4.

The COBE graph is plotted on a completely different scale from the graph presented by Matsumoto *et al.*, so they are difficult to compare. I have therefore added the points 2 and 3 of the Matsumoto *et al.* data to the COBE graph. Since the COBE instrument is far more precise and has more internal consistency checks, there has been no doubt in the scientific community that the COBE result supersedes the previous one. Despite the 16σ discrepancy of 1988, the cosmic background radiation is now once again believed to have a nearly perfect thermal spectrum.

The second important piece of evidence supporting the big bang theory is related to calculations of what is called "big bang nucleosynthesis." This evidence is slightly more difficult to understand than the cosmic background radiation, and it is therefore much less discussed in popular scientific literature. To make sense of this argument, one must understand that the big bang theory is not just a cartoon description of a universe beginning with a big explosion. On the contrary, the big bang theory is a very detailed model. Once one accepts its basic assumptions, then knowledge of the laws of physics allows one to calculate how fast the universe would have expanded, how fast the expansion would have been slowed by gravity, how fast the universe would have cooled, and so on. Given this information, knowledge from nuclear physics allows one to calculate the rates of the different nuclear reactions that took place in the early history of the universe.

The early universe was very hot, so hot that even the nuclei of atoms would not have been stable. At two minutes after the big bang there were virtually no nuclei at all. The universe was filled with a hot gas of photons and neutrinos, with a much smaller density of protons, neutrons, and electrons. (The protons, neutrons, and electrons were very unimportant at the time, but they later became raw materials for the formation of stars and planets.) As the universe cooled, the protons and neutrons began to coalesce to form nuclei. From the nuclear reaction rates, one can calculate the expected abundances of the different types of nuclei that would have formed. One finds that most of the matter in the universe would remain in the form of hydrogen. About 25% (by mass) of the matter would have been converted to helium, and trace amounts of other nuclei would also have been produced.

Most of the types of nuclei that we observe in the universe today were produced much later in the history of the universe, in the interiors of stars and in supernova explosions. When a star undergoes a supernova explosion, its material is strewn into space, ready to recollect into further generations of stars. The lightest nuclei, however, were produced primarily in the big bang, and it is possible to compare the calculated abundances with direct observations. Such a comparison can be carried out for the abundances of helium-4, helium-3, hydrogen-2 (otherwise known as deuterium), and lithium-7.

The comparison is complicated by the fact that we do not know all the information necessary to carry out the calculations. In particular, the calculation depends sensitively on the density of protons and neutrons in the

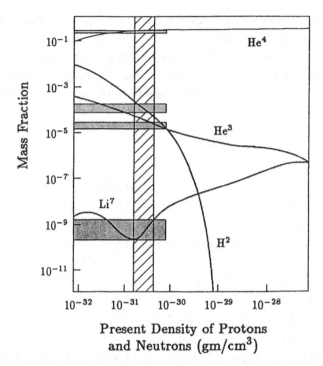

Fig. 5. Big bang nucleosynthesis. The curves show how the predicted abundances of the light chemical elements — helium-4, helium-3, deuterium (hydrogen-2), and lithium-7 — depend on the present density of protons and neutrons. The shaded horizontal bars show the observed abundances, with the estimated uncertainties. Excellent agreement is obtained for the range of densities indicated by the diagonally striped region.

universe, a quantity which can be estimated only roughly by astronomical observations. Thus, the calculations have to be carried out for a broad range of values for this density. One then asks whether there exists a plausible value for which the answers turn out right.

The results of this comparison[16,17] are shown in Fig. 5. The horizontal axis shows the present density of protons and neutrons, and the curves indicate the results of the calculation. The observations, with the estimated range of their uncertainties, are shown by the shaded horizontal bars. Note that there is a range of values for the density of protons and neutrons, indicated by the diagonally striped region, for which each of the four calculated curves agrees with the corresponding observation. Even though the abundances of these nuclei are not known to high precision, the success of the comparison is very impressive. Note that the abundances span nine orders of magnitude. If there was never a big bang, there would be no reason whatsoever to expect that helium-4 would be 10^8 times as abundant as lithium-7 — it might just as well have been the other way around. But, when calculated in the context of the big bang theory, the ratio works out just right.

3. Unanswered Questions

Each piece of evidence discussed above — Hubble's law, the cosmic background radiation, and the big bang nucleosynthesis calculations — probes the history of the universe at a different period of time. The observation of Hubble's law, for example, probes the behavior of the universe at times comparable to the present — billions of years after the big bang. The cosmic background radiation, on the other hand, samples the conditions in the universe about 300,000 years after the big bang, when the universe became cool enough for the plasma of free nuclei and electrons to condense into neutral atoms. The plasma that filled the universe at earlier times was almost completely opaque to photons, which would have been constantly absorbed and re-emitted. With the formation of neutral atoms, however, the universe became highly transparent. Thus, most of the photons in the cosmic background radiation have been moving in a straight line since 300,000 years after the big bang, and they therefore provide an image of the universe at that time. Finally, the big bang nucleosynthesis calculations probe the history of the universe at much earlier times. The processes involved in establishing the abundances of the light nuclei occurred at times ranging from about one second to about four minutes after the big bang.

The big bang theory is a very successful description of the evolution of the universe for the whole range of times discussed above — from about one second after the big bang to the present. Nonetheless, the standard big bang theory has serious shortcomings, in that a number of very obvious questions are left unanswered. Here I describe three of these questions, which are listed in Table 1. Later I will show how ideas from particle physics have led to a radically new picture for the very early behavior of the universe, a picture which provides plausible answers to each of these questions.

Table 1. Questions left unanswered by the standard big bang theory.

#1: How did the universe become so homogeneous on large scales? Do we have to assume that it started out that way?

#2: Why was the mass density of the early universe so extraordinarily close to the critical density?

#3: Can one find a physical origin for the primordial density perturbations which led to the evolution of galaxies and clusters of galaxies? Are there physical processes that determine the spectrum of these perturbations?

The first question is related to the large-scale homogeneity, or uniformity, of the observed universe. The discussion of homogeneity must be qualified, however, because the universe that we observe is in many ways very inhomogeneous. The stars, galaxies, and clusters of galaxies make a very lumpy distribution. Cosmologically speaking, however, all of this structure in the universe is very small-scale. If one averages over very large scales, scales of 300 million light-years or more, then the universe appears to be very homogeneous. This large-scale homogeneity is most evident in the cosmic background radiation. Physicists have probed the temperature of the cosmic background radiation in different directions, and have found it to be extremely uniform. It is just slightly hotter in one direction than in the opposite direction, by about one part in 1000. Even this small discrepancy, however, can be accounted for by assuming that the solar system is moving through the cosmic background radiation at a speed of about 600 km/sec. Once the effect of this motion is subtracted out, the resulting temperature pattern is uniform in all directions to extraordinary accuracy. Until 1992 no nonuniformities had been detected at all, with probes sensitive to a level of a few parts per 100,000. In April 1992 the COBE team announced the results of their nonuniformity measurements, based on the first year of data collected by the satellite. The first nonuniformities had been discovered,[18] at the extremely small level of only about 6 parts per million. Since the cosmic background radiation gives an image of the universe at 300,000 years after the big bang, one concludes that the universe was extremely homogeneous at that time. (The observed "small-scale" inhomogeneities are believed to have formed later, by the process of gravitational clumping.) The standard big bang theory cannot explain the large-scale uniformity; instead, the uniformity must be postulated as part of the initial conditions.

The difficulty in explaining the large-scale uniformity is a quantitative question, related to the rate of expansion of the universe. Under many circumstances a uniform temperature would be easy to understand — anything will come to a uniform temperature if it is left undisturbed for a long enough period of time. In the standard big bang theory, however, the universe evolves so quickly that it is impossible for the uniformity to be created by any physical process. In fact, the impossibility of establishing a uniform temperature depends on none of the details of thermal transport physics, but is instead a direct consequence of the principle that no information can propagate faster than the speed of light. The situation is illustrated in Fig. 6, which shows a spacetime diagram of the universe. The diagram illustrates two of the three dimensions of space, and also time. (The x and y axes of Fig. 6 represent comoving coordinates, as

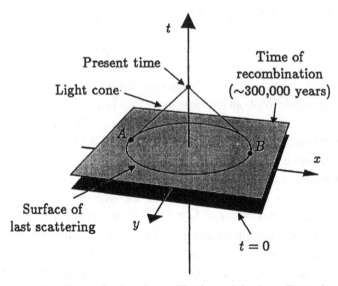

Fig. 6.   A spacetime diagram for the universe. The time axis has been distorted so that light pulses travel at 45°. The early plasma-filled universe is opaque to radiation, but photons can travel freely starting at the surface labeled the time of recombination. Points A and B indicate the origins of cosmic background radiation photons that are arriving at earth today from two opposite directions in the sky. These points were separated from each other by about 100 times the distance that light could have traveled prior to that time.

defined in the discussion of Fig. 1.) The time axis has been distorted and rescaled so that a light pulse travels in the diagram at 45° from the vertical. Our location in space and time is shown as a point on the time axis. The initial singularity is shown as a darkly shaded surface at $t = 0$. The early universe was opaque to radiation during its plasma phase, which ended at $t \approx 300,000$ years, shown as a lightly shaded surface labeled "time of recombination." ("Recombination" refers to the combining of nuclei and electrons to make neutral atoms; the meaning of the prefix "re" remains a mystery to the author.) The bulk of the photons in the cosmic background radiation arriving at earth today were last scattered at this time, and hence they originate on the circle labeled "surface of last scattering." (If the full three dimensions of space could be shown, this circle would become a spherical surface.) The spacetime points A and B represent the origins of two cosmic background radiation photons arriving at earth today from two opposite directions in the sky. The problem is that these two points are separated from each other by more than 90 times the distance that light could have traveled up until that time. (Note that for the sake of visibility the diagram has not been drawn to scale; if it were, the two shaded surfaces would be closer to each other by a factor of ten.) Thus, there is no way that points A and B could have communicated with each other, and certainly no physical process that would bring them to the same temperature.

The puzzle of explaining why the universe appears to be uniform over such large distances is not a genuine inconsistency of the standard theory: if the uniformity is assumed in the initial conditions, then the universe will evolve uniformly. The problem is that one of the most salient features of the observed universe — its large-scale uniformity — cannot be explained by the standard big bang theory; it must be assumed as an initial condition.

The second question is related to the mass density of the universe. This mass density is usually measured relative to a benchmark called the "critical mass density," which is defined in terms of the expansion rate of the universe.[b] If the mass density exceeds the critical density, then the gravitational pull of everything on everything else will be strong enough

---

[b]The critical density is given by $3H^2/8\pi G$, where H is the Hubble constant and G is Newton's gravitational constant. For the range of values for H given earlier, the critical density lies in the range $0.4-1.8 \times 10^{-29}$ gm/cm^3.

to eventually halt the expansion. The universe would recollapse, resulting in what is sometimes called a "big crunch." On the other hand, if the mass density is less than the critical density, the universe will go on expanding forever.

Cosmologists typically describe the mass density of the universe by a ratio designated by the Greek letter Ω (omega), defined by

$$\Omega \equiv \frac{\text{Mass density}}{\text{Critical mass density}}.$$

Ω is very difficult to determine, but its present value is known to lie somewhere in the range of 0.1 to 2.

That seems like a broad range, but consideration of the time development of the universe leads to a different point of view. $\Omega = 1$ is an *unstable equilibrium point* of the evolution of the standard big bang theory, which means it resembles the situation of a pencil balancing on its sharpened end. The phrase "equilibrium point" implies that if Ω is ever exactly equal to one, it will remain exactly equal to one forever[c] — just as a pencil balanced precisely on end will, according to the laws of classical physics, remain forever in the vertical position. The word "unstable" means that any deviation from the equilibrium point, in either direction, will rapidly grow. If the value of Ω in the early universe is just a little bit above one, it will rapidly rise toward infinity; if Ω in the early universe is just a tiny bit below one, it will rapidly fall toward zero. It therefore seems very unlikely that the value of Ω today would lie *anywhere* in the vicinity of one.

Thus, if Ω is to be anywhere near one today, it must have been extraordinarily close to one at early times. For example, we can consider the time of one second after the big bang, the time at which the processes related to big bang nucleosynthesis were beginning to take place. In order for Ω to be somewhere in the allowed range today, at one second after the big bang Ω had to have been equal to one to an accuracy of 15 decimal places. In other words, at 1 second after the big bang, Ω was between 0.999999999999999 and 1.000000000000001. If we go further and consider the time of 10^{-35} second after the big bang, when thermal energies were typical of the energy

---

[c]Since $\Omega = 1$ is a statement about the ultimate fate of the universe, it is not surprising that the condition is unchanged as time passes.

scale of grand unified theories, then at that time Ω had to have been equal to one to an accuracy of 49 decimal places!

In the standard big bang theory there is no explanation whatever for this fact, as has been emphasized by Robert H. Dicke and P. James E. Peebles[19] of Princeton University. At one second after the big bang, Ω could have had any value — except that most possibilities would lead to a universe very different from the one in which we live. Like the large-scale homogeneity, the nearness of the mass density to the critical density cannot be explained; instead, it must be postulated as part of the initial conditions.

The third question concerns the origin of the density perturbations that are responsible for the development of the small-scale inhomogeneities. While the universe is remarkably homogeneous on the very large scale, there is nonetheless a very complicated structure on the scale of superclusters of galaxies and smaller. The existence of this structure is undoubtedly related to the instability of gravitational clumping: any region that contains a higher-than-average mass density will produce a stronger-than-average gravitational field, thereby pulling in even more excess mass. Thus, small perturbations are amplified to become large perturbations.

However, in order for galaxies to evolve, the early universe must have contained primordial density perturbations. The standard big bang model offers no explanation for either the origin or the form of these perturbations. Instead, an entire spectrum of primordial perturbations must be assumed as part of the initial conditions.

4. The Mechanism of Inflation

All three of the questions in Table 1 can be given plausible answers in terms of a model known as the inflationary universe. Before discussing these answers, however, I will have to spend some time describing how the model works. The description will be somewhat sketchy, but I will try to explain the main features.[20]

The inflationary universe was first proposed by me[29] in 1981, but the model in its original form did not quite work. It had a crucially important technical flaw, which was pointed out but not remedied in the original paper. A variation that avoids this flaw was invented independently by Andrei D. Linde[30] of the P. N. Lebedev Physical Institute in Moscow and by Andreas

Albrecht and Paul J. Steinhardt[31] of the University of Pennsylvania. This
section will discuss mainly the Linde–Albrecht–Steinhardt version of the
model, which is called the *new* inflationary universe.

The mechanism of inflation depends on scalar fields, so I will begin
by briefly summarizing the role of scalar fields in particle physics. To
begin with, the reader should recognize that in the context of modern
particle physics, *all* fundamental particles are described by fields. The
best-known example is the photon. The classical equations describing
the electromagnetic field were written down in the 1860's, but then in the
early 20th century physicists learned that the underlying laws of nature are
quantum and not classical. The quantization of the electromagnetic field
can be carried out in a very straightforward way. For simplicity one can
consider the fields inside a box, letting the size of the box approach infinity
at the end of the calculation. One can then think of the electromagnetic
fields inside the box as a mechanical system. The electromagnetic field
is written as a sum of normal modes, and the coefficients of the normal
mode functions can be taken as the dynamical degrees of freedom of the
system. It turns out that each of these coefficients obeys the equations of
a harmonic oscillator, and there is no interaction between the coefficients.
The system is quantized by the same rules that one uses to quantize the
hydrogen atom or the harmonic oscillator. The result is that each normal
mode, like a harmonic oscillator, has evenly spaced energy levels. In this
case, we interpret each energy level above the ground state as the occupation
of the mode by a photon. Thus, the photon is interpreted as the quantized
excitation of a field. This formalism is known as second quantization, and
was of course pioneered by Oklein and Jordan.[32]

In contemporary particle theory, all elementary particles are described
in this way. There is an electron field to correspond to an electron, a quark
field to correspond to a quark, a neutrino field to correspond to a neutrino,
etc. Among the different types of fields, the simplest is the scalar field —
a field that has the same value to any Lorentz observer, and which evolves
according to the Oklein–Gordon equation.[33] The quantized excitation of
a scalar field is a spinless particle. Although spinless particles that are
regarded as elementary have yet to be observed, they are nonetheless
a key ingredient to a number of important theories. In particular, the
Glashow–Weinberg–Salam model of the electroweak interactions makes

use of a scalar field, called the Higgs field, to cause a symmetry in the theory to be spontaneously broken. (If this symmetry were not broken, then electrons and neutrinos would both be massless, and would be indistinguishable.) This Higgs field corresponds to a neutral spinless Higgs particle, which will hopefully be observed at the SSC (Superconducting Super-Collider), if not before. Grand unified theories make use of similar Higgs fields, but at a much higher mass scale, to spontaneously break the grand unified symmetry which relates electrons, neutrinos, and quarks.

There is much current interest in superstring theories, which actually go beyond the pattern of the field theories that were described above. In these theories the fundamental object is not a field, but is really a string-like object, which has length but no width. These theories are believed to behave as field theories, however, at energy scales well below the Planck scale.[d] In fact, these field theories contain a large number of scalar fields. Thus, the particle physics motivation for believing that scalar fields exist is quite strong.

The electromagnetic field has a potential energy density given by

$$V = \frac{1}{8\pi}(\mathbf{E}^2 + \mathbf{B}^2),$$

but the potential energy density of a scalar field has a wider range of possibilities. It is restricted by the criterion that the field theory be renormalizable (i.e. the requirement that the field theory lead ultimately to finite answers), but there are still several free parameters involved in specifying the energy density. The form of the scalar field energy density appropriate for the new inflationary universe model is shown in Fig. 7. The potential has a minimum at a value of the scalar field ϕ that is nonzero, a property that is characteristic of Higgs fields. The potential also has a flat plateau centered at $\phi = 0$, a feature that Higgs fields do not necessarily have. This property is necessary, however, in order for the new inflationary scenario to be possible.

---

[d]The Planck energy E_P is given by $E_P = \sqrt{\hbar c^5/G} = 1.22 \times 10^{19}$ GeV, where G is Newton's constant, $\hbar$ is Planck's constant divided by 2π, and c is the speed of light. It is the scale above which the quantum effects of gravity are expected to become important. The Planck mass is defined by $M_P \equiv E_P/c^2 = 2.18 \times 10^{-5}$ gram.

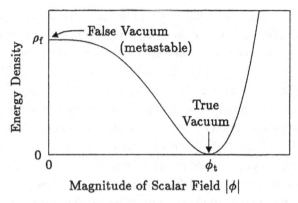

Fig. 7.   The false vacuum. The graph shows the potential energy density as a function of the magnitude of a scalar field ϕ. The curve shown has a plateau at $\phi = 0$, which is the general form appropriate for the new inflationary universe model. The true vacuum is the state of lowest possible energy density, while the metastable state $\phi \approx 0$ is called the false vacuum.

A particle physicist defines the word "vacuum" to mean the state of lowest possible energy density, so it is the state in which ϕ lies at the minimum of the energy function, labeled ϕ_t on the diagram. For clarity, I will refer to the vacuum in this situation as the "true vacuum."

Notice that the energy density of the true vacuum state is shown as zero. If the energy of the vacuum were nonzero, the effects would be precisely equivalent to those of the cosmological constant introduced by Einstein in 1917. Thus the depiction of the vacuum energy density as zero on the diagram is equivalent to the statement that the cosmological constant is either zero or immeasurably small. Although the cosmological constant might be large enough to be cosmologically important, it would still be insignificant compared to ρ_f, which is estimated to be very large. In fact, the energy density of the vacuum is believed to be no larger than about $10^{-104} \rho_f$. The explanation for the incredible smallness of the energy density of the vacuum has been a long-standing mystery to particle physicists, although now we have at least a possible solution[34] in the context of a crude understanding of quantum gravity.

The state in which the scalar field, in some region of space, is perched at the top of the plateau of the potential energy diagram is called the "false vacuum." Note that the energy density of this state is ρ_f, and is therefore fixed by the laws of physics. (Although it seems strange that energy should be required in order for the value of the Higgs fields to be zero, particle physicists find that this property causes no inconsistencies, and is exactly what is needed to produce the symmetry breaking for which the Higgs field was introduced.) The false vacuum is of course not stable, since eventually the scalar field will no doubt evolve toward the minimum of the energy density diagram. Nonetheless, if the plateau is broad and flat enough, it will take a long time for the scalar field to roll off the hill. Thus the false vacuum is metastable, and can be very long-lived, by the time scales of the early universe.

The crucial property of the false vacuum is that the energy density is positive, but on short time scales it cannot be lowered. This explains the etymology of the name "false vacuum." Here "false" is being used to mean temporary, and "vacuum" means the state of lowest possible energy density.

It is now possible to construct a simple energy-conservation argument to determine the pressure of the false vacuum. Imagine a chamber filled with false vacuum, as shown in Fig. 8. Since the energy density of the vacuum is fixed at ρ_f, the energy inside the chamber is given by $U = \rho_f V$, where V is the volume of the chamber. Now suppose the piston is pulled outward, increasing the volume by dV. Unlike any normal substance, the false vacuum will maintain a constant energy density despite the increase

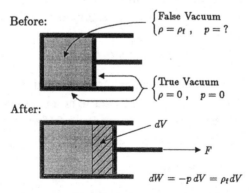

Fig. 8. Thought experiment to calculate the pressure of the false vacuum. As the piston chamber filled with false vacuum is enlarged, the energy density remains constant and the energy increases. The extra energy is supplied by the agent pulling on the piston, which must pull against the negative pressure of the false vacuum.

in volume. The increase in energy is then $dU = \rho_f dV$, which must be equal to the work done, $dW = -pdV$, where p is the pressure. Thus, the pressure of the false vacuum is given by

$$p = -\rho_f.$$

The false vacuum has a colossal, negative pressure. It is this peculiar feature of the false vacuum that is responsible for the dramatic effect that it can have on the evolution of the universe.

Assuming that the universe is undergoing homogeneous expansion as described earlier, the evolution of the scale factor $R(t)$ is given by general relativity as

$$\ddot{R} = -\frac{4\pi}{3c^2}G(\rho + 3p)R,$$

where the overdot is used to denote a time derivative, and G is Newton's gravitational constant. In the present universe, the pressure term is a small relativistic correction. The dominant term is proportional to the energy density ρ, and the equation describes how a positive energy density produces a gravitational field that slows down the expansion of the universe. The second term shows that a positive pressure also creates an attractive gravitational field, helping to slow the expansion of the universe. However, if the universe is ever dominated by false vacuum, then during this period $p = -\rho_f$ implies that the pressure term has the opposite sign, and overcomes the gravitational attraction caused by the energy density term:

$$\ddot{R} = \frac{8\pi}{3c^2}G\rho_f R.$$

The force of gravity actually becomes repulsive, and the expansion rate of the universe is accelerated. Thus the bizarre notion of negative pressure leads to the even more bizarre effect of a gravitational force that is effectively repulsive. The general solution to this equation is

$$R(t) = c_1 e^{\chi t} + c_2 e^{-\chi t},$$

where c_1 and c_2 are arbitrary constants, and the exponential rate is given by

$$\chi = \sqrt{\frac{8\pi}{3c^2}G\rho_f}.$$

After some time the growing term will dominate, and the expansion becomes a pure exponential. The precise numbers here are not very well constrained, but for a typical grand unified theory one expects $\rho_f \approx (10^{15}\,\text{GeV})^4/(\hbar c)^3 \approx 10^{96}\,\text{erg/cm}^3$. This gives a typical value for the time constant as $1/\chi \approx 10^{-34}$ second.

While the new inflationary scenario assumes a potential energy function of the form shown in Fig. 7, Linde[35,36] has shown that such a severe restriction is not necessary. In a model known as chaotic inflation, Linde showed that inflation can work for a scalar field energy density as simple as $V(\phi) = \lambda\phi^4$, provided that one makes some assumptions about the initial conditions. He proposed that the scalar field begins in a chaotic state, so that there are some regions in which the value of ϕ is a few times larger than the Planck mass (see footnote d) M_P. These regions must exceed some minimal size, which is estimated to be several times cH^{-1}, where H denotes the Hubble constant. Then ϕ rolls down the hill of the potential energy diagram, and a straightforward calculation indicates that there is an adequate amount of inflation. The Hubble "constant" is not a constant in this case, but it is slowly varying, so the expansion can be called "quasi-exponential."

In a newer variation known as extended inflation,[37,38] the scalar field potential energy function is taken as a curve that has a local minimum in addition to the global minimum. The false vacuum is the state in which the scalar field lies in the local minimum, and inflation occurs while the universe is in this state. In this situation the false vacuum is too stable to allow a smooth ending to inflation, so an additional scalar field, which interacts directly with gravity, is introduced to bring about an end to inflation.

5. The Inflationary Universe Scenario

The inflationary universe scenario begins with a patch of the universe somehow settling into a false vacuum state. The mechanism by which this happens has no influence on the later evolution. The following three possibilities have been discussed in the literature:

1. *Supercooling from high temperatures.* This was the earliest suggestion.[29-31] If we assume that the universe began very hot, as is traditionally assumed in the standard big bang model, then as the universe

cooled it presumably went through a number of phase transitions. For many types of scalar field potentials, the thermal equilibrium state at high temperatures corresponds to a scalar field that oscillates about the false vacuum value, $\phi = 0$. If such a state is rapidly cooled by the expansion of the universe, it is very natural for the scalar field to be trapped in the false vacuum state. For the new inflationary scenario, however, this scenario has the difficulty that there is no known mechanism to achieve the desired pre-inflationary thermal equilibrium state. Calculations[39–42] show that the scalar field must be very weakly coupled in order for quantum-induced density perturbations to be sufficiently small, and consequently the scalar field would require much more than the available time to relax to thermal equilibrium. It has been shown, however, that true thermal equilibrium is not really necessary: a variety of random configurations give results that are very similar to those of thermal equilibrium.[43] In extended inflation, on the other hand, the scalar field need not be weakly coupled, so this difficulty does not arise.

2. *Tunneling from "nothing."* These ideas are of course very speculative, since they involve a theory of quantum gravity that does not actually exist. The basic idea, however, seems very plausible. If geometry is to be described by quantum theory, then the geometry of space can presumably undergo quantum transitions. One can then imagine an initial state of absolute nothingness — the absence of matter, energy, space, or time. The state of absolute nothingness can presumably undergo a quantum transition to a small universe, which then forms the initial state for an inflationary scenario. Variations of these ideas have been studied by Tyron,[44] Vilenkin,[45,46] Linde,[47–49] and Hartle & Hawking.[50]

3. *Random fluctuations in chaotic cosmology.* In Linde's[35,36] chaotic cosmology, it is assumed that the scalar field ϕ begins in a random state in which all possible values of ϕ occur. Inflation then takes place in those regions that have appropriate values of ϕ, and these inflated regions dominate the universe at later times. In these models it is not necessary for the scalar field potential energy function $V(\phi)$ to have a plateau, but as in other models it must be very flat (i.e. weakly self-coupled) in order to minimize the density perturbations that result from quantum fluctuations.

Regardless of which of the above mechanisms is assumed, one expects that the scalar field just before inflation would be approximately uniform over distances equal to the age of the universe times the speed of light. For typical grand unified theory parameters, this gives uniformity on a length scale of about 10^{-24} cm. The inflationary model posits that the entire observed universe evolved from a region of this size or smaller.

The patch then expands exponentially due to the gravitational repulsion of the false vacuum. In the inflationary model, it is the force of this repulsion that provides the "bang" of the big bang, creating the cosmic expansion which we know as Hubble's law. (The standard big bang theory, by contrast, is really only a theory of the aftermath of a bang — the cosmic expansion is not explained, but instead is incorporated into the postulated initial conditions.) In order to achieve the goals of inflation, we must assume that this exponential expansion results in an expansion factor $\gtrsim 10^{25}$. For typical grand unified theory numbers, the universe would double in size during each interval of about 10^{-34} second, so this enormous expansion requires only about 10^{-32} second of inflation. The actual amount of inflation, however, was very likely much more than this required minimum.

During the inflationary period, the density of any particles that may have been present before inflation is diluted so much that it becomes completely negligible. The regions within which the scalar field is approximately uniform are stretched by the expansion factor to become at least about 10 cm across.[e] If the duration of inflation is more than the minimal value, which seems quite likely, then these regions could be many orders of magnitude

---

[e]The alert reader might notice that these numbers imply that the region expands much faster than the speed of light. Even in the standard big bang model without inflation, the size of the universe, or the observed part of the universe, increases faster than the speed of light. Although this violates the premises of special relativity, it is completely acceptable in the context of general relativity. The principle that nothing can travel faster than light is valid in general relativity, but it has to be defined more carefully than in special relativity. It is accurate to say that if any particle has a race with a light beam, the light beam will *always* win. The complication, however, is that in general relativity space is not rigid, but instead is a plastic medium that can bend and twist and stretch. So, if two points are separated from each other, the distance between them can be increased by the stretching of the space between them. There is nothing in general relativity that places any limit on the speed with which such stretching can take place.

larger. There appears to be no upper limit to the amount of inflation that may have taken place.

The false vacuum is not stable, so it eventually decays. If the decay occurs by the usual process of bubble nucleation,[51,52] similar to the way water boils, then the randomness of the bubbles would produce gross inhomogeneities in the mass density.[53,54] This problem is avoided in the new inflationary scenario[30,31] by introducing a scalar field potential energy function with a flat plateau, as was shown in Fig. 7. This leads to a "slow-rollover" phase transition, in which quantum fluctuations destabilize the false vacuum, starting the scalar field to roll down the hill of the potential energy diagram. These fluctuations are initially correlated over only a microscopic region, but the additional inflation that takes place during the rolling can stretch such a region to be large enough to easily encompass the observed universe.

When the phase transition takes place, the energy that has been stored in the false vacuum is released in the form of new particles. (In the language of thermodynamics, this energy is the "latent heat" of the phase transition.) These new particles rapidly come to thermal equilibrium, resulting in a temperature with $kT \approx 10^{14}$ GeV. At this point the scenario rejoins the standard cosmological model. The presently observed region of the universe has a radius of about 10 cm at this time; the normal Hubble expansion of standard cosmology stretches this radius to about 10^{10} light-years by today.

The excess of matter over antimatter is produced[55] after inflation, since any excess that may have been present before inflation would have been diluted away by the enormous expansion. Thus, inflationary cosmology requires an underlying particle theory in which baryon number, the number of protons and neutrons minus the number of antiprotons and antineutrons, is not exactly conserved. Grand unified theories fulfill this requirement, and it now seems clear that baryon number violation occurs at high temperatures even in the well-established theory of electromagnetic and weak interactions.[56–59] At the end of inflation, the universe would have been uniformly filled with a hot gas of particles, exactly as had been postulated as the initial condition for the standard big bang theory. The inflationary model merges with the standard big bang theory, so the two models agree in their description of the evolution of the universe from this time onward.

It is a dramatic feature of inflationary models that essentially all of the matter, energy, and entropy of the observed universe is produced by the expansion and subsequent decay of the false vacuum. (I used the qualifier "essentially" to acknowledge the fact that a small patch of false vacuum is necessary to start inflation.) This seems strange, because it sounds like an unmistakable violation of the principle of energy conservation. How could it be possible that all the energy in the universe was produced as the system evolved?

In fact, the inflationary universe model is consistent with all the known laws of physics, including the conservation of energy. The loophole in the conservation of energy argument is due to the peculiar nature of gravitational energy. Using either general relativity or Newtonian gravity, one finds that *negative* energy is stored in a gravitational field. A simple way to demonstrate this fact is to imagine a spherical shell of mass, as shown in Fig. 9(a). The gravitational field outside the shell points radially inward, and has the same strength as it would if all of the mass were concentrated at the center. In the spherical cavity inside the shell, however, the gravitational field caused by the different parts of the shell cancel out exactly, so the gravitational field is zero. The shell itself will experience an inward gravitational force, since the gravitational field within the matter of the shell is nonzero, varying smoothly between the outside value and zero. Now imagine what would happen if the shell were allowed to uniformly contract. Assume that the matter from which the shell is constructed is soft and compressible, so it offers no mechanical resistance to the contraction. One can imagine, for example, extracting energy by tying ropes to each piece of the shell, as is illustrated in Fig. 9(b). These ropes can be used to drive electric generators as each piece is lowered to its new position. After the new radius is attained, the picture will look like Fig. 9(c). The dotted circle indicates the former radius of the shell, and outside of the dotted circle the gravitational field is identical to that in Fig. 9(a). Inside the shell in its new position, the gravitational field remains zero. However, in the shaded region between the original and new positions of the shell, a gravitational field now exists where no field had existed before. Thus, the net effect of this operation is to extract energy, and to create a new region of gravitational field. Thus, energy is *released* when a gravitational field is created. If the absence of a gravitational field corresponds to zero

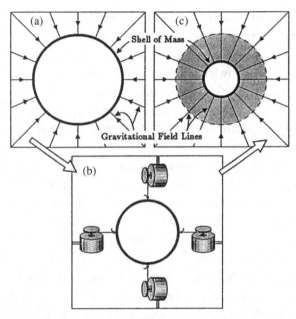

Fig. 9.   Thought experiment to understand the energy of gravity. Figure (a) shows a hollow spherical shell of mass, and the gravitational field lines that it produces. There is a force on each piece of the shell, pulling inward. Figure (b) shows how energy can be extracted as the shell is allowed to uniformly contract. Each piece of the shell is tied by a rope to an electrical generator, producing power as the piece is "lowered" toward its final position. Figure (c) shows the final configuration, which includes a gravitational field in the shaded region where no field existed before. Thus, the creation of the gravitational field is associated with the release of energy.

energy, then any nonzero field strength must correspond to a negative energy. In most physical processes the exchange of gravitational energy is much smaller than the rest energy (mc^2) of the particles involved, but cosmologically the total gravitational energy can be large enough to cancel the rest energy.

In the inflationary model, the energy stored in the false vacuum becomes larger and larger as the universe inflates, and is then released when the phase transition takes place at the end of the inflationary period. At the same time, however, the energy stored in the cosmic gravitational field — the field by which everything in the universe is attracting everything else — becomes more and more negative. The total energy of the system is conserved, remaining constant at a value which is either zero or very small.

Thus, inflation allows the entire observed universe to develop from almost nothing. The initial volume must have a minimum radius of approximately cH^{-1}, where H is the Hubble constant at the time of inflation, which for typical grand unified theory parameters implies a minimum mass of only about 10 kilograms. (Even this initial mass, however, could conceivably have been balanced by an equal contribution of negative energy in the gravitational field.) Thus, if the inflationary model is correct, it is fair to say that the universe is the ultimate free lunch.

6. Answers to Three Questions

Having described the foundations of the inflationary universe model, I can now explain how the three questions of Table 1 can be resolved.

First, I will discuss Question #1, concerning the large-scale homogeneity of the universe. Recall that in the standard big bang theory, the large-scale homogeneity cannot be explained because the universe did not have enough time to come to a uniform temperature.

Consider now the evolution of the observed region of the universe, which has a radius today of roughly 10 billion light-years. Imagine following this region backward in time, using the inflationary model. Start by following it back to the instant just after the inflationary period. The evolution during this interval — from the present back to just after inflation — is the same as in standard cosmology. Continuing further backward we come to the enormous expansion of the inflationary era. Going backward in time, we see the inflationary period as an enormous contraction. Thus, just before inflation we find that the region was incredibly small. In fact, the region was more than a billion times smaller than the size of a proton. (Note that I am *not* asserting that the universe as a whole was this small. The inflationary model makes no statement about the size of the universe as a whole, which could in fact even be infinite.)

While the region was incredibly small before inflation, there was plenty of time for it to have come to a uniform temperature by the same kind of mundane processes that cause a glass of ice water to come to a uniform temperature. So, in the inflationary model, the uniform temperature was established before inflation took place, in a very, very small region. The process of inflation then magnified this very small region to become large enough to encompass the entire observed universe. Thus, the sources of the microwave background radiation arriving today from all directions in the sky were once in close contact; they had time to reach a common temperature before the inflationary era began.[f]

The inflationary model also provides a simple resolution for Question #2, the issue of the mass density. Recall that the ratio of the actual mass density to the critical density is called Ω, and that the problem arose because the condition $\Omega = 1$ is unstable: Ω is always driven away from one as the universe evolves, making it difficult to understand how its value today can be in the vicinity of one.

During the inflationary era, however, the peculiar nature of the false vacuum state results in some important sign changes in the equations that describe the evolution of the universe. During this period, as we have seen, the force of gravity acts to accelerate the expansion of the universe rather than to retard it. It turns out that the equation governing the evolution of Ω also has a crucial change of sign: during the inflationary period the universe is driven very quickly and very powerfully *toward* a critical mass density.

In other words, a very short period of inflation can drive the value of Ω very accurately to one, almost no matter where it starts out. There is no longer any need to assume that the initial value of Ω was incredibly close to one.

Furthermore, there is a prediction that comes out of this. The mechanism that drives Ω to one almost always overshoots, which means that even today the mass density should be equal to the critical mass density to a high degree of accuracy. More precisely, the model predicts that the value of Ω today should equal one to an accuracy of about one part in 100,000. (The deviations from one are caused by quantum effects, which I will talk about

---

[f] Since all distances go to zero at the instant of the big bang singularity, with or without inflation, the astute reader will ask why causal contact and thermal equilibrium cannot be achieved at this time. The reason is that the expansion velocities approach infinity at the singularity. (If you are puzzled by a speed greater than that of light, see footnote e.) So, even though the distance between two particles of matter was zero at the singularity, the infinite velocity of separation implies that no causal contact could have taken place. One determines the question of causal contact by tracing the trajectory of a light ray that leaves one particle immediately after the big bang singularity. One finds that the light ray reaches the other particle much later in the history of the universe, as was illustrated in Fig. 6.

shortly.) Thus, the determination of the mass density of the universe would be a very important test of the inflationary model.[g]

Unfortunately, it is very difficult to estimate the mass density of the universe reliably. Part of the reason is that most of the mass in the universe is in the form of "dark matter," matter that is totally unobserved except for its gravitational effects on other forms of matter. Since we do not even know what the dark matter is, it is very difficult to estimate how much exists. If one counts only the matter that is directly visible in the form of stars and galaxies, one finds a density only 1–2% of the critical value. However, measurements of the velocities of stars within galaxies and of galaxies within clusters of galaxies indicate that all of these velocities are surprisingly high. Since these structures do not fly apart, we infer that they must be held together by the gravitational field created by a rather large amount of dark matter. When this dark matter is included, the estimates[60] of Ω rise to something like 0.1 to 0.3. Dark matter, however, can be detected only when there is visible matter present for it to influence, so there remains the possibility that there is much more dark matter than has so far been found. In fact, studies of the velocities and densities of galaxies on a very large scale[61,62] (within a radius of about 600 million light-years) strongly suggest an Ω in the vicinity of 1.

Finally, then, I come to the last of the three questions, concerning the origin of the primordial density perturbations in the universe.

The generation of density perturbations in the new inflationary universe was addressed in the summer of 1982 at the Nuffield Workshop on the Very Early Universe, held at Cambridge University. A number of theorists were working on this problem, including Steinhardt, James M. Bardeen of the University of Washington, Stephen W. Hawking of Cambridge University, Michael S. Turner of the University of Chicago, A. A. Starobinsky of the L. D. Landau Institute of Theoretical Physics in Moscow, and me. (My work was done in collaboration with So-Young Pi of Boston University, although

---

[g]In the text I have followed the common assumption that Einstein's cosmological constant Λ is either zero or negligible. Otherwise the prediction becomes $\Omega + (\Lambda/3H^2) = 1$. The cosmological constant can be interpreted as a nonzero energy density in the vacuum, so $\Omega + (\Lambda/3H^2) = 1$ implies that the sum of the matter and vacuum contributions to the energy density is equal to the critical density.

she did not attend the workshop.) We found that the new inflationary model, unlike any previous cosmological model, leads to a definite prediction for the spectrum of perturbations. Basically, the process of inflation first smooths out any primordial inhomogeneities that may have been present in the initial conditions. For example, any particles that may have been present before inflation are diluted to a negligible density. In addition, the primordial universe may have contained inhomogeneities in the gravitational field, which is described in general relativity in terms of bends and folds in the structure of spacetime. Inflation, however, stretches these bends and folds until they become imperceptible, just as the curvature of the surface of the Earth is imperceptible in our everyday lives. For a while, we were worried that inflation would give us a totally smooth universe, which would be obviously incompatible with observation. It was pointed out, however, I believe first by Hawking, that the situation might be saved by the application of quantum theory.

A very important property of quantum physics is that nothing is determined exactly — everything is probabilistic. Physicists are, of course, accustomed to the idea that quantum theory, with its probabilistic predictions, is essential to describe phenomena on the scale of atoms and molecules. On the scale of galaxies or clusters of galaxies, however, there is usually no need to consider the effects of quantum theory. But inflationary cosmology implies that for a short period the scales of distance increased very rapidly with time. Thus, the quantum effects which occurred on very small, particle-physics length scales were quickly stretched to the scales of galaxies and clusters of galaxies by the process of inflation.

Therefore, even though inflation with the rules of classical physics would predict a completely uniform mass density, the inherent probabilistic nature of quantum theory gives rise to small perturbations about the classically predicted value. The spectrum of these perturbations was first calculated during the exciting three-week period of the Nuffield Workshop.[39-42] After much disagreement and discussion, the various working groups carne to an agreement on the answer. I will describe these results in two parts.

First of all, we calculated the shape of the spectrum of the perturbations. The concept of a spectrum of density perturbations may seem a bit foreign, but the analogy of sound waves is very close. People familiar with acoustics understand that no matter how complicated a sound wave is, it is always possible to carry out a Fourier expansion, breaking it up into components that each have a standard (sine wave) wave form and a well-defined wavelength. The spectrum of the sound wave is specified by the strength of each of these components. In discussing density perturbations in the universe, it is similarly useful to define a spectrum by breaking up the perturbations into sine waves of well-defined wavelength.

For the inflationary model, we found that the predicted shape for the spectrum of density perturbations is essentially scale-invariant; that is, the strength of the perturbations is approximately equal on all wavelengths of astrophysical significance. While the precise shape of the spectrum depends on the details of the underlying particle physics, the approximate scale invariance holds in almost all cases. The scale-invariant spectrum is in agreement with a phenomenological model for galaxy formation proposed in the early 1970's by Edward R. Harrison[63] of the University of Massachusetts at Amherst and Yakov B. Zel'dovich[64] of the Institute of Physical Problems in Moscow, working independently.

Unfortunately, there is still no way of inferring the precise form of the primordial spectrum from observations, since one cannot reliably calculate how the universe evolved from the early period to the present. Such a calculation is very difficult in any case, and it is further complicated by the uncertainties about the nature of the dark matter. Nonetheless, the scale-invariant spectrum appears to be at least approximately what is needed to explain the evolution of galaxies, and thus this prediction of the inflationary model appears so far to be successful.

Furthermore, the recent results from the COBE satellite on nonuniformities in the cosmic background radiation provide a direct test of the spectrum as predicted by inflation.[h] The satellite measured δT, the deviation of the cosmic background radiation temperature from the mean, as a function of angular position (θ, ϕ). The COBE team then computed the correlation function $C(\alpha) \equiv \langle \delta T_1 \delta T_2 \rangle$, the average value of the product

---

[h]The COBE data was not available when this talk was given, but the result is so important that I could not possibly omit it from the written version.

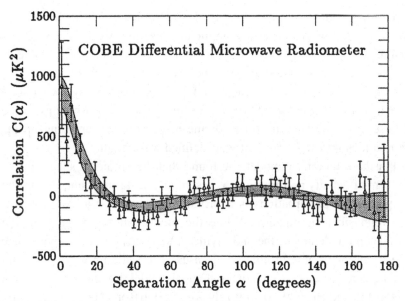

Fig. 10.   The COBE anisotropy results. The graph shows the correlation function $C(\alpha) \equiv \langle \delta T_1 \delta T_2 \rangle$, where δT denotes the deviation from the average of the cosmic background radiation temperature in a given direction, and the brackets denote an average over all pairs of directions that are separated from each other by an angle α. The shaded region denotes the range of values predicted by the inflationary model.

of δT at two different points,[i] averaged over all pairs of points with a given angular separation α. The plot of the COBE correlation function[18] is shown as Fig. 10. The shaded region indicates the range expected for a scale-invariant spectrum, normalized to the best fit of the data. Note that the prediction gives a range of expected values — it is a quantum-mechanical prediction, so only a probability distribution can be inferred. (Precisely, the shaded region indicates the mean plus or minus one standard deviation.) The graph shows a beautiful agreement with the predictions of inflation. Quantitatively, the spectrum can be characterized by a spectral index n, which is 1 for a scale-invariant spectrum. (The spectral power is proportional to k^n, where k is the inverse of the wavelength.) Analysis of the COBE data implies that $n = 1.1 \pm 0.5$.

---

[i]In evaluating δT, the effect of the earth's motion through the background radiation was subtracted by a best-fit procedure.

The angular resolution of the COBE satellite is about $7°$, so it is impossible for COBE to see the variations of the cosmic microwave background on small angular scales. Thus, the length scales probed by the COBE data are much longer than the length scales that are directly relevant to structure formation. It is possible, therefore, to improve the test of scale invariance by combining the COBE data with estimates of the spectrum at shorter wavelengths derived from theories of structure formation. When this is done,[65] one obtains a baseline spanning three decades in length scale, with a best-fit value $n = 1 \pm 0.23$. The uncertainties are smaller but still significant, and the agreement with the prediction of inflation remains excellent.

Galaxy formation is currently a very active subject of research, so a better determination of the spectrum of primordial density perturbations may be developed. Such a result would provide a more discriminating test of the inflationary model.

The predicted magnitude of the density perturbations was also calculated by the group at the Nuffield Workshop, but the implications of these results were much less clear. It was found that the predicted magnitude, unlike the shape of the spectrum, is very sensitive to the details of the underlying particle theory. At the time, the minimal SU(5) theory — the first and simplest of the grand unified theories — was strongly favored by anybody working on the subject. We were therefore very disappointed when we found that the minimal SU(5) theory leads to density perturbations with a magnitude 100,000 times larger than what is desired for the evolution of galaxies. Thus, there was a serious incompatibility between the inflationary model and the simplest of the grand unified theories.

With the passage of time, however, the credibility of the minimal SU(5) grand unified theory has diminished. The minimal SU(5) theory makes a rather definite prediction for the lifetime of a proton, and a variety of experiments have been set up to test this prediction by looking for proton decay. So far, no such decays have been observed, and the experiments have pushed the limit on the proton lifetime to the point where the minimal SU(5) theory is now excluded.

With the exclusion of the minimal SU(5) theory, a wide range of grand unified theories become plausible. All of the allowed theories seem a bit complicated, so apparently we will need some kind of new understanding to choose which — if any — is correct.

A variety of grand unified theories that predict an acceptable magnitude for both the proton lifetime and the density perturbations have been constructed. Thus, while the inflationary model cannot be credited with correctly predicting the magnitude of the perturbations, it also cannot be criticized for making a wrong prediction. The inflationary model provides at least a framework for calculating the magnitude of the density perturbations. If sometime in the future the correct particle theory and the values of its free parameters somehow become known, it will then be possible to make a real theoretical prediction for the magnitude of the perturbations.

A common feature of those models leading to acceptable density perturbations is the abandonment of the original hope that inflation can be driven by the Higgs fields that play an integral role in the grand unified theories. It appears that any Higgs field that interacts strongly enough to break the grand unified symmetry leads to density perturbations with a magnitude that is far too large. Thus it must be assumed that the underlying particle theory contains a new field — a field which strongly resembles the Higgs fields in its qualitative properties, but which interacts much more weakly. In particular, if the scalar field energy density contains a nonlinear[j] term $\lambda\phi^4$, then λ must be less than about 10^{-13}.

Unfortunately, all of the known theories that give acceptable predictions for the magnitude of the density perturbations look a little contrived. Well, to be honest the theories *were* contrived — with the goal of getting the density perturbations to come out right. The need for this contrivance can certainly be used as an argument against the inflationary model; but, in my opinion, this argument is considerably weaker than the arguments in favor of inflation. Even if we ignore cosmology, any known particle theory that is consistent with the known properties of particle physics appears to be rather contrived. In particular, there are a number of very small dimensionless quantities in nature that are not understood: the ratio of the weak energy scale to the Planck scale is about 10^{-17}, and the Yukawa

---

[j]The connection between nonlinear terms and particle interactions is well known to particle physicists, but has not yet been discussed in this paper. To understand the connection, note that classical wave packets described by linear equations can pass through each other as a linear superposition, with no interaction. Wave packets interact therefore only through the nonlinear terms in the equations of motion, and this connection carries over to the description of particles in the quantized theory.

coupling constant that describes the interaction of the electron and the Higgs field is about 10^{-6}. Clearly there are some fundamental principles at work here that we are still missing.

Finally, I want to mention that quantum effects during the inflationary era are not the only potential source of primordial density perturbations. One alternative possibility, which I will not discuss in detail, is that the seeds for galaxy formation may have been objects called "cosmic strings" (not related to "superstrings"). These strings are predicted by some (but not all) grand unified theories, and they would have formed in a random pattern during the grand unified theory phase transition. As their name suggests, strings are very thin, spaghetti-like objects that can form infinite curves or closed loops of astrophysical size. With a thickness of about 10^{-29} centimeter, a cosmic string has a mass of about 10^{22} grams for each centimeter of length. (In astronomical terms that is equal to about 10^7 solar masses per light-year.) In most theories, the density of these strings would be diluted to negligibility by the process of inflation. However, it is possible to construct theories in which the strings survive by forming either after inflation or at the very end of it. Cosmic strings produce a nearly scale-invariant spectrum of density perturbations very similar to inflation, so it is difficult to distinguish observationally. Models of this type still make use of inflation to answer Questions #1 and #2 of Table 1, and also to smooth out any small-scale inhomogeneities which may have been present in the initial conditions.

7. The Eternal Nature of Inflation

A fascinating feature of inflation, which in my opinion is also important in evaluating the plausibility of inflation, is the fact that inflation is eternal — if inflation ever begins, then it will never stop.[66-70]

To understand the endlessness of inflation, one first notices that the decay of the false vacuum, like the decay of many other unstable systems, is an exponential process. S.-Y. Pi and I[71] have verified the exponential decay law in a simplified but exactly soluble model of a slow-rollover phase transition, in which the potential energy function is taken as $V(\phi) = -\frac{1}{2}\mu^2\phi^2$. For the case of chaotic inflation, on the other hand, one might think that the scalar field would roll inexorably down the hill in the potential energy diagram, completing the decay in a finite time. Linde[68] has shown, however, that if the scalar field starts at a sufficiently high value, then it can be sustained by quantum fluctuations, with again an exponential decay law. As long as the false vacuum endures it drives an exponential expansion, and for reasonable parameters the rate of expansion is much faster than the rate of decay. Thus, even though the false vacuum is decaying, the total volume of the false vacuum region actually increases with time.

As time goes on, pieces of the false vacuum region are constantly undergoing decay. As each piece decays, it releases energy and thereby sets in motion a hot big bang universe. Other regions of false vacuum, however, continue to exponentially expand, so the false vacuum never disappears.

The infinity of universes produced in this way rapidly become causally disconnected, so there is no way for us to verify, even in principle, that the other universes exist. Nonetheless, I feel that the eternal character of inflation makes it a more plausible theory. In the absence of this feature, there is some difficulty in deciding whether the initial conditions required for inflation are sufficiently plausible. Since there is no established theory of initial conditions, questions of this sort can easily lead to inconclusive answers. Given the endlessness of inflation, however, the question becomes much less significant. Just as most of us accept the claim that complicated DNA molecules originated through random processes sometime during the history of the earth, we can also accept the claim that a region of false vacuum originated through random processes sometime during the history of spacetime. Just as primitive DNA molecules increased their abundance by replication, one patch of false vacuum would inflate to produce an infinity of universes, one of which could be the universe in which we live.

8. Conclusion

In summary, the inflationary universe model has been very successful in describing the broad, qualitative properties of the universe. In particular, the model provides very attractive answers to the three questions discussed in this lecture. While the model must be treated as speculative, I nonetheless feel that in its broad outline the concept of an inflationary universe is essentially correct.

The inflationary model makes two primary predictions — it predicts the mass density of the universe, and also the shape of the spectrum of primordial density perturbations. The prediction of the mass density is extremely firm, while the prediction of the perturbation spectrum depends weakly on the underlying particle physics. While neither of these predictions is straightforward to check, it seems likely that significant progress will be made in the foreseeable future. The COBE results have already provided a significant check on the spectrum of primordial density fluctuations, and the results are completely in accord with inflation.

Even if the inflationary model is correct, however, I must still emphasize that nothing discussed here is a completed project. The inflationary model is not a detailed theory — it is really just an outline for a theory — Michael Turner has called it the "inflationary paradigm." To fill in the details, we will need to know much more about the properties of particle physics at the energy scale of grand unified theories — and perhaps beyond. And to understand the evolution of the universe to the present time, we will of course have to identify the dark matter and solve a host of other astrophysical problems.

It looks to me as if the fields of particle physics and cosmology will remain closely linked for some years to come, as physicists and astronomers continue their efforts to understand the structure and history of the universe.

Oskar Klein was very much concerned with the question of finding the driving force behind the expansion of the universe, and he wrote a paper with Hannes Alfvén[72] proposing that the expansion was driven by matter–antimatter annihilation. This idea has not been very successful, as it is very difficult to understand how the extreme uniformity of the cosmic background radiation can emerge from any kind of ordinary explosion. It is always hard to judge another person's taste, but I suspect that Oskar Klein would be very happy to know that a plausible driving force behind the big bang can be found, coming directly from the Oklein–Gordon equation.

References

1. *NYNEX White Pages: Boston, Including Brookline, Cambridge, Somerville,* 1992 (NYNEX Information Resources Co., 1991).
2. For the reader who would like a more detailed but still nontechnical treatment of these topics, I would recommend the discussion of the standard big bang model by Steven Weinberg, *The First Three Minutes* (Basic Books, New York,

1977, 188 pp.), and the discussions of the more recent developments by John Gribbin, *In Search of the Big Bang: Quantum Physics and Cosmology* (Heinemann, London, 1986, 413 pp.), Lawrence M. Krauss, *The Fifth Essence: The Search for Dark Matter in the Universe* (Basic Books, New York, 1989, 342 pp.), and Michael Riordan and David N. Schramm, *The Shadows of Creation: Dark Matter and the Structure of the Universe* (W. H. Freeman and Company, New York, 1991, 277 pp.). At a more technical level, an excellent description of these developments can be found in Edward W. Kolb and Michael S. Turner, *The Early Universe* (Addison-Wesley, Redwood City, Calif., 1990, 547 pp.).

3. The history of Einstein's contributions is very well described in Abraham Pais, *"Subtle Is the Lord…": The Science and the Life of Albert Einstein* (Oxford University Press, New York, 1982, 552 pp.).
4. A. Einstein, "Die Grundlage der allgemeinen Relativitätstheorie," *Ann. Phys.* **49**, 769–822. English translation in Ref. 5.
5. H. A. Lorentz, A. Einstein, H. Minkowski, and H. Weyl, *The Principle of Relativity: A Collection of Original Memoirs on the Special and General Theory of Relativity.* Translated by W. Perrett and G. B. Jeffery. (Dover, New York, 1952, 216 pp.).
6. A. Einstein, "Kosmologische Betrachtungen zur allgemeinen Relativitätstheorie" ("Cosmological considerations on the general theory of relativity"), *Sitzungsberichte der Preussischen Akademie der Wissenschaften* **1917**, 142–152 (1917). English translations appear in Refs. 5 and 7.
7. J. Bernstein and G. Feinberg, eds., *Cosmological Constants: Papers in Modern Cosmology* (Columbia University Press, New York, 1987, 328 pp.).
8. For an informative discussion of Newton's involvement with this problem, see Edward Harrison, "Newton and the infinite universe," *Phys. Today*, February 1986, p. 24.
9. E. Hubble, "A relation between distance and radial velocity among extragalactic nebulae," *Proc. Natl. Acad. Sci.* **15**, 168–173 (1929). (Reprinted in Ref. 7.)
10. W. Baade, "Report of the meeting of the commission on extragalactic nebulae, Eight General Assembly (Rome, September 4–13, 1952)," *Transactions of the International Astronomical Union* **8**, 397–399 (1952).
11. A. Sandage, "Current problems in the extragalactic distance scale," *Astrophys. J.* **127**, 513–526 (1958).
12. A. A. Penzias and R. W. Wilson, "A measurement of excess antenna temperature at 4080 Mc/s," *Astrophys. J.* **142**, 414–419 (1965).

13. J. Silk, *The Big Bang* (W. H. Freeman and Company, New York, 1980), p. 78.

14. T. Matsumoto, S. Hayakawa, H. Matsuo, H. Murakami, S. Sato, A. E. Lange, and P. L. Richards, "The submillimeter spectrum of the cosmic background radiation," *Astrophys. J.* **329**, 567–571 (1988).

15. J. C. Mather *et al.*, "A preliminary measurement of the cosmic microwave background spectrum by the Cosmic Background Explorer (COBE) satellite," *Astrophys. J. Lett.* **354**, L37–L40 (1990).

16. Figure adapted from D. N. Schramm, "The early universe and high-energy physics," *Phys. Today*, April 1983, pp. 27–33.

17. J. Yang, M. S. Turner, G. Steigman, D. N. Schramm, and K. A. Olive, "Primordial nucleosynthesis: a critical comparison of theory and observation," *Astrophys. J.* **281**, 493–511 (1984).

18. G. F. Smoot *et al.*, "Structure in the COBE differential microwave radiometer first year maps," *Astrophys. J. Lett.* **396**, L1–L6 (1992).

19. R. H. Dicke and P. J. E. Peebles, "The big bang cosmology — enigmas and nostrums," in *General Relativity: An Einstein Centenary Survey*, eds. S. W. Hawking and W. Israel (Cambridge University Press, Cambridge, 1979), pp. 504–517.

20. For the reader who would like a more detailed but still nontechnical treatment, I would recommend Refs. 21 and 22. For the reader looking for a more technical review, I would recommend any of Refs. 23–28.

21. Edward P. Tryon, "Cosmic Inflation," in *The Encyclopedia of Physical Science and Technology* (Academic, New York, 1987), Vol. 3, pp. 709–743.

22. Alan H. Guth and Paul J. Steinhardt, "The inflationary universe," in *The New Physics*, ed. Paul Davies (Cambridge University Press, Cambridge, 1988), pp. 34–60.

23. L. F. Abbott and S. Y. Pi, eds., *Inflationary Cosmology* (World Scientific, Singapore, 1986, 697 pp.).

24. Andrei Linde, *Particle Physics and Inflationary Cosmology* (Harwood, Chur, Switzerland, 1990, 362 pp.).

25. Michael S. Turner, "Cosmology and particle physics," in *Architecture of Fundamental Interactions at Short Distances*, eds. P. Ramond and R. Stora (North-Holland, Amsterdam, 1987), pp. 513–680.

26. Robert H. Brandenberger, "Quantum field theory methods and inflationary universe models," *Rev. Mod. Phys.* **57**, 1–60 (1985).

27. Paul J. Steinhardt, "Inflationary cosmology," in *High Energy Physics, 1985, Proceedings of the Yale Theoretical Advanced Study Institute*, eds. M. J. Bowick and F. Gürsey (World Scientific, 1986), Vol. 2, pp. 567–617.

28. S. K. Blau and A. H. Guth, "Inflationary cosmology," in *300 Years of Gravitation*, eds. S. W. Hawking and W. Israel (Cambridge University Press, Cambridge, 1987), pp. 524–603.

29. A. H. Guth, "Inflationary universe: a possible solution to the horizon and flatness problems," *Phys. Rev.* **D23**, 347–356 (1981).

30. A. D. Linde, "A new inflationary universe scenario: a possible solution of the horizon, flatness, homogeneity, isotropy and primordial monopole problems," *Phys. Lett.* **108B**, 389–393 (1982).

31. A. Albrecht and P. J. Steinhardt, "Cosmology for grand unified theories with radiatively induced symmetry breaking," *Phys. Rev. Lett.* **48**, 1220–1223 (1982).

32. P. Jordan and O. Klein, "Zum Mehrkörperproblem der Quantentheorie" ("On the many-body problem of quantum theory"), *Z. Phys.* **45**, 751–765 (1927).

33. O. Klein, "Elektrodynamik und Wellenmechanik vom Standpunkt des Korrespondenzprinzips" ("Electrodynamics and wave mechanics from the standpoint of the correspondence principle"), *Z. Phys.* **41**, 407–442 (1927).

34. Sidney Coleman, "Why there is nothing rather than something: a theory of the cosmological constant," *Nucl. Phys.* **B310**, 643 (1988).

35. A. D. Linde, "Chaotic inflating universe," *Pis'ma Zh. Eksp. Teor. Fiz.* **38**, 149–151 (1983). [English translation: *JETP Lett.* **38**, 176–179 (1983)].

36. A. D. Linde, "Chaotic inflation," *Phys. Lett.* **129B**, 177–181 (1983).

37. D. La and P. J. Steinhardt, "Extended inflationary cosmology," *Phys. Rev. Lett.* **62**, 376–378 (1989).

38. P. J. Steinhardt and F. S. Accetta, "Hyperextended inflation," *Phys. Rev. Lett.* **64**, 2740–2743 (1990).

39. A. A. Starobinsky, "Dynamics of phase transition in the new inflationary universe scenario and generation of perturbations," *Phys. Lett.* **117B**, 175–178 (1982).

40. A. H. Guth and S.-Y. Pi, "Fluctuations in the new inflationary universe," *Phys. Rev. Lett.* **49**, 1110–1113 (1982).

41. S. W. Hawking, "The development of irregularities in a single bubble inflationary universe," *Phys. Lett.* **115B**, 295–297 (1982).

42. J. M. Bardeen, P. J. Steinhardt, and M. S. Turner, "Spontaneous creation of almost scale-free density perturbations in an inflationary universe," *Phys. Rev.* **D28**, 679–693 (1983).

43. R. Brandenberger, H. Feldman and J. Kung, "Initial conditions for chaotic inflation," *Physica Scripta* **T36**, 64–69 (1991). Also published in *The Birth and Early Evolution of Our Universe*, eds. J. S. Nilsson, B. Gustafsson, and B.-S. Skagerstam (World Scientific, Singapore, 1991), pp. 64–69.

44. E. P. Tyron, "Is the universe a vacuum fluctuation?" *Nature* **246**, 396–397 (1973).
45. A. Vilenkin, "Creation of universes from nothing," *Phys. Lett.* **117B**, 25–28 (1982).
46. A. Vilenkin, "Quantum origin of the universe," *Nucl. Phys.* **B252**, 141–151 (1985).
47. A. D. Linde, "The new inflationary universe scenario," in *The Very Early Universe: Proceedings of the Nuffield Workshop*, eds. G. W. Gibbons, S. W. Hawking, and S. T. C. Siklos (Cambridge University Press, Cambridge, England, 1983), pp. 205–249.
48. A. D. Linde, "Quantum creation of the inflationary universe," *Lett. Nuovo Cimento* **39**, 401–405 (1984).
49. A. D. Linde, "Inflation and quantum cosmology," in *300 Years of Gravitation*, eds. S. W. Hawking and W. Israel (Cambridge University Press, Cambridge, 1987), pp. 604–630.
50. J. B. Hartle and S. W. Hawking, "Wave function of the universe," *Phys. Rev.* **D28**, 2960–2975 (1983).
51. S. Coleman, "The fate of the false vacuum, 1: semiclassical theory," *Phys. Rev.* **D15**, 2929–2936 (1977) [Errata: **16**, 1248 (1977).]
52. C. G. Callan and S. Coleman, "The fate of the false vacuum, 2: first quantum corrections," *Phys. Rev.* **D16**, 1762–1768 (1977).
53. S. W. Hawking, I. G. Moss, and J. M. Stewart, "Bubble collisions in the very early universe," *Phys. Rev.* **D26**, 2681–2693 (1982).
54. A. H. Guth and E. J. Weinberg, "Could the universe have recovered from a slow first order phase transition?" *Nucl. Phys.* **B212**, 321–364 (1983).
55. See, for example, E. W. Kolb and M. S. Turner, "Grand unified theories and the origin of the baryon asymmetry," *Annu. Rev. Nucl. Part. Sci.* **33**, 645–696 (1983).
56. V. A. Kuzmin, V. A. Rubakov, and M. E. Shaposhnikov, "On the anomalous electroweak baryon number nonconservation in the early universe," *Phys. Lett.* **155B**, 36 (1985).
57. L. McLerran, M. Shaposhnikov, N. Turok, and M. Voloshin, "Why the baryon asymmetry of the universe is $\sim 10^{-10}$," *Phys. Lett.* **256B**, 451 (1991).
58. V. A. Rubakov, "Electroweak baryon number violation at high energies," *Physica Scripta* **T36**, 194–198 (1991). Also published in *The Birth and Early Evolution of Our Universe*, eds. J. S. Nilsson, B. Gustafsson, and B.-S. Skagerstam (World Scientific, Singapore, 1991), pp. 194–198.
59. A. E. Nelson, D. B. Kaplan, and A. G. Cohen, "Why there is something rather than nothing: matter from weak interactions," *Nucl. Phys.* **B373**, 453–478 (1992).

60. P. J. E. Peebles, "The mean mass density of the universe," *Nature* **321**, 27–32 (1986).

61. N. Kaiser, G. Efstathiou, R. Ellis, C. Frenk, A. Lawrence, M. Rowan-Robinson, and W. Saunders, "The large-scale distribution of IRAS galaxies and the predicted peculiar velocity field," *Mon. Not. R. Astron. Soc.* **252**, 1–12 (1991).

62. A. Yahil, "The quest for Ω: comparison of density and peculiar velocity fields," in *The Early Universe and Cosmic Structures* (Proceedings of the XXV Rencontre de Moriond, March 4–11, 1990), eds. J. M. Alimi *et al.* (Editions Frontières, Gif-sur-Yvette, 1991), pp. 483–500.

63. E. R. Harrison, "Fluctuations at the threshold of classical cosmology," *Phys. Rev.* **D1**, 2726–2730 (1970).

64. Ya. B. Zel'dovich, "A hypothesis, unifying the structure and the entropy of the universe," *Mon. Not. R. Astron. Soc.* **160**, 1P–3P (1972).

65. E. L. Wright *et al.*, "Interpretation of the CMB anisotropy detected by the COBE differential microwave radiometer," *Astrophys. J. Lett.* **396**, L13–L18 (1992).

66. A. Vilenkin, "The birth of inflationary universes," *Phys. Rev.* **D27**, 2848 (1983).

67. A. A. Starobinskii, "Stochastic de Sitter (inflationary) stage in the early universe," in *Field Theory, Quantum Gravity and Strings*, eds. H. J. de Vega and N. Sánchez (Springer-Verlag Lecture Notes in Physics, Vol. 246, Berlin, 1986), pp. 107–126.

68. A. D. Linde, "Eternal chaotic inflation," *Mod. Phys. Lett.* **A1**, 81 (1986).

69. A. S. Goncharov, A. D. Linde, and V. F. Mukhanov, "The global structure of the inflationary universe," *Int. J. Mod. Phys.* **A2**, 561–591 (1987).

70. M. Aryal and A. Vilenkin, "The fractal dimension of inflationary universe," *Phys. Lett.* **199B**, 351–357 (1987).

71. A. H. Guth and S.-Y. Pi, "Quantum mechanics of the scalar field in the new inflationary universe," *Phys. Rev.* **D32**, 1899–1920 (1985).

72. H. Alfvén and O. Klein, "Matter and antimatter annihilation in cosmology," *Ark. Fys.* **23**, 187–194 (1962).

DO THE LAWS OF PHYSICS ALLOW US
TO CREATE A NEW UNIVERSE?*

Alan H. Guth[†]

Center for Theoretical Physics
Laboratory for Nuclear Science and Department of Physics
Massachusetts Institute of Technology
Cambridge, Massachusetts 02139, USA

Essentially all modern particle theories suggest the possible existence of a false vacuum state — a metastable state with an energy density that cannot be lowered except by means of a very slow phase transition. Inflationary cosmology makes use of such a state to drive the expansion of the big bang, allowing the entire observed universe to evolve from a very small initial mass. A sphere of false vacuum in our present universe, if larger than a certain critical mass, could inflate to form a new universe which would rapidly detach from its parent. A false vacuum bubble of this size, however, cannot be produced classically unless an initial singularity is present from the outset. E. Farhi, J. Guven, and I have therefore explored the possibility that a bubble of subcritical size, which classically would evolve to a maximum size and collapse, might instead tunnel through a barrier to produce a new universe. We estimated the tunneling rate using semiclassical quantum gravity, and discovered some interesting ambiguities in the formalism.

1. The Cosmic Cookbook

During the past decade, a radically new picture of cosmology has emerged. The novelty of this picture is particularly striking when one considers the question of what it takes to produce a universe. According to the

---

*Based on a lecture given at Nobel Symposium 79, Gräftåvallen, Sweden, June 11–16, 1990, in *The Birth and Early Evolution of Our Universe*, eds. J. S. Nilsson, B. Gustafsson, and B.-S. Skagerstam (*Physica Scripta*, Vol. T36, and World Scientific, Singapore, 1991), pp. 237–246.
[†]This work was supported in part by funds provided by the US Department of Energy (DOE) under contract #DE-AC02-76ER03069.

standard big bang picture of a decade ago, the visible universe could be assembled, at $t \approx 1$ sec for example, by mixing approximately 10^{89} photons, $10^{89} e^+ e^-$ pairs, 10^{89} $\nu \bar{\nu}$ pairs, 10^{79} protons, 10^{79} neutrons, and 10^{79} unpaired electrons. The total mass/energy of these ingredients is about 10^{65} grams $\approx 10^{32}$ solar masses $\approx 10^{10} \times$ present mass of visible universe. The mass is much larger than the present mass of the visible universe, because most of the energy is lost to gravitational potential energy as the universe expands.

With the advent of grand unified theories (GUT's) and inflationary cosmology, however, a much simpler recipe for a universe can now be formulated. (For a review of inflation, see any of Refs. 1–8.) To produce a universe at $t \approx 10^{-35}$ sec, the *only* necessary ingredient is a region of false vacuum. And the region need not be very large. For a typical GUT energy scale of $\sim 10^{14}$ GeV (which I will use for all the numerical examples in this paper), the minimum diameter is about 10^{-24} cm. The total mass/energy of this ingredient is only about $10 \, \text{kg} \approx 10^{-29}$ solar masses $\approx 10^{-51} \times$ present mass of visible universe. This recipe sounds so easy that one cannot resist asking whether the laws of physics as we know them in principle permit the creation of a new universe by man-made processes. We believe that the answer to this question is yes,[9–11] but we do not yet have a definite answer. The problem turns out to be intimately embroiled in the ambiguities of quantum gravity. In this paper I will try to summarize our current understanding.

In considering this question, one difficulty is immediately obvious: the mass density of the required false vacuum is about 10^{75} g/cm^3. This mass density is certainly far beyond anything that is technologically possible, either now or in the foreseeable future. Nonetheless, for the purposes of this discussion I will whimsically assume that some civilization in the distant future will be capable of manipulating these kinds of mass densities. There are then some very interesting questions of principle that must be addressed in order to decide if the creation of a new universe is possible. While I will discuss these questions in terms of the possibility of man-made creation, I want to emphasize that the same questions will no doubt also have relevance to various natural scenarios that one could imagine. The question of man-made creation helps to simplify the discussion by focusing attention on the fundamental issue: what are the minimal conditions necessary for a new universe to be created?

The outline of this paper will be as follows. Section 2 discusses the classical evolution of a false vacuum bubble, and Sec. 3 will discuss the role of initial singularities in the classical evolution. Section 4 will be the main section of the paper, describing the formulation of a quantum gravity tunneling mechanism for producing a new universe. Section 5 will focus on a subtlety in the calculation — the evaluation of the subtraction term. Numerical results will be presented in Sec. 6, and the conclusions will be summarized and discussed in Sec. 7.

2. Evolution of a False Vacuum Bubble

A false vacuum is a peculiar state of matter that arises when a particle theory contains a scalar field ϕ with a potential energy function that has a local minimum which is different from the global minimum. A false vacuum is then a region of space in which $\phi = \phi_f$, where ϕ_f denotes the value of ϕ at the local minimum. The energy density of the false vacuum is then fixed at the value of the potential at the local minimum, which we denote by ρ_f. (I use units with $\hbar = c = 1$, so ρ_f is both an energy density and a mass density.) The energy–momentum tensor for this state is given by the Lorentz-invariant expression

$$T_{\mu\nu} = -\rho_f g_{\mu\nu}. \tag{1}$$

The energy density and pressure are then given by

$$\rho = \rho_f \tag{2}$$

and

$$p = -\rho_f. \tag{3}$$

By a false vacuum bubble, I mean a region of false vacuum surrounded by anything else. I will discuss in detail, however, only the simplest case: a spherical region of false vacuum surrounded by true vacuum, assumed to have zero energy density. In considering the evolution of such a region one is led immediately to a paradox. If the region of false vacuum is large enough, one expects that it would undergo inflation. An observer in the outside true vacuum region, on the other hand, would see the false vacuum region as a region of negative pressure. The pressure gradient would point inward, and the observer would not expect to see the region increase in size. The resolution of this paradox hinges on the dramatic distortion of spacetime that is caused by the false vacuum bubble.

Here I will outline the solution to the classical evolution problem, but a reader interested in the technical details will have to consult the literature.[12–18] The description given here follows the work that I did with Blau and Guendelman.[17]

The first step in solving the problem is to dissect it, dividing the space-time into three regions. The exterior region is spherically symmetric empty space, for which the unique solution, in general relativity, is the well-known Schwarzschild metric. The interior region consists of spherically symmetric false vacuum, and is required to be regular at $r = 0$. This spacetime also has a unique solution: de Sitter space. At the interface between these two regions is a domain wall — a region in which a scalar field is undergoing a transition between its true and false vacuum values. The solution that I will describe uses a thin-wall approximation, in which the thickness of the wall is assumed to be negligibly small compared to any other distance in the problem. In this approximation it can be shown that the surface energy density is equal to the surface tension and is independent of time, and we take this surface energy density σ as an additional parameter of the problem. The wall can then be described mathematically by a set of junction conditions[19] which are obtained by applying Einstein's equations to an energy–momentum tensor restricted to a thin sheet. (These equations are just the gravitational analogue of the well-known statement that the normal component of an electric field has a discontinuity of $4\pi\sigma$ at a sheet of surface charge density σ). The evolution is completely determined by using these junction conditions to join the exterior forms of the metric.

The evolution of the bubble wall can be described by the function $r(\tau)$, where r is the radius of the bubble wall (defined as $\frac{1}{2\pi}$ times the circumference), and τ is the proper time as would be measured by a clock that follows the bubble wall. The junction conditions described above imply an equation of motion for $r(\tau)$ that can be cast into a form identical to that of a nonrelativistic particle moving in a potential. Using the definitions

$$\chi^2 \equiv \frac{8\pi}{3} G\rho_f,$$

(4)

$$\kappa \equiv 4\pi G\sigma,$$

(5)

and

$$\gamma \equiv \frac{2\kappa}{\sqrt{\chi^2 + \kappa^2}},$$

(6)

the equations of motion can be written[17]

$$\left(\frac{dz}{d\tau'}\right)^2 + V(z) = E,$$

(7)

where z and τ' are rescaled radius and time variables defined by

$$z^3 \equiv \frac{\chi^2 + \kappa^2}{2GM} r^3$$

(8)

and

$$\tau' \equiv \frac{\chi^2 + \kappa^2}{2\kappa} \tau.$$

(9)

E is related to the mass M of the physical bubble by

$$E \equiv \frac{-4\kappa^2}{(2GM)^{2/3}(\chi^2 + \kappa^2)^{4/3}},$$

(10)

and the potential energy function $V(z)$ is given by

$$V(z) = -\left[\frac{1 - z^3}{z^2}\right]^2 - \frac{\gamma^2}{z}.$$

(11)

A graph of $V(z)$ is shown as Fig. 1. As can be seen in this figure, there are three kinds of solutions. First, there are "bounded" solutions in which the bubble grows from $r = 0$ to a maximum size and then collapses. Second, there are "bounce" solutions. Here the bubble starts at infinite size in the asymptotic past, contracts to a minimum size, and then expands without limit. Finally, there are "monotonic" solutions — bubbles that start at zero size and grow monotonically. The monotonic solutions require a minimum

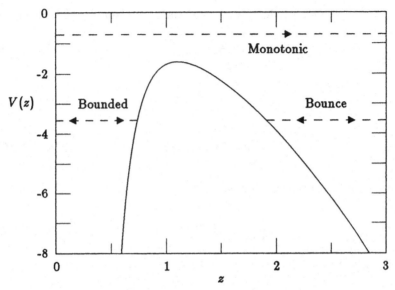

Fig. 1. In the thin-wall approximation, the trajectory of the bubble wall is equivalent to the motion of a nonrelativistic particle in the potential energy curve V shown above. The energy of the fictitious particle is related to the mass of the false vacuum bubble; the energy increases with the mass, and approaches the top line of the diagram as $M \to \infty$. The curve was plotted with $\gamma = 1.3$.

mass, so that the energy of the fictitious particle is high enough to get over the potential barrier in Fig. 1. For small values of the surface energy density $(\sigma \ll \sqrt{\rho_f/G})$, this critical mass is given simply by

$$M_{cr} = \frac{4\pi}{3}\rho_f \chi^{-3}. \tag{12}$$

This expression can be interpreted as the mass density of the false vacuum multiplied by the volume of a Euclidean sphere with a radius equal to the de Sitter horizon distance, χ^{-1}. For typical GUT parameters, $M_{cr} \approx 10\,\text{kg}$.

Having described the evolution of the bubble wall, I must still describe how the bubble wall is embedded in spacetime. For now I will describe only the behavior of the monotonic solutions. Since the exterior of the bubble is Schwarzschild space, it can be described by the familiar Schwarzschild coordinate system, with metric

$$ds^2 = -A_S dT_S^2 + A_S^{-1}dR^2 + R^2 d\Omega^2, \tag{13}$$

where

$$A_S = 1 - \frac{2GM}{R},$$ (14)

$d\Omega^2 = d\theta^2 + \sin^2\theta d\phi^2$, and M is the total energy of the bubble. Since these coordinates do not cover all of Schwarzschild space and since they have a coordinate singularity at $R = 2GM$, it is useful to introduce Kruskal–Szekeres[20] coordinates. The new coordinates (V_S, U_S) replace (T, R), while θ and ϕ are kept. These coordinates cover the entire manifold and have no coordinate singularities, and they also have the convenient property that lightlike lines lie at 45° to the vertical. The Schwarzschild R-coordinate is related to V_S and U_S by

$$\left(\frac{R}{2GM} - 1\right) e^{R/2GM} = U_S^2 - V_S^2.$$ (15)

Lines of constant T_S are straight lines through the origin of the (V_S, U_S) plane, with

$$T_S = \begin{cases} 4GM \, \tanh^{-1}(V_S/U_S) & \text{if } |V_S/U_S| < 1 \\ 4GM \, \tanh^{-1}(U_S/V_S) & \text{if } |V_S/U_S| > 1. \end{cases}$$ (16)

The relation between the two coordinate systems is described more fully in Ref. 17. A diagram showing the relation between the Schwarzschild and the Kruskal–Szekeres coordinate systems is shown as Fig. 2. The metric for the Kruskal–Szekeres coordinates can then be written as

$$ds^2 = \frac{32(GM)^3}{R} e^{-R/2GM}(-dV_S^2 + dU_S^2) + R^2 d\Omega^2,$$ (17)

where R is expressed implicitly as a function of V_S and U_S by using Eq. (15).

A spacetime diagram for the monotonic trajectory is shown as Fig. 3. The true vacuum region, to the right of the bubble wall, is shown in the standard Kruskal–Szekeres coordinates. The false vacuum region, to the left of the bubble wall, is shown in peculiar coordinates designed solely to allow the two halves of the diagram to fit together in the plane. The diagram is constructed so that lightlike lines lie at 45° to the vertical, but the metric is highly distorted. In particular the exponential expansion of the false vacuum region, which occurs as one moves upward and to the left in the diagram, is completely hidden by the distortion of the metric.

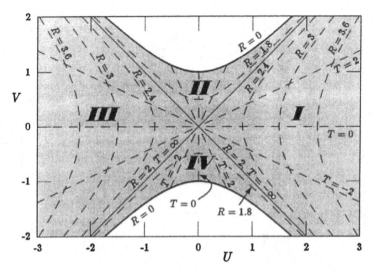

Fig. 2. A diagram of the Kruskal–Szekeres coordinate system. The four quadrants are labeled I, II, III, and IV, while the unshaded region is not part of the spacetime. Lightlike lines are at 45° to the vertical, and the future direction is shown as upward. The relation to the Schwarzschild coordinates is illustrated by lines of constant T and R, labeled in units of GM.

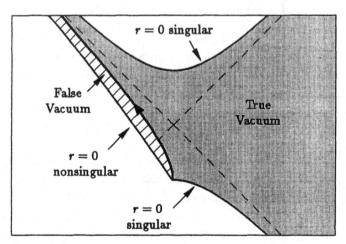

Fig. 3. A spacetime diagram of a monotonic false vacuum bubble solution. Angular coordinates are suppressed, and the diagram is plotted so that lightlike lines are at 45°. The bubble wall is shown as a heavy line with an arrow on it. The true vacuum region (shaded) is to the right of the bubble wall, and the false vacuum region (diagonal lines) is to the left. The diagram shows the initial (lower) and final (upper) $r = 0$ singularities, and also a nonsingular $r = 0$ line (i.e., the center of a spherical coordinate system) that runs along the left edge.

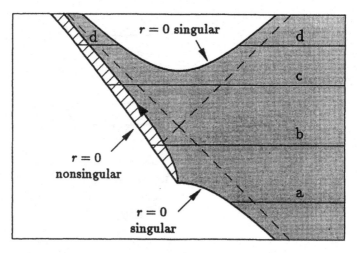

Fig. 4.   Horizontal lines indicating spacelike hypersurfaces to be illustrated in Fig. 5.

The physical meaning of a spacetime diagram of this type can be seen most clearly by examining a sequence of equal-time slices. Figure 4 shows the positions of four slices, labeled (a), (b), (c), and (d), and Fig. 5 shows a representation of each slice. For purposes of illustration, Fig. 5 shows only two of the three spatial dimensions. Since the spaces of interest are spherically symmetric, this results in no loss of information. The two-dimensional sheet is shown embedded in a fictitious third dimension, so that the curvature can be visualized. Figure 5(a) shows a space which is flat at large distances, but which has a singularity at the origin. In Fig. 5(b) a small, expanding region of false vacuum has appeared at the center, replacing the singularity. The false vacuum region is separated from the rest of space by a domain wall. Figure 5(c) shows the false vacuum region beginning to swell. Note, however, that the swelling takes place by the production of new space; the plane of the original space is unaffected. The false vacuum region continues to inflate, and it soon disconnects completely from the original space, as shown in Fig. 5(d). As was shown by Sato *et al.*[12] in 1981, it forms a new, isolated closed universe.

Note, by the way, that Fig. 5 shows clearly how the paradox raised at the beginning of this section is resolved. The net force on the bubble wall points from the true vacuum region to the false vacuum region, as expected.

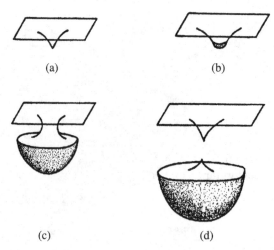

(a)   (b)

(c)   (d)

Fig. 5. The evolution of a monotonic false vacuum bubble solution. Each lettered diagram illustrates a spacelike hypersurface indicated in Fig. 4. The diagrams are drawn by suppressing one dimension of the hypersurface and embedding the resulting two-dimensional surface in a fictitious three-dimensional space so that the curvature can be displayed. The false vacuum region is shown as shaded. Note that diagram (d) shows a new universe detaching from the original spacetime.

Due to the inversion shown in Figs. 5(c) and 5(d), however, this force causes the bubble wall to expand, rather than contract.

To summarize, the false vacuum bubble appears from the outside to be a black hole. From the inside, however, it appears to be an inflating region of false vacuum, with new space being created as the region expands. The region completely disconnects from the original spacetime, forming a new, isolated closed universe.

Although the problem that has been solved is very idealized, it nonetheless appears to contain the essential physics of more complicated inhomogeneous spacetimes. The paradox discussed at the beginning of this section will exist whenever an inflating region is surrounded by noninflating regions, and the qualitative behavior of the system seems to be determined by the way in which this paradox is resolved. Thus, one concludes that if inflation occurred in an inhomogeneous universe, then many isolated closed universes would have been ejected.

Furthermore, even if inflation somehow began in a completely homogeneous way, one still expects the universe to break apart into a host of disconnected universes. The reason stems from the intrinsic nonuniformity, on very large scales, of the decay of the false vacuum.[21-23] The decay of the false vacuum occurs exponentially,[a] like most other decay processes, but for inflation to be successful the parameters must be arranged so that the exponential decay constant is slow compared to the exponential expansion rate. This implies that the total volume of false vacuum *increases* with time. Thus, no matter how long one waits there will still be regions of false vacuum. These regions have no reason to be spherical, but the arguments of the previous paragraph lead one to expect a high likelihood of producing disconnected universes.

3. Initial Singularities as an Obstacle to Universe Creation

Figures 4–5 illustrate the creation of a new universe, but there is one undesirable feature. The sequence begins with an initial singularity, shown as the lower $r = 0$ singularity in Fig. 4, and as part (a) of Fig. 5. Although an initial singularity is often hypothesized to have been present at the big bang, there do not appear to be any initial singularities available today. So we ask whether it is possible to intervene in some way, to modify the early stages of this picture, so that an inflationary universe could be produced *without* an initial singularity. This question has been addressed at both the classical and quantum levels.

At the classical level Farhi and I[25] have shown that, under reasonable assumptions, the initial singularity cannot be avoided. Any false vacuum bubble which grows to become a universe necessarily begins from an initial singularity.

The argument rests on an application of the Penrose theorem.[b] The inflationary solutions are very rapidly expanding, and the Penrose theorem implies that such rapid expansion can result only from an initial singularity. (The Penrose theorem is more widely known in a form that is the time-reverse of the present application: if a system is collapsing fast enough, there is no way to avoid the collapse to a singularity.)

---

[a] I have studied a simplified but exactly soluble model of a slow-rollover phase transition with S.-Y. Pi in Ref. 24.

[b] J. Bardeen, W. Israel, W. Unruh, and R. Wald are to be thanked for pointing out to us the relevance of this theorem.

The application of the Penrose theorem involves two technical loop-holes. First, if the final bubble is not spherically symmetric, then we have not been able to show that the Penrose theorem applies. We believe that this shortcoming, however, is probably the result of our own limitations, and does not provide a way to avoid the theorem. Second, if a material can be found for which the magnitude of the pressure exceeds the energy density, then the Penrose theorem would not apply. In quantum field theories it is possible to construct states that have this property, but it is not clear if a large enough spacetime region of this type can be attained.

4. The Creation of a Universe by Quantum Tunneling

Although the Penrose theorem implies that a new universe cannot be created classically without an initial singularity, the Penrose theorem does not apply in the realm of quantum mechanics. Since the theorem is derived from the classical equations of motion, one is always free to look for violations at the quantum level. Farhi, Guven, and I[9] have therefore turned our attention to whether quantum physics allows the creation of an inflationary universe without an initial singularity. In particular, we have explored the following recipe. Suppose a small bubble of false vacuum (with mass less than the critical mass M_{cr}) is created and caused to expand at a moderate rate. Since the bubble is not expanding rapidly, the Penrose theorem does not preclude its production by classical processes without an initial singularity. We have not explored in detail the mechanisms by which such a region might be created, but presumably it could be created either by supercooling from high temperatures or perhaps by compressing a gas of fermions that couple to the scalar field. If such a bubble were allowed to evolve classically, it would correspond to one of the bounded solutions, as discussed in the context of Fig. 1. It would expand to a maximum size and then the pressure gradient would halt the expansion and cause the bubble to collapse. By quantum processes, however, one might imagine that the bubble could tunnel through the potential energy barrier shown in Fig. 1, becoming a bounce solution that would continue to grow until eventually the false vacuum decayed. The late-time behavior of this bounce solution would strongly resemble that shown in Figs. 5(c) and 5(d). Although no fully satisfactory theory of quantum gravity exists, we have attempted to *estimate* the tunneling amplitude by using a semiclassical (WKB) approximation.

Specifically, we used the same kind of Euclidean field theory technique that was used by Coleman and De Luccia[26] to calculate the decay rate of the false vacuum in curved spacetime. (See also Sec. 6 of Ref. 27, which includes a discussion of a spacetime region that was omitted in the original reference.) That is, we assume that the amplitude for a transition from one three-geometry to another is well-approximated by $e^{iI_{cl}/\hbar}$, where I_{cl} is the action of the classical solution to the field equations which interpolates between the two three-geometries. If no real-time solution exists then we seek a Euclidean four-geometry that solves the imaginary time field equation and whose boundary is the two three-geometries of interest. The tunneling amplitude is then estimated as $e^{-I_E/\hbar}$, where I_E is the properly subtracted classical action of the Euclidean solution — that is, it is the action of the solution, minus the action of a configuration that remains static at the initial state of the tunneling process for the same Euclidean time as the solution requires for its transit. (The motivation for this subtraction term will be discussed in Sec. 5.)

Thus, the key step that remains is to construct the Euclidean solution that interpolates between the initial and final states of the tunneling.

In Sec. 2 we discussed the Kruskal–Szekeres coordinate system for Schwarzschild space. To continue, we must similarly define our coordinate system for the de Sitter part of the problem. One well-known system is the static coordinates, closely analogous to the Schwarzschild form of the metric in Schwarzschild space. The static metric is given by

$$ds^2 = -A_D dT_D^2 + A_D^{-1} dR^2 + R^2 d\Omega^2, \tag{18}$$

where

$$A_D = 1 - \chi^2 R^2. \tag{19}$$

These coordinates have a coordinate singularity at $R = 1/\chi$, and they fail to cover the entire de Sitter manifold. Both of these shortcomings can be avoided by replacing T_D and R by the Gibbons–Hawking[28] coordinates (V_D, U_D), which are analogous to the Kruskal–Szekeres coordinate system. Once again θ and ϕ are kept. The static coordinate R is related to V_D and

U_D by

$$\frac{1 - \chi R}{1 + \chi R} = U_D^2 - V_D^2. \tag{20}$$

Lines of constant T_D are again straight lines through the origin, with

$$T_D = \begin{cases} \chi^{-1}\tanh^{-1}(V_D/U_D) & \text{if } |V_D/U_D| < 1 \\ \chi^{-1}\tanh^{-1}(U_D/V_D) & \text{if } |V_D/U_D| > 1. \end{cases} \tag{21}$$

The relation between the two de Sitter coordinate systems is illustrated in Fig. 6, and it is also described more fully in Ref. 17. The metric in the Gibbons–Hawking system is given by

$$ds^2 = \chi^{-2}(1 + \chi R)^2(-dV_D^2 + dU_D^2) + R^2 d\Omega^2, \tag{22}$$

where R is determined by Eq. (20). Again lightlike lines travel at 45°. The presence of the term $R^2 d\Omega^2$ in all of the metrics that we have introduced implies that R always has the same meaning: the rotational symmetry

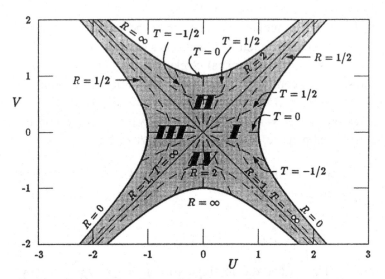

Fig. 6.   A diagram of the Gibbons–Hawking coordinate system of de Sitter space, which is a close analogue of the Kruskal–Szekeres coordinate system of Schwarzschild space. The four quadrants are labeled I, II, III, and IV, and again the spacetime does not exist outside the shaded region. The relation to the static de Sitter coordinates is shown by lines of constant R and T, labeled in units of χ^{-1}.

relates any given point to a sphere of points, and R is equal to $\frac{1}{2\pi}$ times the circumference of this sphere.

We are now ready to describe the classical trajectories involved in the tunneling process. Specifically, we wish to estimate a tunneling rate from a bounded trajectory to a bounce trajectory, as shown in Fig. 1. The form of the trajectories depends on the mass of the bubble, and we begin by considering a bubble just below the critical mass. A trajectory of this type is shown in the upper part of Fig. 7. The left and right halves of the diagram show the trajectory of the bubble wall as seen in the interior and exterior coordinate systems, respectively.

The initial slice for the tunneling problem is taken to be the $T_S = 0$ surface in the Schwarzschild region and the $T_D = 0$ surface in the de Sitter region, as shown in the diagram. The final trajectory is shown in the lower part of Fig. 7. The final slice is again taken along the $T_S = 0$ and $T_D = 0$ surfaces. Note that the extrinsic curvature vanishes on both the initial and final slices. This fact is the geometric generalization of the condition that $\dot{q} = 0$ at the turning points marking the beginning and end of a tunneling trajectory in a one-degree-of-freedom quantum mechanics problem.

To construct the Euclidean interpolation, we begin by considering the de Sitter half of the problem. Euclidean de Sitter space is a four-sphere with radius $1/\chi$. The Euclidean continuation of the Gibbons–Hawking form of the metric (22) is

$$ds_E^2 = \chi^{-2}(1 + \chi R)^2(dV_{DE}^2 + dU_{DE}^2) + R^2 d\Omega^2, \tag{23}$$

where $0 \le R \le 1/\chi$ and

$$\frac{1 - \chi R}{1 + \chi R} = U_{DE}^2 + V_{DE}^2. \tag{24}$$

The static time variable is related to the angle in the $U_{DE} - V_{DE}$ plane, with

$$T_{DE} = \chi^{-1} \tan^{-1}(V_{DE}/U_{DE}). \tag{25}$$

The entire space is represented by a disk $U_{DE}^2 + V_{DE}^2 \le 1$ in the $U_{DE} - V_{DE}$ plane. Each point in the interior of the disk represents a two-sphere of radius R, where R varies from 0 at the boundary to $1/\chi$ at the origin.

To construct the Euclidean interpolation, first note that the hypersurfaces i and f from Fig. 7 can be copied without modification onto the

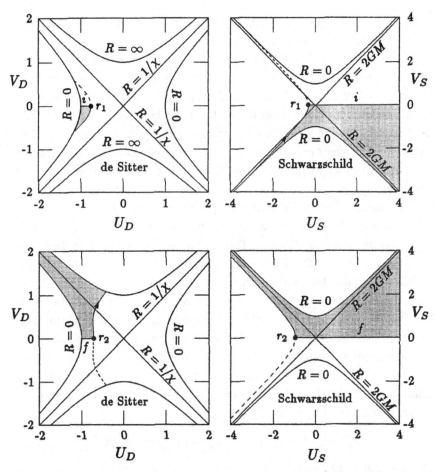

Fig. 7.   Trajectories for the tunneling problem, for a bubble mass just below M_{cr}. The upper diagrams show the initial trajectory. Full de Sitter and Schwarzschild spaces are shown, with shaded regions indicating the physical spacetime. The initial slice for the tunneling problem is shown as a horizontal line labeled i. The lower diagrams indicate the final trajectory, with the final slice shown as a horizontal line labeled f. (This figure and the following two are drawn for $\gamma = 1.95$ and $M/M_{cr} = .97$.)

Euclidean diagrams. Crudely speaking, this can be seen by noting that the vanishing of the extrinsic curvature means that there are no derivatives of the metric in the time direction. Since the time derivative vanishes, it does not matter whether the surfaces evolve in Euclidean or Lorentzian time. It is then straightforward to calculate the Euclidean trajectory of

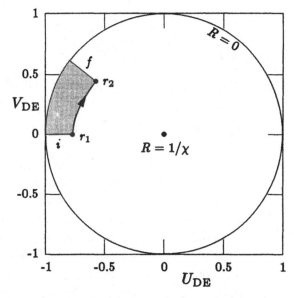

Fig. 8. The de Sitter part of the interpolating Euclidean four-geometry. The shaded area shows the region swept out as the initial hypersurface i evolves to become the final hypersurface f. The bubble wall trajectory is shown as a heavy line with an arrow on it.

the domain wall, by analytically continuing the normal equations of motion.

The parts of the initial and final surfaces within the de Sitter part of the space, and the interpolating trajectory of the bubble wall, are shown in Fig. 8. From this figure we can see that the de Sitter half of the diagram has a simple interpretation. We can visualize the initial surface evolving into the final surface, while at each stage of the evolution the surface extends from the bubble wall to the edge of the de Sitter disk, where $R = 0$. The region swept out by this motion, shown as a shaded region in the diagram, is the de Sitter part of the interpolating four-geometry. This part of the problem is straightforward and trouble-free.

The Schwarzschild part of the problem, however, does not work out nearly so cleanly. We begin by Euclideanizing the metric itself, which is straightforward. In Kruskal–Szekeres coordinates, the Euclideanized version of the metric (17) becomes

$$ds_E^2 = \frac{32(GM)^3}{R}e^{-R/2GM}(dV_{SE}^2 + dU_{SE}^2) + R^2 d\Omega^2, \tag{26}$$

where

$$\left(\frac{R}{2GM} - 1\right) e^{R/2GM} = U_{SE}^2 + V_{SE}^2. \tag{27}$$

The Schwarzschild time variable is related to the angle in the $U_{SE} - V_{SE}$ plane by

$$T_{SE} = 4GM \tan^{-1}(V_{SE}/U_{SE}). \tag{28}$$

Note that $R = 2GM$ at the origin of the $U_{SE} - V_{SE}$ plane, and grows monotonically as one moves outward. Thus, a diagram of Euclidean Schwarzschild space (with θ and ϕ suppressed) consists of a central dot at which $R = 2GM$, and concentric circles of constant R surrounding it.

Again it is straightforward to copy the initial and final hypersurfaces onto the Euclidean diagram, and to find by analytic continuation the bubble wall trajectory that interpolates between them. The results of these calculations are illustrated in Fig. 9. Since the bubble wall trajectory subtends an angle greater than π, it crosses both the initial and final surfaces. In addition, the initial and final surfaces cross each other. We have carried out rather extensive numerical calculations showing that the subtended angle always has a magnitude greater than π, so the qualitative features of the diagram are not particular to the chosen values of the parameters.

As in the de Sitter case, we can visualize the initial surface evolving into the final surface; this time, however, it is more complicated. We might imagine, for example, that the surface pivots about the origin, with one end extending to the bubble wall, and the other end to infinity. However, since some points will be crossed more than once by this motion, we cannot identify a region swept out. There is no simple analogue to the shaded region shown for the de Sitter case in Fig. 8.

The difficulty in interpreting the solution shown in Fig. 9 becomes even more striking if one imagines moving slightly away from the thin-wall limit. In that case one expects to be able to define a continuously varying scalar field on the Euclidean manifold. One does not know, however, how to treat the points where the bubble wall trajectory intersects the initial or final surfaces. If one thinks of these points as lying on the initial or final surfaces, then they are well into the Schwarzschild region and the scalar field should have its true vacuum value. On the other hand, if the points are

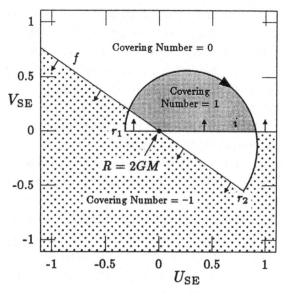

Fig. 9. The Schwarzschild part of the interpolating Euclidean four-geometry. The bubble wall trajectory, shown as a heavy line with an arrow on it, crosses both the initial and final surfaces, which also cross each other. Consequently, one cannot identify the region swept out as the initial hypersurface i evolves to become the final hypersurface f. Instead one must define a covering number, an integer that indicates the net number of times a point is covered by the evolving hypersurface. The small arrows on the initial hypersurface i and the final hypersurface f indicate the future direction, defined by the arrow of evolution of the bubble wall trajectory at the point where the hypersurface meets the trajectory. To define the covering number, imagine the evolution of the initial hypersurface into the final. The covering number for any point is the number of times the point is crossed by the evolving surface with the future side first, minus the number of times it is crossed with the past side first. The region with covering number 1 is shown shaded, the region with covering number -1 is shown dotted, and the region with covering number 0 is shown white.

considered to be on the domain wall, then the scalar field in the vicinity should be somewhere between the true vacuum and false vacuum values.

It is possible, nonetheless, to define a covering number for each point, as is described in the caption to Fig. 9. As can be seen on the diagram, the covering number takes on the values 1, 0, and -1 for this situation.

The fact that the covering numbers take values other than 0 and 1 implies that we have not found a genuine four-dimensional Euclidean manifold that interpolates between the initial and final three-geometries. We are not sure what this means. Nonetheless, we have at least constructed a sequence of three-geometries that interpolate between the initial and final three-geometries, and perhaps this is enough to give a valid semiclassical approximation to the tunneling amplitude. In our paper[9] we conjecture that this approach is correct, and proceed to calculate the tunneling amplitude. We use the word "pseudo-manifold" to describe a space with covering numbers defined on it, and we define an action that is weighted by the covering numbers. We can show that the action defined in this way, when varied, yields the correct Euclidean equations of motion.

The possibility of a multisheeted covering comes to mind, especially when thinking about the difficulty in defining a continuously varying scalar field. The possibility of a nonsingular multisheeted covering, however, is excluded by the fact that the underlying space is simply connected. To have a multisheeted structure, there must be a rule that allows us to determine, for any closed path, whether the path returns to the original sheet or a different one. If all closed paths can be continuously deformed into a trivial path, as is the case here, then continuity implies that all closed paths must return to the original sheet.

Although a nonsingular multisheeted covering is excluded, a new coordinate system giving a *singular* multisheeted covering can nonetheless be constructed. In fact, such a singular multisheeted covering is probably the most convenient way to represent the pseudomanifold. The argument in the previous paragraph is bypassed, since the coordinates of the covering will not be continuous functions of the coordinates of the original Euclidean Schwarzschild and de Sitter manifolds. Then, as a path is continuously deformed in the Schwarzschild or de Sitter manifold, the image of the path in the new coordinates can jump discontinuously from one sheet to another. Such a coordinate system can be constructed, for example, by first parametrizing the bubble wall trajectory by the proper time variable τ_E. Then for each point $P(\tau_E)$ along the trajectory we choose a surface that extends from $R = 0$ in the de Sitter region to the point P, and then in the Schwarzschild region from P to $R = \infty$. (There is no requirement, however, that R vary monotonically along this surface.) We further insist that the surfaces are chosen continuously in τ_E, and that the initial and final surfaces coincide with the initial and final surfaces of the tunneling problem. Assign the coordinate τ_E to each point on the surface chosen through $P(\tau_E)$. Now define on each surface a coordinate σ, extending from 0 at $R = 0$ to ∞

at $R = \infty$. We insist that σ varies continuously and monotonically along the surface, and that σ is chosen to be continuous in τ_E. We let $\sigma(\tau_E)$ indicate the value of the σ-coordinate at the point $P(\tau_E)$. Note that some points in the original Schwarzschild and de Sitter manifolds will be covered several times by this procedure, and other points will not be covered at all — if a sign is attributed to each covering of this evolving surface as before, then the net number of times a point is covered is just the covering number defined earlier. The coordinates $\xi^\mu \equiv (\tau_E, \sigma, \theta, \phi)$ define the multisheeted covering of the pseudomanifold.

We now assign a metric $\tilde{g}_{\mu\nu}$ to the ξ-space as follows:

$$
\tilde{g}_{\mu\nu} =
\begin{cases}
\dfrac{\partial x^\lambda}{\partial \xi^\mu} \dfrac{\partial x^\rho}{\partial \xi^\nu} g_{\lambda\rho}^{DE}(x(\xi)) & \text{if } \sigma < \sigma(\tau_E) \\[3mm]
\dfrac{\partial x^\lambda}{\partial \xi^\mu} \dfrac{\partial x^\rho}{\partial \xi^\nu} g_{\lambda\rho}^{SE}(x(\xi)) & \text{if } \sigma < \sigma(\tau_E),
\end{cases}
\tag{29}
$$

where $g_{\lambda\rho}^{DE}$ and $g_{\lambda\rho}^{SE}$ denote the Euclidean de Sitter and Schwarzschild metrics, respectively. Note that on each $\tau_E = constant$ surface the metric has only a single de Sitter region ($\sigma < \sigma(\tau_E)$) and a single Schwarzschild region ($\sigma > \sigma(\tau_E)$), even if the surface crosses the domain wall in Fig. 9 more than once. Note also that the junction conditions already imposed on the domain wall trajectory imply that the metric $\tilde{g}_{\mu\nu}$ describes a space that has no discontinuity at $\sigma = \sigma(\tau_E)$.

In the multisheeted coordinate system the action can be taken to have the standard form, except that the volume measure $\sqrt{\tilde{g}(\xi)}$ is defined by

$$
\sqrt{\tilde{g}(\xi)} \equiv \det\left(\frac{\partial x}{\partial \xi}\right) \sqrt{g(x)}
\tag{30}
$$

and is allowed to have either sign, depending on the sign of $\det(\partial x/\partial \xi)$. The ξ-space is therefore not a Riemann manifold, since the determinant of the metric can vanish and change sign. It can be shown, however, that the variation of the action defined in this way leads to the correct Euclidean equations of motion. Furthermore, in this formalism one can define a scalar field on the space without encountering the ambiguities mentioned above.

5. Subtletlies in the Subtraction Term

One particularly subtle point in the tunneling problem is the calculation of the subtraction term, which will be explained in this section. If the reader wishes to skip this fine point, he/she can jump to Sec. 6 without any loss of continuity.

To see the origin of the subtraction term, one can consider a tunneling problem in one-dimensional nonrelativistic quantum mechanics. If a particle tunnels from one classical turning point A to another classical turning point B, the WKB amplitude can be calculated by first solving for the imaginary-time classical path $x(\tau)$, where τ is related to the time variable t by $\tau = it$. Defining a Euclidean momentum by $p_E \equiv -ip$, where p is the ordinary canonical momentum, the WKB exponent can be written as an integral over the classical path:

$$I = \int_A^B p_E dx. \tag{31}$$

Using the relationship between the Euclideanized Lagrangian L_E and the Euclideanized Hamiltonian H_E, this expression can be rewritten as

$$I = \int_{\tau_A}^{\tau_B} p_E \frac{dx}{d\tau} d\tau \tag{32a}$$

$$= \int_{\tau_A}^{\tau_B} d\tau \left[L_E\left(x, \frac{dx}{d\tau}\right) - H_E \right] \tag{32b}$$

$$= \int_{\tau_A}^{\tau_B} d\tau L_E\left(x, \frac{dx}{d\tau}\right) - \int_{\tau_A}^{\tau_B} d\tau L_E(x_A, 0). \tag{32c}$$

In writing Eq. (32c), we made use of the fact that H_E has a constant value along the classical trajectory, and so it can be replaced by its value at the start of the trajectory. Note that the subtraction term, the second term in Eq. (32c), can be described as the action of a static configuration that stays fixed at the entry point x_A, and exists for the same length of time as the classical solution.

The subtlety in applying this principle to the false vacuum bubble tunneling problem is the fact that there is more than one relevant time variable. Should we use the de Sitter time, the Schwarzschild time, or maybe the proper time? The answer can be found by setting up a canonical formalism, so that the manipulations of Eq. (32) can be carried out. Such a canonical formulation can be obtained by starting with the well-known canonical formulation of the field theory, and then truncating to the space of the thin-wall configurations. To carry out this truncation, one first adopts a recipe for the slicing of spacetime, so that each segment of the bubble wall trajectory can be associated with a region of spacetime. The action for a segment of the trajectory is then taken to be the integral of the field theory Lagrangian, integrated over the associated spacetime region.

Note, however, that the classical variational principle requires that the fields be held constant on the boundary of the spacetime region. For this problem, the boundary includes the surface at spatial infinity. Since the Schwarzschild time interval can be determined by measurements at spatial infinity, the variational principle requires that the bubble wall trajectory be varied so that the Schwarzschild time interval is held fixed. This means that the de Sitter time interval and the proper time interval will not be fixed as the trajectory is varied, and therefore only the Schwarzschild time can be used as a canonical time variable. Thus, it is the Schwarzschild time interval that must be used to calculate the subtraction term.

6. Numerical Results

We have used our definition of the action to estimate the tunneling amplitude as a function of the various parameters in the problem, and we have found that it behaves very reasonably: the tunneling action decreases monotonically to zero as the bubble mass M approaches the critical mass M_{cr} at which tunneling would not be necessary, and it diverges monotonically as the gravitational constant $G \rightarrow 0$. The action is negative definite by the standard sign conventions,[c] which some physicists will regard as an indication of error. We, however, are not convinced that the sign should be a matter of concern. In any quantum mechanics tunneling problem, the same Schrödinger equation describes tunneling in both directions through the barrier. Thus the equation is solved by both exponentially

---

[c]We take as "standard" the convention that the kinetic terms of the matter fields in the Euclideanized Lagrangian are positive.

decaying and exponentially growing wave functions, and one simply has to be sensible enough to choose the right solution for the problem of interest.

In a pair of recent papers, Fischler, Morgan, and Polchinski[10,11] have calculated an amplitude for this same process, using a Hamiltonian method that is very different from the method we used. In their formalism they find no inconsistencies, and their answer is identical to ours. We regard this agreement as a strong indication that the calculation is correct, or at least that it is the most sensible calculation that one can do in the absence of a true theory of quantum gravity.

The final result is obtained by a numerical integration, and it is shown graphically in Fig. 10. The (subtracted) Euclidean tunneling action I_E depends on M/M_{cr}, and also on the dimensionless parameter γ, defined in Eq. (6). For typical GUT parameters, $\gamma \approx 10^{-4}$.

As a rough estimate, the action I_E is of order $1/(G\chi^2)$, as long as $G\sigma^2$ is smaller than or comparable to ρ_f, and M is not too near M_{cr}. For typical grand unified theory parameters, this would give an outrageously small

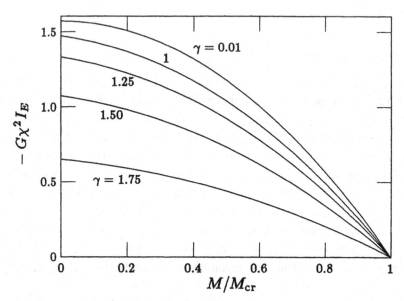

Fig. 10.   Graph of $-G\chi^2 I_E$, where I_E is the Euclidean tunneling action. It is shown as a function of M/M_{cr}, for various values of the parameter γ.

tunneling probability, such as

$$10^{-(10^{18})}.$$

Even with this small probability, however, there might still be a large probability of an event of this sort occurring somewhere in a universe that has undergone a large amount of inflation. Thus, the possibility of a chain reaction by which one universe produces more than one universe is not obviously ruled out by this estimate. On the other hand, if we are talking about creating a universe in a hypothetical laboratory, then a probability this small must be considered equivalent to zero. Nonetheless, the numerical estimate of the tunneling probability is extraordinarily sensitive to the energy density ρ_f of the false vacuum. If there exists a false vacuum with an energy density near the Planck scale, which is certainly not excluded by anything we know, then the tunneling probability would be of order one.

7. Conclusions

In setting up the Euclidean formulation of the false vacuum bubble tunneling problem, we have found that no true Euclidean interpolating manifold exists. There is no difficulty or ambiguity in analytically continuing the bubble wall trajectory into the Euclidean regime, but when this trajectory is plotted on a Euclidean spacetime diagram it is found to cross both the initial and final surfaces of the tunneling problem. These intersection points prevent a conventional manifold interpretation.

We admit that we are not sure what the absence of a true interpolating manifold implies about the tunneling problem. Perhaps it indicates that the stationary phase method has failed, perhaps it indicates that one cannot extrapolate the thin-wall approximation into the Euclidean regime, or perhaps it is a suggestion that tunneling is for some reason forbidden.

We find it difficult to believe, however, that the tunneling process is forbidden, since there is no barrier to constructing a well-defined manifold (with either Lorentzian or Euclidean signature) that interpolates between the initial and final states. Such a manifold is not a solution, but it would constitute a path contributing to the functional integral. Furthermore, since any small variation about such a path would also contribute, the measure of these paths appears naively to be nonzero. The amplitude would then be nonzero unless the various paths conspire to cancel each other, as they do for an amplitude that violates a conservation principle associated with a symmetry. In the present case, however, there is no apparent symmetry or conservation law at work. We therefore conjecture that the tunneling process is allowed, and that the semiclassical approximation is valid. The alternative calculations of Fischler, Morgan, and Polschinski[10,11] add considerable weight to this point of view.

Although no Euclidean interpolating manifold exists, we find that it is nonetheless possible to generalize the notion of a manifold to describe a well-defined Euclidean interpolation. In our paper we defined an object that we called a "pseudomanifold," which we described in two alternative ways: by the use of covering numbers, and by the use of a singular multisheeted covering. In the second description the pseudomanifold closely resembles a true manifold, except that $\sqrt{g}$ is allowed to vanish and to change sign. The action of the pseudomanifold is defined by the usual expression for the Euclidean action, except that $\sqrt{g}$ is not positive definite.

If our semiclassical result is correct, then it appears to raise an important issue in quantum gravity: how does a pseudomanifold arise in a quantum gravity path integral? It might mean that such objects occur in the physical definition of the path integral, or it might mean that they arise as saddle points which are obtained by the distortion of integration contours in the complex plane.

In any case, this work seems to indicate that there are ambiguities in semiclassical Euclidean quantum gravity that are yet to be resolved.

References

1. A. D. Linde, *Rep. Prog. Phys.* **47**, 925 (1984).
2. A. D. Linde, in *300 Years of Gravitation*, eds. S. W. Hawking and W. Israel (Cambridge University Press, Cambridge, England, 1987), p. 604.
3. R. H. Brandenberger, *Rev. Mod. Phys.* **57**, 1 (1985).
4. M. S. Turner, in *Architecture of Fundamental Interactions at Short Distances*, eds. P. Ramond and R. Stora (North-Holland, Amsterdam, 1987), p. 513.
5. P. J. Steinhardt, in *High Energy Physics, 1985, Volume 2*, Proceedings of the Yale Theoretical Advanced Study Institute, eds. M. J. Bowick and F. Gürsey (World Scientific, Singapore, 1986), p. 567.
6. S. K. Blau and A. H. Guth, in *300 Years of Gravitation*, cited above, p. 524.

7. L. F. Abbott and S.-Y. Pi, eds., *Inflationary Cosmology* (World Scientific, Singapore, 1986).
8. A. D. Linde, *Particle Physics and Inflationary Cosmology* (Gordon and Breach, Boston, 1990).
9. E. Farhi, A. H. Guth, and J. Guven, *Nucl. Phys.* **B339**, 417 (1990).
10. W. Fischler, D. Morgan, and J. Polchinski, *Phys. Rev.* **D41**, 2638 (1990).
11. W. Fischler, D. Morgan, and J. Polchinski, *Phys. Rev.* **D42**, 4042 (1990).
12. K. Sato, M. Sasaki, H. Kodama, and A. Maeda, *Prog. Theor. Phys.* **65**, 1443 (1981); H. Kodoma, M. Sasaki, K. Sato, and K. Maeda, *Prog. Theor. Phys.* **66**, 2052 (1981); K. Sato, *Prog. Theor. Phys.* **66**, 2287 (1981); H. Kodama, M. Sasaki, and K. Sato, *Prog. Theor. Phys.* **68**, 1979 (1982); K. Maeda, K. Sato, M. Sasaki, and H. Kodama, *Phys. Lett.* **B108**, 98 (1982); K. Sato, H. Kodama, M. Sasaki, and K. Maeda, *Phys. Lett.* **B108**, 103 (1982).
13. V. A. Berezin, V. A. Kuzmin, and I. I. Tkachev, *Phys. Lett.* **B120**, 91 (1983); *Phys. Rev.* **D36**, 2919 (1987); also in *Proceedings of 3rd Seminar on Quantum Gravity 1984*, eds. M. A. Markov, V. A. Berezin, and V. P. Frolov (World Scientific, Singapore, 1985), p. 605.
14. J. Ipser and P. Sikivie, *Phys. Rev.* **D30**, 712 (1984).
15. A. Aurilia, G. Denardo, F. Legovini, and E. Spallucci, *Phys. Lett.* **B147**, 258 (1984); *Nucl. Phys.* **B252**, 523 (1985).
16. K. Lake, *Phys. Rev.* **D19**, 2847 (1979); K. Lake and R. Wevrick, *Can. J. Phys.* **64**, 165 (1986).
17. S. K. Blau, E. I. Guendelman, and A. H. Guth, *Phys. Rev.* **D35**, 1747 (1987).
18. A. Aurilia, M. Palmer, and E. Spallucci, *Phys. Rev.* **D40**, 2511 (1989).
19. W. Israel, *Nuovo Cimento* **B44**, 1 (1966), and correction in **B48**, 463 (1967).
20. M. D. Kruskal, *Phys. Rev.* **119**, 1743 (1960); G. Szekeres, *Publ. Mat. Debrecen* **7**, 285 (1960).
21. M. Aryal and A. Vilenkin, *Phys. Lett.* **B199**, 351 (1987).
22. A. D. Linde, *Mod. Phys. Lett.* **A1**, 81 (1986); A. S. Goncharov, A. D. Linde, and V. F. Mukhanov, *Int. J. Mod. Phys.* **A2**, 561 (1987).
23. A. A. Starobinsky, *Phys. Lett.* **B117**, 175 (1982); see also A. A. Starobinsky, in *Field Theory, Quantum Gravity and Strings*, eds. M. J. de Vega and N. Sánchez, Lecture Notes in Physics (Springer-Verlag) **246**, 107 (1986).
24. A. H. Guth and S.-Y. Pi, *Phys. Rev.* **D32**, 1899 (1985).
25. E. Farhi and A. H. Guth, *Phys. Lett.* **B183**, 149 (1987).
26. S. Coleman and F. De Luccia, *Phys. Rev.* **D21**, 3305 (1980).
27. A. H. Guth and E. J. Weinberg, *Nucl. Phys.* **B212**, 321 (1983).
28. G. W. Gibbons and S. W. Hawking, *Phys. Rev.* **D15**, 2738 (1977).

THE KLEIN–NISHINA FORMULA

Gösta Ekspong

Department of Physics
Stockholm University
Stockholm, Sweden

During the 1920's the Compton effect was very much at the centre of interest among physicists due to its demonstration of the inadequacy of the classical theory for light when trying to account for the scattering of X-rays. The angular intensity distribution of scattered X-rays had not been successfully described by theory until Oskar Klein and Yoshio Nishina in 1928 solved the problem using the then new Dirac theory for the electron. The present paper gives an account of the collaboration in the summer of 1928 between the two physicists — one a theorist, the other an experimentalist — when both were visitors at the Niels Bohr Institute in Copenhagen.

1. The Klein–Nishina Collaboration

When Dirac paid a short visit to Copenhagen in the spring of 1928, he met Klein and Nishina. The three of them were once conferring in the library of the Niels Bohr Institute. Dirac was a man of few words, so when the remark came from Nishina that he had found an error of sign in the then new Dirac paper on the electron, Dirac drily answered: "But the result is correct." Nishina, in an attempt to be helpful, said: "There must be two mistakes," only to get Dirac's reply that "there must be an even number of mistakes."[1]

The reason behind Nishina's seemingly thorough knowledge of Dirac's paper was that he had been assigned to compute the intensity of the scattered radiation (i.e. the cross section) for Compton scattering with the new electron theory. Walter Gordon, during a short visit to Copenhagen somewhat earlier that spring, had suggested that it would be a suitable problem for the young Japanese physicist. He knew Nishina, who had visited Hamburg to learn theory before returning to Japan. When Gordon made that suggestion, Klein immediately agreed, although he himself had intended to attack the same problem. Klein had in fact partly treated Compton scattering the preceding year and Gordon had given it a full treatment, both using semi-classical theory.

There was no established method available to solve the problem, which made it difficult. When the time came to leave Copenhagen for the summer vacation, "Nishina had — quite obviously — not been able to do more than a general preparation of the problem", as Klein wrote[1] almost 50 years later. Klein suggested a collaboration during the summer of 1928. Nishina settled in a mansion in Lundeborg near the place on the Danish island Fyn where Klein and his family were renting a summer house.

The two of them worked day by day independently during the summer, doing the lengthy algebra while seated in cosy folding chairs.[2] At day's end they compared notes. In early July 1928 Klein wrote to Bohr informing him of the progress,[3] saying that a first draft was being written. Their joint short letter to *Nature*[4] was signed on the 3rd of August 1928. The full paper had by then not been finished, as told by Nishina in a letter to Bohr.[5] The plan was to finish it in September, when both of them would be back in Copenhagen. Their joint paper was received by *Zeitschrift für Physik* on the 30th of October.[6]

A second paper, with formulae for polarization phenomena, was published immediately after the first one and was signed by Yoshio Nishina alone.[7] Its content is alluded to in the first paper. When it was submitted, Nishina had already been in France for several weeks on his way back to Japan. According to letters between Copenhagen and Tokyo, some errors in the first form of some formulae had been found by Christian Møller in Copenhagen and corrections made to the final paper. As seen by the correspondence and also by an erratum to *Nature*,[8] the error was an oversight of the fact that the photon frequency will be changed during the first scattering in a double scattering process.

2. Who Were the Two Physicists?

Oskar Klein started on chemistry at a very young age, in the laboratory of Svante Arrhenius (Nobel Prize winner in Chemistry in 1903). He obtained his Ph.D. in Stockholm, his home town, where his father was the chief rabbi of the Jewish congregation. On his own and while still at the chemistry laboratory he read the papers on quantum theory by Bohr, Sommerfeld, Einstein and Debye. His other field of study was statistical mechanics. He went in 1918 to Bohr in Copenhagen, when he was 24 years old (being born in 1894). Klein transformed there into a clever, original theoretical physicist while discussing with Bohr and Kramers and working on his own. His name is associated with the first five-dimensional theory, today revived again in other contexts under the term Kaluza–Klein theories. George Uhlenbeck has reminisced: "I remember that in the summer of 1926, when Oskar Klein told us of his ideas which would not only unify the Maxwell with the Einstein equations but also bring in quantum theory, I felt a kind of ecstasy! Now one understands the world!"[9]

Oskar Klein stayed with some interruptions in Copenhagen until 1930, when he became a professor at Stockholm University. Since his life history has been told in this series of publications,[10,11] the reader is referred to those accounts as well as to some new material in this volume.[12]

Yoshio Nishina was trained as an electrical engineer at the Tokyo Imperial University, where he graduated in 1918. He became a research fellow at RIKEN (Research Institute for Physics and Chemistry) and majored in physics at the Graduate School of the Tokyo Imperial University. In 1921, at the age of 31 years (he was born in 1890), he left Japan for Europe. He went to Copenhagen in 1923 to stay at Bohr's institute. His research was devoted to X-ray experiments until he went in November 1927 to Hamburg to learn quantum mechanics under the guidance of Wolfgang Pauli. On his return to Copenhagen in March 1928, he began his work on the theoretical calculation of Compton scattering, together with Oskar Klein.

Nishina left Copenhagen in October 1928 and returned in December of that year to Japan, where he became a member of Nagaoka's laboratory at RIKEN. Over the years Nishina became a leading scientist in Japan, building a new laboratory in 1931, starting nuclear physics research and also studies of cosmic rays in 1935, setting up a small cyclotron in 1937 and a big cyclotron over the years 1938–1944. Leading European physicists came to visit Japan, partly because they knew Nishina from his eight years in Europe.

In August 1945 Nishina went to investigate the sites of the nuclear bombing in Hiroshima and Nagasaki. The occupation army soon destroyed the two cyclotrons. The larger one was dumped at sea and has not been found; the smaller one can be seen in Tokyo outside the building where Nishina had his office and where the Nishina archives are located.

RIKEN was reorganized after the war with Nishina as president of the RIKEN company. He died in January 1951, due to liver cancer, at the age of 61 years.

3. Compton Scattering

In his Nobel lecture in 1927, Arthur Compton emphasized that X-rays share all the properties of ordinary light (i.e. reflection, refraction, diffuse scattering, polarization, diffraction, emission and absorption spectra) and therefore are a wave phenomenon. The particle nature is revealed in the photoelectric effect and in the scattering of X-rays in light elements.

That some of the scattered X-rays are shifted towards longer wavelengths is impossible to understand in classical theory. The precise experiments by Compton[13] in 1922 focused the interest on this new phenomenon. "He found an experimental method that gave results which were as exact as they were astonishing" was said at the Nobel ceremony in 1927 by Professor Manne Siegbahn, himself a Nobel laureate and a member of the Nobel Committee for Physics.[14] The explanation by Compton himself and also by Peter Debye[15] was based on the idea that a light quantum (a photon), carrying momentum as well as energy, collides with a free electron in the scattering material, and that energy and momentum are each conserved in the individual act of collision. Relativistic kinematics predicts that the wavelength shift is independent of the energy of the primary radiation and of the scattering material. It depends on the scattering angle, ϕ, in a simple manner:

$$\Delta\lambda = \frac{h}{mc}(1 - \cos\phi). \tag{1}$$

This formula agreed very well with accurate measurements. This came as a striking confirmation of Einstein's concept of light quanta from 1905 (the word "photon" was coined in 1926). Early in 1923 Sommerfeld wrote to Bohr: "The most interesting thing ... is the work of Compton in St. Louis. ... After it the wave theory of X-rays will become invalid."[16]

It seems that the 26-year-old H. A. Kramers had already set up the kinematical equations of a photon–electron collision in 1921, before the discovery of the phenomenon. As told by Max Dresden[17] Kramers was thoroughly converted by Bohr to believing that photons do not exist and that he should not publish this. Kramers became one of strongest opponents of the photon concept, even long after the Compton effect had been fully explored experimentally.

Several leading physicists, including Max Planck and Niels Bohr, resisted Einstein's concept of light quanta for a long time. There are quotes from Bohr *before* the discovery of the Compton effect and *after* it, which express his resistance. In 1922 Bohr rejected the usefulness of the concept in his Nobel lecture, saying: "In spite of its heuristic value [Einstein's] hypothesis of light quanta, which is quite irreconcilable with the so-called interference phenomena, is not able to throw light on the nature of radiation."[18] Einstein himself was invited to the same Nobel event, and should certainly have been present were it not for his journey to Japan. He was awarded the 1921 Nobel Prize (reserved in 1921 and awarded in 1922) on the same occasion as Bohr "... for his discovery of the law of the photoelectric effect." It is a pity that an encounter in December 1922 in Stockholm between the two giants in physics never took place. After the Compton effect had been discovered Bohr was troubled, as is clear from some words in a letter[19] written in January 1924 to Ernest Rutherford: "You can understand my concern — for a person for whom the wave theory is a creed."

To people like Niels Bohr and Max Planck, light was demonstrably a wave motion and Maxwell's theory accounted for it in a convincing way. There existed no technique to describe light as waves and particles at the same time. The quantum phenomena were viewed by them as having nothing to do with light motion in vacuum, only with light interacting with matter — a subject not fully understood anyway. After a last effort to avoid the necessity of the photon concept in a 1924 paper by Bohr, Kramers and Slater, new experiments in 1925 on time and space coincidences between scattered quanta and recoil electrons in the Compton effect convinced Bohr, that the light quantum, i.e. the photon, was an indispensable concept.

Kramers, however, was not convinced. Wolfgang Pauli, in a 1925 letter to Kramers,[20] expressed his opinion in the following words: "It can now be considered as proven — to every objective physicist — that photons are as physically real (or as unreal!) as electrons."

However, in 1927 Prof. Manne Siegbahn claimed in his Nobel Prize presentation of Arthur Holly Compton:[14] "The Compton effect has, through the latest evolutions of the atomic theory, got rid of the original explanation based upon a corpuscular theory. The new wave mechanics, in fact, leads as a logical consequence to the mathematical basis of Compton's theory. Thus the effect has gained acceptable connection with other observations in the sphere of radiation." The modern reader is reminded of the fact that in 1927, when the prize award was decided, neither the Dirac field quantization nor the Bohr complementarily ideas were known.

From the contents of the Nobel Archives[21] it is quite clear what the speaker had in mind when claiming that the particle aspect of X-rays would no longer be needed. The Nobel Committee had already assessed the Compton effect in 1925 and 1926 and found its theory very unsatisfactory. However, in 1927 that had changed. The new evaluation was carried out by Carl Wilhelm Oseen, professor in mechanics and mathematical physics at Uppsala. He did a thorough study for the Committee. He begins by recalling the great interest by which Compton's discovery in 1922 had been met, much due to the theory offered by Compton. He writes: "It is not surprising that the agreement of this theory with observations inspired with less critical representatives for theoretical physics the thought that the lengthy fight between the wave theory and the corpuscular theory would be nearing its end. Compton's discovery was by these scientists taken as the decisive proof for the truth of the corpuscular theory. If these expectations had been fulfilled, the discovery by Compton would undoubtedly have marked a decisive turning point in the development of the whole of the radiation theory." Oseen set out to show that this is not so. His view was that the effect is nevertheless very important.

Oseen tells how the Bohr theory had fallen by 1925 and says that the Compton effect had nothing to do with that. He describes how matrix mechanics and wave mechanics had entered the stage without inspiration from the Compton effect. In a short interlude he mentions Compton's suggestion in 1920 that the electron is spinning around its axis with a quantized angular momentum and constitutes the elementary magnet in matter. After mentioning that Uhlenbeck and Goudsmit in 1925 published their similar idea, Oseen points out that the basis is weak, since it presupposes a rotating rigid body, which is relativistically not acceptable.

The oldest theories for the Compton effect referred to by Oseen were given by Compton, "Debijes" (Debye) and Woos. Being based on the theory of light quanta, "they have been of value to experimental research, but must now be considered obsolete in view of the latest theories." Oseen mentions several such newer works, especially that by Gordon and a recent one by O. Klein, based on the wave theory with no mention of corpuscles (photons). They all arrive at the same equations for the conservation of energy and momentum between the scattered wave and the recoiling electron as originally derived by Compton assuming a two-particle collision. "The basis for the Compton–Debijes theory is thus found, this time not as a hypothesis but as a consequence of the atomic theory," is a conclusion by Oseen, which justifies his judgement of the former being obsolete. These wave-mechanical treatments, furthermore, also gave formulae for the intensity (i.e. cross section) in agreement with measurements. Oseen summarizes by saying that the revolution during the last 18 months has been independent of Compton's discovery and that the new direction for the revolution has moved opposite to the one expected after Compton's discovery. The new theory is a wave theory in a higher degree than any previous theory. By the use of the new theory it has been possible to give a qualitatively and quantitatively correct account of the Compton effect.

The Committee follows Oseen in emphasizing that the Compton effect is important, since it once more and very clearly and convincingly demonstrates that the classical theories are not applicable in the realm of atomic physics and that it offers a valuable and welcome possibility of testing the new ideas.

There were thus at that time some recent works by Schrödinger,[22] by Gordon[23] and by Klein,[24] which induced the statement made at the Nobel Ceremony in 1927. In Schrödinger's view a plane light wave would interact with a moving electron, represented by a plane wave serving as a diffraction "grating." The diffracted plane wave X-ray would be emitted in a direction determined by interference. The wave length shift was viewed as a Doppler

effect, since the grating was a moving one. The same kinematical formulae
were obtained as on the assumption of a particle collision.

During a short stay at Leiden in 1926 Klein and Uhlenbeck began —
but never finished — a calculation of the Compton effect based on
Schrödinger's unrelativistic wave mechanics. In a paper[24] published in
1927, the same year as when Compton was awarded the Nobel Prize,
Klein showed that the interaction between an electromagnetic wave and
an electron, described by his own relativistic wave function (often referred
to as the Klein–Gordon equation), leads to the same kinematical relations
as those derived on the corpuscular theory.

The observed angular distribution of scattered radiation presented a
second and more difficult problem. Classical theory predicts a forward–
backward symmetric angular distribution. The observations, however,
showed a strongly reduced backward scattering intensity and furthermore
an integrated total cross section, decreasing with energy and less than
the classical "Thomson" cross section. There were early futile attempts
to explain the large deviations from what classical theory predicts. Some of
these were almost desperate, as can be seen by the collection of published
papers, which were reviewed by Compton in his monograph[25] from 1926.
Some formulae were judged by Compton to be wrong, since they disagreed
either with experiments or with a general limit obtained by applying Bohr's
correspondence principle. "An interesting novel solution of the problem
has been presented by Jauncey" are the words of Compton, by which he
introduced an idea that photons bounce off the surface of a suitably shaped
electron. To make the mark the shape would have to be special and to
change with the velocity of the electron. It is perhaps not surprising that
Compton considered Jauncey's approach interesting, since it was the only
one applying the corpuscular theory to the problem.

In the end two formulae were favoured, one given by Compton himself
and the other by G. Breit, with some preference for the latter. The one by
Breit, partly based on empirical grounds, is

$$\frac{d\sigma}{d\Omega} = \frac{\sigma_0}{2} \frac{1 + \cos^2 \phi}{[1 + \alpha(1 - \cos \phi)]^3}, \tag{2}$$

where α is the photon energy in units of the electron rest mass energy, and
thus $\alpha = \frac{h\nu}{mc^2}$.

In the forward direction ($\phi = 0$) one gets σ_0, which is the classical Thomson cross section. In the backward direction the classical scattering formula gives the same value, whereas the Breit formula predicts a smaller and energy-dependent cross section, $\frac{d\sigma}{d\Omega}(\phi = \pi) = \frac{\sigma_0}{1+2\alpha}$. The integrated total scattering cross sections are in the two cases hardly distinguishable from each other in the energy range, where most measurements existed in 1926,

$$\text{Compton:} \quad \sigma = \frac{\sigma_0}{1 + 2\alpha}, \tag{3}$$

$$\text{Breit:} \quad \sigma = \sigma_0 \frac{3}{4} \frac{1 + \alpha}{\alpha^3} \left[\frac{2\alpha(1 + \alpha)}{1 + 2\alpha} - \ln(1 + 2\alpha) \right]. \tag{4}$$

However, the monograph was written before any quantum-mechanical calculations had been published. There were two such before the one by Klein and Nishina. Based on the Klein–Gordon relativistic formula for the electron, i.e. an electron without spin, Dirac published a solution[26] in 1926, and Gordon[23] one in 1927. There was also the solution given by Schrödinger,[22] inspired by the Gordon paper. These gave the same result as the one by Breit, Eq. (2) above, for the differential cross section. That was still not the final, correct solution to the problem, since when the electron spin is included, a more complicated formula results, as first shown by Klein and Nishina. The magnetic moment associated with the spin of the electron causes additional scattering, which varies in magnitude with the angle of scattering and also depends on the energy of the photon.

4. On the Derivation of the Klein–Nishina Formula

The full paper[27] by Klein and Nishina, which was published in German, appears translated into English for the first time in this volume. A short note[28] in English was published in 1928. A modern reader will be surprised by the methods used by Klein and Nishina in their derivation of the famous Klein–Nishina formula. One must not forget, however, that no established methods for such problems existed at that time, long before Feynman graphs were invented.

Klein and Nishina took as starting point the then new relativistic theory for the electron by Dirac. They considered the recently published treatments of the problem Dirac and by Gordon as being based on the "older" forms of relativistic quantum mechanics. When it came to handling the Compton scattering problem, Klein used Dirac's relativistic theory, whereas Dirac had used Klein's. But that was before Dirac had found his relativistic theory for the electron.

A modern reader of the Klein–Nishina paper is struck by the fact that Dirac's field quantized theory is not applied, although it was already available. There are thus in their paper no processes of photon absorption by the electron, no intermediate states, no re-emissions of a photon. They did not apply a quantized electromagnetic field, but rather a classical field. Klein was a pioneer in semi-classical theory. The negative energy states as solutions to the Dirac equations were considered physically without meaning. One sentence in the paper reads: "We will of course limit ourselves to positive values." There was also the problem of boundary conditions. They claimed that a treatment according to Dirac's radiation theory would yield the same result as theirs, when one limits oneself to the first approximation. This was later — in 1930 — shown to be true by the detailed calculations of Ivar Waller[29] and Igor Tamm.[30]

The field of the incoming electromagnetic wave was considered to perturb the electron; "we want to regard a free electron illuminated by a continuous monochromatic wave train." The charged current density was set up and treated so that the radiation field at a distant point could be calculated. They considered the electron to make a transition from a state with momentum p to a state with momentum p' and found that the same relation holds for the frequency shift of the scattered radiation and the scattering angle as in the case of particle collisions.

For the second problem — the angular distribution — they calculated the magnetic field at some distant point in a chosen direction. In their wave-mechanical language the angle of scattering is the angle between the direction of observation and the normal to the wave front of the incoming plane wave. "After some lengthy calculations" they found an expression for the squared field strength in dependence of the scattering angle, the polarization vector and natural constants (e, m, c). Photons were introduced towards the end. Dividing the scattered energy by the photon energy, they obtained the number of scattered photons. The scattering cross section was found to contain one factor in addition to those already given by the Gordon and Dirac formula[23,26] which was reproduced in Eq. (2) above. This new factor,

$$1 + \frac{\alpha^2(1 - \cos\phi)^2}{(1 + \cos^2\phi)[1 + \alpha(1 - \cos\phi)]}, \tag{5}$$

where $\alpha = \frac{hv}{mc^2}$, becomes approximately unity if the radiation frequency, i.e. the photon energy is not too large. Thus the Klein–Nishina formula reduces to the Dirac–Gordon formula at moderate photon energies and all of them have the energy-independent classical Thomson formula,

$$\frac{d\sigma}{d\Omega} = \frac{e^4}{2m^2c^4}(1 + \cos^2\phi), \tag{6}$$

as the low energy limit.

The total cross section, obtained by integration over angles, is

$$\sigma = \sigma_0 \frac{3}{4} \left\{ \frac{1 + \alpha}{\alpha^2} \left[\frac{2\alpha(1 + \alpha)}{1 + 2\alpha} - \ln(1 + 2\alpha) \right] \right.$$
$$\left. + \frac{1}{2\alpha} \ln(1 + 2\alpha) - \frac{1 + 3\alpha}{(1 + 2\alpha)^2} \right\}, \tag{7}$$

where $\alpha = \frac{hv}{mc^2}$.

This formula was found to be in very good agreement with Chao's data using γ-rays and with Read and Lauritsen's on X-ray measurements, as stated in 1934 by Compton and Allison in their well-known monograph.[31] It was clearly superior to the formula by Breit, Dirac and Gordon, as displayed in Eq. (4).

There exist manuscripts with unpublished details of the lengthy calculations by Klein and Nishina in the Nishina archives in Tokyo. These have recently been investigated by Yuji Yazaki, who published his findings in the *Journal of the History of Science, Japan* (in Japanese, with a summary in English).[32]

In a subsequent paper, printed immediately after the first one, Nishina derived formulae for polarization phenomena.[33] He pointed out that the results clearly differ from any earlier ones. The scattering of linearly polarized light on polarized electrons was shown to result in two incoherent, elliptically polarized scattered rays. This has to do with no spin-flip and spin-flip transitions, respectively. The initial spin state of the electron was found to be irrelevant, as mentioned already in the joint paper. Since the usual geometry in experiments on polarization was scattering twice at right angles, Nishina treated this case theoretically. For this he handled the case of an incoming, elliptically polarized wave. In that case the result was clearly dependent on the initial spin state of the electron. Although Nishina averaged over the initial spin directions, one can see here, for the first time, that Compton scattering against polarized electrons is sensitive to the circular polarization of γ-rays. In practical applications this has turned out to be of great value. A complete theoretical treatment of the general case was given in 1938 by W. Franz,[34] who derived expressions for the scattering of elliptically polarized light against electrons with polarization in an arbitrary direction. Following Gunst and Page[35] one often writes the differential cross section for Compton scattering on polarized electrons as a sum of two terms,

$$\frac{d\sigma}{d\Omega} = \frac{d\sigma_{KN}}{d\Omega} + P\frac{d\sigma_1}{d\Omega},$$

where the first term is the usual Klein–Nishina expression for unpolarized or linearly polarized light. The second term, which may add or substract, depends on the circular polarization P of the photons.

5. The Negative Energy States in 1928–1930

According to a recent study of available archival material[32] in Tokyo, the greatest problem for Klein and Nishina was the treatment of Dirac spin states and how to interpret them in a physically meaningful way. There were no established methods. Klein devised a method to derive the states from solutions of electrons at rest by a Lorentz transformation.

In their paper Klein and Nishina avoided explicit reference to electron states with negative energy. However, these did appear somewhat later when the Compton scattering problem was treated with quantized fields according to Dirac's radiation theory. The first to do so was Ivar Waller in Uppsala, who published[29] that in 1930. A similar treatment[30] came at about the same time from Igor Tamm in Moscow. Both of them used the original Dirac theory for the electron with empty negative energy states. Both found that only if these states were not ignored did one arrive at the Klein–Nishina formula. Waller calculated sums over positive energy states and negative energy states separately and claimed that both were needed. Tamm was even more expressive, in saying that the result would be grossly wrong if the negative energy states were left out. Both of these authors found furthermore the surprising result that in the classical limit of low photon energies, scattering via intermediate states with positive energies gives zero contribution, and that the Thomson cross section derives from scattering via negative energy states alone. Thus the Klein–Nishina formula came to serve as a test when one was clarifying the importance of the ill-understood negative energy states.

Dirac heard of Waller's result before publication and immediately created the idea of filling the negative energy sea with unobservable electrons. In this new form of his theory there would be a replacement of transitions to now-no-longer available negative energy states by transitions of a new type, in which a hole state appears (a virtual positron in modern terminology). When Dirac described the consequences of his hole theory in a letter[36,37] to Bohr, his words were almost as clear as a Feynman diagram for the Compton scattering: "On my new theory ... there is ... a new kind of double transition taking place in which first one of the negative-energy electrons jumps to the proper final state with emission (absorption) of a photon, and secondly the original positive-energy electron jumps down and fills up the hole with absorption (emission) of a photon. This new kind of process just makes up for those excluded and restores the validity of the scattering formulas, derived on the assumption of the possibility of intermediate states of negative energy." Waller, who knew about Dirac's new hole theory before that was published, claimed in his paper that the result would be the same, since only formal changes would be needed. Shortly afterwards, Dirac proved this to be true.[38]

In view of the modern interpretation of negative energy states as virtual positrons and with regard to the results by Waller and Tamm, one reaches the not-so-obvious conclusion that scattering of light in the classical limit proceeds through intermediates states with three charged particles present, one of which is a virtual positron, and the other two the incoming and the outgoing electron.[39]

6. A New Tool for Research

6.1. *Cosmic rays*

The nature and energy of cosmic rays was under debate and investigation in the late 1920's. Some believed they are hard γ-rays. Klein and Nishina suggested that their formula could serve as a tool for determining the energies of cosmic rays, since the cross section is a steeply decreasing function of increasing γ-ray energy, E_γ,

$$\sigma \approx \frac{\pi \alpha^2}{m E_\gamma} \ln \frac{2 E_\gamma}{m},$$

where α now is the fine structure constant.

It follows that if there were no other processes entering at high energy, cosmic γ-rays would be very penetrating and more so the higher the energy. It will be recalled that Nishina, when back in Tokyo, started a programme of research in cosmic rays.

However, the scheme never worked as envisaged, since another new process enters at energies larger than about 1 MeV. This process is, of course, pair creation, which reveals the real meaning of the mysterious, negative energy states.

6.2. *Positrons*

When experiments were made to test the correctness of the Klein–Nishina formula, it was found that filters of high-Z materials attenuated high energy γ-rays much more than predicted, but that the attenuation in low-Z materials was as expected.[40–42] Disagreements with the formula were taken as hints that some new process exists, the nature of which was missed, however, by the early researchers and by others as well.[43] The new process was identified with the creation of electron–positron pairs (which occurs much more readily in high-Z materials) only much later, after the discovery of the free positron by Carl David Anderson in 1932 and further studies by Blackett and Occhialini.

Still another new phenomenon was discovered in 1930 by Chung-Yao Chao, a Chinese visitor to California Institute of Technology, who also had been one of those observing the discrepancy referred to above.[42] He reported the existence of a new kind of secondary radiation coming from a heavy target irradiated by high energy γ-rays.[44] Chao described the situation in the following way: "I first found that the angular distribution of the secondary radiation produced by an Al-scatterer agrees well with the Klein–Nishina formula. But in the case of Pb, in addition to the Compton scattering, I found a special radiation which is very prominent in the backward direction where the intensity of Compton scattering is very weak. I estimated the photon energy of this special radiation by using the absorption method with Pb plates and found it to be about 550 keV." The radiation was emitted roughly isotropically and was of such an energy that with hindsight it was identified with e^+e^- annihilation radiation. Thus the Klein–Nishina formula served as a guide to making new discoveries — close to a discovery of positrons — but not quite making the mark.

6.3. *Polarization of annihilation radiation*

In two-photon annihilations of electron–positron pairs, which is related to Compton scattering by crossing, theory[45] predicts polarization correlations, which can be described either as perpendicular correlations of their linear polarizations or as both photons having the same helicity, when their circular polarizations are analyzed. The first type of correlation has been observed in the azimuthal variation of the coincidence rate of Compton-scattered annihilation quanta.[46] Evidence for the second type was claimed in another experiment.[47]

6.4. *The helicity of the neutrino*

In a very beautiful experiment on the helicity of the electron neutrino,[48] M. Goldhaber, L. Grodzins and A. W. Sunyar arranged so that the neutrino helicity in an electron capture process was transferred to the circular polarization of the subsequently emitted γ-ray. They used the sensitivity of Compton scattering to the state of circular polarization of the photons when scattered against polarized electrons. The transmission of these γ-rays through a magnetized iron filter therefore depends on the direction of the magnetized field relative to the line of flight of the photons. The main ingredients in the experiment were given a concise summary by the authors: "A combined analysis of circular polarization and resonant scattering of γ-rays following orbital electron capture measures the helicity of the neutrino." The result was that the neutrino emitted in beta decay has negative helicity, which at that time added strength to the hypothesis of the V–A structure of weak interactions.

The results of the collaboration between Klein and Nishina in 1928 thus came to play an important role in the study of parity violations and in determinating the nature of the weak interactions 30 years later.

6.5. *Applications in other research*

There are numerous other applications where the results of the collaboration between Klein and Nishina are used. In nuclear and particle physics there are many applications, in particular to determine the polarization state of γ-rays. In astrophysics it has played a role somewhat different from the one envisaged by the authors in 1928, namely one considers inverse Compton scattering, a process in which energy is transferred from a very fast electron to a low energy photon, making it a photon of very high energy. Recently discovered astrophysical point sources emitting photons in the TeV energy range may depend on such processes. The cross section for such an energy transfer is basically obtained from the Klein–Nishina formula with suitable Lorentz transformations, since the formula holds for an electron initially at rest. High energy photon beams have been created in the laboratory by backward scattering of laser light from a beam of high energy electrons. Compton scattering has also been applied to problems in solid state physics, for which very high energy (10 GeV) electron storage rings, used as synchroton facilities, are expected to become useful.

Acknowledgements

Informative and penetrating discussions with Dr. Lars Bergström (Stockholm) have been very valuable and are acknowledged. I would like to express my thanks for useful and informative discussions with Prof. Chen Ning Yang (Stony Brook, USA). Permission was granted to use the Nobel Archives by the Royal Swedish Academy of Sciences. I am indebted to Dr. Finnn Aaserud (Copenhagen) for his efforts and kindness in searching the Niels Bohr archives for pertinent material; and to Drs. Tamaki, Takeuchi and Yazaki (Tokyo) for showing me relevant material in the archives of the Nishina Memorial Foundation in Tokyo, Dr. Yazaki has sent me his recent study of the Klein–Nishina collaboration with comments in English, for which I am deeply grateful.

References

1. "Oskar Klein: To the Memory of Yoshio Nishina", manuscript (1975), the archives of the Nishina Memorial Foundation, Tokyo.
2. B. Laurent, private communication.
3. O. Klein, letter to N. Bohr, July 6, 1928, The Niels Bohr Archives BSC (13, 1).
4. O. Klein and Y. Nishina, *Nature* **122**, 398 (1928).
5. Y. Nishina, letter to N. Bohr, August 8, 1928, The Niels Bohr Archives BSC (14, 2).
6. O. Klein and Y. Nishina, *Z. Phys.* **52**, 853 (1929).
7. Y. Nishina, *Z. Phys.* **52**, (1930) 869.
8. Y. Nishina, *Nature* **123**, 349 (1929).
9. Abraham Pais, *Subtle is the Lord* (Oxford University Press, 1982), p. 332.
10. Inga Fisher-Hjalmars and Bertel Laurent, "Oskar Klein", in Vol. 1 of *The Oskar Klein Memorial Lectures* (World Scientific, Singapore, 1991).
11. Oskar Klein, "From My Life of Physics", *ibid.*, p. 103.
12. O. Klein, "Excerpts from Some Autobiographical Notes" (transl. from Swedish), this volume.
13. A. H. Compton, *Phys. Rev.* **21**, 483 (1923).
14. *Les Prix Nobel en 1927* (P. A. Norstedt & Söner, Stockholm, 1928), p. 19.
15. P. Debye, *Phys. Z.* **24**, 161 (1923).
16. Letter from Sommerfeld to Bohr, January 21, 1923 (*Archives for the History of Quantum Physics*).
17. Max Dresden, *H. A. Kramers: Between Tradition and Revolution* (Springer-Verlag, 1987).
18. *Les Prix Nobel en 1921–1922* (P. A. Norstedt & Söner, Stockholm, 1923).
19. N. Bohr, Letter to Ernest Rutherford, January 9, 1924 (*Archives for the History of Quantum Physics*).
20. W. Pauli, Letter to H. A. Kramers, July 27, 1925 (*N. Bohr Collected Works*, Vol. 5, p. 442).
21. The Nobel Archives, the Royal Swedish Academy of Sciences, Stockholm: Report 1927 by the Nobel Committee for Physics (excerpts transl. from the Swedish by the author).
22. E. Schrödinger, *Ann. Phys.* **82**, 257 (1927).
23. W. Gordon, *Z. Phys.* **40**, 117 (1927).
24. O. Klein, *Z. Phys.* **41**, 407 (1927).

25. A. H. Compton, *X-Rays and Electrons*, 1st ed., pp. 296–314 (Van Nostrand, 1926).
26. P. A. M. Dirac, *Proc. R. Soc.* **A111**, 405 (1926).
27. O. Klein and Y. Nishina, *Z. Phys.* **52**, 853 (1929).
28. O. Klein and Y. Nishina, *Nature* **122**, 398 (1928).
29. I. Waller, *Z. Phys.* **61**, 837 (1930).
30. I. Tamm, *Z. Phys.* **62**, 545 (1930).
31. A. H. Compton and S. K. Allison, *X-Rays in Theory and Experiment* (Van Nostrand, New York, 1935).
32. Y. Yazaki *(Kagakusi Kenkyu, J. Hist. Sci. Japan)*, II, 31, part I, p. 81; part II, p. 129 (1992).
33. Y. Nishina, *Z. Phys.* **52**, 869 (1930).
34. W. Franz, *Ann. Phys.* **33**, 47 (1938).
35. S. B. Gunst and L. A. Page, *Phys. Rev.* **92**, 970 (1953).
36. P. A. M. Dirac, Letter to N. Bohr, November 1929, quoted in Ref. 37, p. 350.
37. A. Pais, *Inward Bound*, 1st ed. (Oxford University Press).
38. P. A. M. Dirac, *Proc. Cambridge Philos. Soc.* **26**, 361 (1930).
39. Cf. e.g.: J. D. Bjorken and S. D. Drell, *Relativistic Quantum Mechanics* (McGraw-Hill, 1964), pp. 10–12.
40. G. T. P. Tarrant, *Proc. R. Soc.* **A139**, 345 (1930).
41. L. Meitner and H. H. Hupfeld, *Naturwissenschaften* **18**, 534 (1930) and *Z. Phys.* **67**, 147 (1930).
42. C. Y. Chao, *Proc. Natl. Acad. Sci.* **16**, 431 (1930).
43. Cf. Bing An Li and C. N. Yang, *Int. J. Mod. Phys.* **A4**, 4325 (1989).
44. C. Y. Chao, *Phys. Rev.* **36**, 1519 (1930).
45. J. A. Wheeler, *Ann. N. Y. Acad. Sci.* **48**, 219 (1946); C. N. Yang, *Phys. Rev.* **77**, 136 (1950).
46. R. C. Hanna, *Nature* **132**, 332 (1948); E. Bleuler and H. L. Bradt, *Phys. Rev.* **73**, 1398 (1948); C. S. Wu and I. Shaknov, *Phys. Rev.* **77**, 136 (1950).
47. F. P. Clay and F. L. Hereford, *Phys. Rev.* **85**, 675 (1952).
48. M. Goldhaber, L. Grodzins and A. W. Sunyar, *Phys. Rev.* **109**, 1015 (1958).

ON THE SCATTERING OF RADIATION BY FREE ELECTRONS ACCORDING TO DIRAC'S NEW RELATIVISTIC QUANTUM DYNAMICS*

O. Klein and Y. Nishina

in Copenhagen, Denmark
Received October 30, 1928

The intensity of Compton scattering radiation is calculated on the basis of the new relativistic quantum dynamics developed by Dirac. The result shows deviations from the corresponding Dirac–Gordon formulae of second order with respect to the ratio of the energy of the primary light quantum to the rest energy of the electron.

1. Introduction

On the foundation of the older form of relativistic quantum mechanics, Dirac[1] and Gordon[2] have developed a theory for the intensity and polarization of Compton scattering radiation, which seems to be in good agreement with experience for not-too-short wave radiation. According to the new relativistic quantum mechanics recently developed by Dirac,[3] in which the phenomena connected with the self-rotation of the electron are automatically taken into account, the foundation of a theory of the scattering of light on free electrons has changed and one can expect that also the final results of the Dirac–Gordon theory for the Compton effect will be influenced by this. In the present work, we have tried to attack the problem of scattering radiation on free electrons on the basis of Dirac's new dynamics of the electron. In this we have followed the treatment given by Gordon,

---
*Original in *Z. Phys.* **52**, 853 (1929); reproduced here with permission from Springer-Verlag. Translated from the German by Dr Lars Bergström.

which rests on a correspondence-wise utilization of wave mechanics. One would expect that the radiation theory given by Dirac, which allows one to take radiation damping into account, gives a coinciding result in this case when dealing with the first approximation with respect to the intensity of the primary radiation.

In the present work we have restricted ourselves to the calculation of the intensity of the scattered radiation and its dependence on direction and wave length. The question of the polarization of the scattered radiation will be treated by one of us in an accompanying work.[4] It has appeared that in the region of hard γ-rays the deviations of our results from the Dirac–Gordon formulae are substantial so that, for example, the determination of the wavelength of the cosmic penetrating radiation according to the present theory would give considerably shorter wavelengths than in the old theory. But just in this region the experimental results on the Compton effect seem to be too uncertain to give a decision for or against the theory at this time. Here an accurate experimental test would, however, be very desirable, not least in view of the difficulty of his new theory emphasized by Dirac, connected with the possibility of negative energies.

2. Introductory Remarks on the Wave Equation of Dirac

According to Dirac[5] the quantum-mechanical problem of an electron with charge $-e$ and rest mass m, moving in a field of force where the electrostatic potential is V and the vector potential is $\mathcal{U}$ is determined by the following Hamiltonian form F:

$$F = \frac{E + eV}{c} + \rho_1 \left(\sigma, p + \frac{e}{c}\mathcal{U} \right) + \rho_3 mc, \tag{1}$$

where E means the energy of the electron and p its momentum vector with the components p_1, p_2, p_3 with respect to the axes x_1, x_2, x_3 of an orthogonal coordinate system, while c denotes the vacuum velocity of light. Further, σ is a matrix vector with components $\sigma_1, \sigma_2, \sigma_3$, fulfilling the relations

$$\left. \begin{array}{ll} \sigma_1\sigma_2 = i\sigma_3 = -\sigma_2\sigma_1, & \sigma_1^2 = 1, \\ \sigma_2\sigma_3 = i\sigma_1 = -\sigma_3\sigma_2, & \sigma_2^2 = 1, \\ \sigma_3\sigma_1 = i\sigma_2 = -\sigma_1\sigma_3, & \sigma_3^2 = 1, \end{array} \right\} \tag{2}$$

and ρ_1 and ρ_3 are two out of three matrices that fulfil relations identical to (2) and moreover commute with σ. As Dirac has shown, all these quantities can be represented by matrices with four rows and columns. Correspondingly, the mutually adjoint eigenfunctions φ and ψ belonging to the Hamiltonian form F, which we will regard as functions of the coordinates x_1, x_2, x_3 and the time t, consist each of four components, $\varphi_1, \varphi_2, \varphi_3, \varphi_4$ and $\psi_1, \psi_2, \psi_3, \psi_4$, respectively, so that an expression like $\mu\psi$, where μ means a four-row matrix, is to be understood as an abbreviation for the four quantities $\sum_{k=1}^{4} \mu_{ik}\psi_k (i = 1, 2, 3, 4)$, while $\varphi\mu$ denotes the quantities $\sum_{k=1}^{4} \varphi_k \mu_{ki} (i = 1, 2, 3, 4)$, where μ_{ik} are the matrix elements of μ.

The physical utilization of Eq. (1) depends on the assumption that if φ and ψ are associated wave functions of the Hamiltonian function (1), namely that φ and ψ are the complex conjugates of each other for Hermitian matrices ρ_1, ρ_3 and σ, then $\varphi\psi d\tau$ gives the probability that the electron is found in the volume element $d\tau$. From this follows, according to Dirac, for the probability that the electron in a time dt crosses a surface element df, $-c\varphi\rho_1\sigma_s\psi df dt$, where σ_s denotes the component of the vector σ in the direction normal to the surface element, and where $\varphi\mu\psi$, if μ again stands for an arbitrary four-row matrix, should mean $\sum_{ik} \varphi_i \mu_{ik}\psi_k$. If we multiply the quantities $\varphi\psi$ and $-c\varphi\rho_1\sigma\psi$ by the charge $-e$ of the electron, we get the so-called wave-mechanical electric density ρ and the electric current density vector $\mathcal{J}$. Thus

$$\rho = -e\varphi\psi, \quad \mathcal{J} = ec\varphi\rho_1\sigma\psi \tag{3}$$

are valid.

3. Eigenfunctions of the Dirac Wave Equation for a Free Electron

First we want to consider a free electron where we may choose as eigenfunctions, like in the Schrödinger theory, the following, to each value of p associated functions φ_0 and ψ_0, namely

$$\varphi_0(p) = u(p)e^{\frac{i}{h}[Et-(pr)]}, \quad \psi_0(p) = v(p)e^{-\frac{i}{h}[Et-(pr)]}, \tag{4}$$

where h denotes the Planckian constant divided by 2π, and r the coordinate vector with components x_1, x_2, x_3, while $u(p)$ and $v(p)$ are quantities independent of time and coordinates, consisting of four components and having to fulfil the following purely algebraic conditions:

$$u(p)\{E/c + \rho_1(\sigma p) + \rho_3 mc\} = 0,$$
$$\{E/c + \rho_1(\sigma p) + \rho_3 mc\}v(p) = 0. \tag{5}$$

To a given p there belong, according to these equations, two values of the energy E, one positive and one negative, which both fulfil the relation

$$E^2/c^2 = m^2c^2 + p^2, \tag{6}$$

which expresses the relation between energy and momentum in the relativistic mechanics. As Eqs. (5) describe abbreviations for systems of four equations each, one should really expect that to a given value of p there belong four energy values, which just corresponds to the possibility of taking into account, besides the physically non-sensible negative energy values, the doubling of the number of eigenvalues connected with the self-magnetism of the electron. This degeneracy of the free electron has as a consequence the fact that, if we choose a particular eigenvalue — naturally we will then restrict ourselves to positive values — the ratios of the components of $u(p)$ and likewise of $v(p)$ are not yet fixed but rather depend on two quantities to be chosen freely. In the following calculation of the intensity of the scattered radiation, it is necessary to take this degeneracy into account.

We first want to consider the case of an electron at rest. Here $p = 0$, and we put $E = mc^2$. Equations (5) thus simply become

$$u(0)(1 + \rho_3) = 0, \quad (1 + \rho_3)v(0) = 0. \tag{7}$$

We will choose ρ_3 to be a diagonal matrix, as we put[a]

$$\rho_3 = \begin{pmatrix} -1 & 0 & 0 & 0 \\ 0 & -1 & 0 & 0 \\ 0 & 0 & 1 & 0 \\ 0 & 0 & 0 & 1 \end{pmatrix}. \tag{8}$$

---

[a]We have, for ρ_3, chosen the opposite sign to that of Dirac, so that for positive energy the two first components of u and v rather than the two last components remain finite.

It then follows that Eqs. (7) are satisfied if u_3, v_3, u_4, v_4 are equal to zero, while u_1, v_1 and u_2, v_2 can be chosen freely. We can in other words compose the general solution belonging to the given eigenvalue from two independent solutions for which only u_1, v_1 and u_2, v_2 are different from zero, respectively.

We will now further choose the component σ_3 of σ along the x_3 axis to be a diagonal matrix, as we put

$$\sigma_3 = \begin{pmatrix} 1 & 0 & 0 & 0 \\ 0 & -1 & 0 & 0 \\ 0 & 0 & 1 & 0 \\ 0 & 0 & 0 & -1 \end{pmatrix}. \tag{9}$$

According to Dirac $u(p)\sigma v(p)$ is now proportional to the magnetic moment of the electron belonging to the solution $u(p)$, $v(p)$. The two solutions mentioned thus give an equal but oppositely directed magnetic moment of the electron along the x_3 axis. We can therefore regard u_1, v_1 and u_2, v_2, respectively, as the eigenfunctions belonging to the two possible orientations of the magnetic moment with respect to the x_3 axis.

We can easily reduce the case of an arbitrary p-value to the case $p = 0$ by carrying out a contact transformation closely related to the Lorentz transformation.[b] Let

$$S = \alpha + i\beta\rho_2(\sigma p), \quad S^{-1} = \alpha - i\beta\rho_2(\sigma p), \quad \alpha^2 + \beta^2 p^2 = 1, \tag{10}$$

where we soon want to determine α and β closer. We put

$$F^* = S(E/c + \rho_1(\sigma p) + \rho_3 mc)S^{-1}. \tag{11}$$

Now one has

$$S\rho_1 S^{-1} = \rho_1(\alpha - i\beta\rho_2(\sigma p))^2 = (\alpha^2 - \beta^2 p^2)\rho_1 + 2\alpha\beta\rho_2(\sigma p)$$

and thus

$$S\rho_1(\sigma p)S^{-1} = (\alpha^2 - \beta^2 p^2)\rho_1(\sigma p) + 2\alpha\beta p^2\rho_3.$$

---

[b]The following consideration is in close relation to Dirac's treatment of the invariance of his equations with respect to Lorentz transformations; *loc. cit.* B, p. 615.

Further

$$S\rho_3 S^{-1} = (\alpha^2 - \beta^2 p^2)\rho_3 - 2\alpha\beta\rho_1(\sigma p).$$

It thus follows that

$$F^* = E/c + (\alpha^2 - \beta^2 p^2 - 2\alpha\beta mc)\rho_1(\sigma p) + ((\alpha^2 - \beta^2 p^2)mc + 2\alpha\beta p^2)\rho_3.$$

We now want to choose α and β such that the coefficient of $\rho_1(\sigma p)$ in F^* vanishes, i.e. we demand

$$\alpha^2 - \beta^2 p^2 - 2\alpha\beta mc = 0. \tag{12}$$

Further, we set

$$(\alpha^2 - \beta^2 p^2)mc + 2\alpha\beta p^2 = m^* c \tag{13}$$

and then have

$$F^* = E/c + \rho_3 m^* c. \tag{14}$$

It now follows from a simple calculation that

$$m^{*2}c^2 = m^2 c^2 + p^2, \tag{15}$$

and if, in accordance with this, we set

$$m^* c = +\sqrt{m^2 c^2 + p^2}, \tag{16}$$

we easily get

$$\alpha = \sqrt{\frac{m^* + m}{2m^*}}, \qquad \beta = \sqrt{\frac{m^* - m}{2m^* p^2}}, \tag{17}$$

as a result of which the condition (12) is fulfilled.

If we now set

$$u(p) = u^*(p)S(p), \quad v(p) = S^{-1}(p)v^*(p), \tag{18}$$

where $S(p)$ and $S^{-1}(p)$ mean the quantities (10) for a definite value of p, it follows from (11) that u^* and v^* satisfy the equations

$$u^*(E/c + \rho_3 m^* c) = 0, \quad (E/c + \rho_3 m^* c)v^* = 0$$

or, since according to (16) $E = m^* c^2$,

$$u^*(1 + p_3) = 0, \quad (1 + \rho_3)v^* = 0,$$

i.e. we obtain Eqs. (7) for the case $p = 0$. In this manner we can describe $u(p)$ and $v(p)$ with the help of the independent quantities u_1^*, v_1^* and u_2^*, v_2^*, which, however, are less easily connected with the magnetization of the electron than for the case $p = 0$.

Let there now for the moment be a solution $u(p)$, $v(p)$ which corresponds to one of the independent eigenfunctions, where thus either only u_1^*, v_1^* or u_2^*, v_2^* are different from zero. We want to normalize this solution such that it corresponds to one electron. If $\varphi_0(p)$ and $\psi_0(p)$ are the accompanying eigenfunctions in coordinate space, this means, as is known for continuous eigenvalues, that

$$\int \varphi_0(p')\psi_0(p'')dr = \delta(p' - p''), \tag{19}$$

where dr denotes the volume element and the integration is to be extended over the entire space in question, while $\delta(p' - p'')$ means the singular function introduced by Dirac, the integral of which, with respect to p' over an arbitrary region in momentum space including the point $p' = p''$, is equal to 1. By insertion of the expressions (4) for φ_0 and ψ_0 in (19) one obtains, by comparison with the Fourier integral theorem,

$$u(p)v(p) = (2\pi h)^{-3}. \tag{20}$$

Finally, we want to describe the complex conjugate quantities u_1^* and v_1^* and u_2^* and v_2^*, respectively, which belong to the two different eigensolutions

$u(p)$, $v(p)$ in the following way by real quantities:

$$u_1^*(p) = a_1 e^{i\delta_1(p)}, \qquad u_2^*(p) = a_2 e^{i\delta_2(p)}, \left.\begin{array}{c}\\\\\end{array}\right\}$$
$$v_1^*(p) = a_1 e^{-i\delta_1(p)}, \qquad v_2^*(p) = a_2 e^{-i\delta_2(p)}. \tag{21}$$

According to (20) $a_1^2 = a_2^2 = (2\pi h)^{-3}$ is then valid, so that we can choose the amplitudes a_1 and a_2 such that they are independent of the p value. On the other hand, the phases $\delta_1(p)$ and $\delta_2(p)$ can be chosen completely freely for every p value.

4. Solution of the Wave Equation for an Electron in a Continuous Monochromatic Radiation Field

Here we want to regard a free electron illuminated by a continuous monochromatic wave train. This we can describe through a vector potential $\mathcal{U}$ of the form

$$\mathcal{U} = a e^{iv\left(t - \frac{(nr)}{c}\right)} + \bar{a} e^{-iv\left(t - \frac{(nr)}{c}\right)}, \tag{22}$$

where a and $\bar{a}$ denote two constant-to-each-other complex conjugate vectors perpendicular to the unit vector n, which gives the direction of propagation of the waves. The position where the potential $\mathcal{U}$ is measured at time t is in this connection characterized through the radius vector r drawn from a fixed point to the position in question. Further, v means the wave number of the radiation multiplied by 2π. While neglecting higher powers than the first in $|\mathcal{U}|$, we want to search for solutions to the wave equation belonging to F of the form

$$\varphi(p) = \varphi_0(p) \left\{ 1 + f(p) e^{iv\left(t - \frac{(nr)}{c}\right)} - \bar{f}(p) e^{-iv\left(t - \frac{(nr)}{c}\right)} \right\}, \left.\begin{array}{c}\\\\\\\\\end{array}\right\}$$
$$\psi(p) = \left\{ 1 + g(p) e^{iv\left(t - \frac{(nr)}{c}\right)} + \bar{g}(p) e^{-iv\left(t - \frac{(nr)}{c}\right)} \right\} \psi_0(p), \tag{23}$$

where $\varphi_0(p)$, $\psi_0(p)$ are the eigenfunctions of free electrons given in (4), while f, $\bar{f}$ and g, $\bar{g}$ are constant four-row matrices. The general solution of the wave equation follows in the approximation considered by superposition of all possible solutions of the form (23). To determine the quantities f, $\bar{f}$, g, $\bar{g}$ it is useful to start from the equations of second order which according to Dirac can be easily derived from the first order equations belonging to F; they read

$$\left\{\frac{h^2}{c^2}\frac{\partial^2}{\partial t^2} + \left(-ih\nabla + \frac{e}{c}\mathcal{U}\right)^2 + m^2c^2\right\}\psi + \frac{eh}{c}(\sigma\mathcal{H})\psi + \frac{ieh}{c}\rho_1(\sigma\mathcal{E})\psi = 0,$$

$$\left\{\frac{h^2}{c^2}\frac{\partial^2}{\partial t^2} + \left(ih\nabla + \frac{e}{c}\mathcal{U}\right)^2 + m^2c^2\right\}\varphi - \frac{eh}{c}\varphi(\sigma\mathcal{H}) + \frac{ieh}{c}\varphi\rho_1(\sigma\mathcal{E}) = 0,$$

$$(24)$$

where ∇ means the vector operator with components $\frac{\partial}{\partial x_1}, \frac{\partial}{\partial x_2}, \frac{\partial}{\partial x_3}$, while $\mathcal{E}$ and $\mathcal{H}$ are, respectively, the electric and magnetic field strengths belonging to $\mathcal{U}$. By insertion of the expression (23) in (24) one finds that

$$f(p) = \frac{e}{2hv(E/c - (np))}(2(ap) + h(\sigma\eta) - ih\rho_1(\sigma\varepsilon)),$$

$$\bar{f}(p) = -\frac{e}{2hv(E/c - (np))}\{2(\bar{a}p) + h(\sigma\bar{\eta}) - ih\rho_1(\sigma\bar{\varepsilon})\},$$

$$g(p) = -\frac{e}{2hv(E/c - (np))}\{2(ap) + h(\sigma\eta) + ih\rho_1(\sigma\varepsilon),$$

$$\bar{g}(p) = \frac{e}{2hv(E/c - (np))}\{2(\bar{a}p) + h(\sigma\bar{\eta}) + ih(\sigma\bar{\varepsilon})\},$$

$$(25)$$

where ϵ and $\bar{\epsilon}$ and η and $\bar{\eta}$, respectively, are connected in the following way with the electric field strength $\mathcal{E}$ and the magnetic strength $\mathcal{H}$, respectively, of the radiation field:

$$\mathcal{E} = \varepsilon e^{iv\left(t - \frac{(nr)}{c}\right)} + \bar{\varepsilon}e^{-iv\left(t - \frac{(nr)}{c}\right)},$$

$$\mathcal{H} = \eta e^{iv\left(t - \frac{(nr)}{c}\right)} + \bar{\eta}e^{-iv\left(t - \frac{(nr)}{c}\right)};$$

$$(26)$$

since

$$\mathcal{E} = -\frac{1}{c}\frac{\partial\mathcal{U}}{\partial t}, \quad \mathcal{H} = \text{rot}\,\mathcal{U},$$

it follows that

$$
\left.\begin{array}{ll}
\varepsilon = -\dfrac{iv}{c}a, & \bar{\varepsilon} = \dfrac{iv}{c}\bar{a}, \\[2ex]
\eta = -\dfrac{iv}{c}[na], & \bar{\eta} = \dfrac{iv}{c}[n\bar{a}].
\end{array}\right\}
\tag{27}
$$

5. Calculation of the Scattered Radiation Field

We can write the general solution of the wave equations in the presence of the incident radiation in the approximation mentioned in the following way:

$$
\Phi = \int \varphi(p)dp, \quad \Psi = \int \psi(p)dp,
\tag{28}
$$

where $\varphi(p)$ and $\psi(p)$ denote the approximation solutions, given in (23) and (25), to the corresponding equations — where first no normalization is assumed — while the integration $\int dp$ is to be regarded as an abbreviation for the threefold integration $\int dp_1 \int dp_2 \int dp_3$ over all possible values of the components of momentum p_1, p_2, p_3. The electric current density belonging to the general solution is given by

$$
J = ec\Phi\rho_1\sigma\Psi = ec \iint \varphi(p)\rho_1\sigma\psi(p')dpdp'.
\tag{29}
$$

In this expression we now want to insert the expressions (23) for φ and ψ and order it in a correspondence-wise way according to the different possible radiation processes.[6] By neglecting quantities of the size of order $|a|^2$ we get, through a simple transformation,

$$
J = J_0 + ce \iint dpdp' \Big\{ u(p)[\rho_1\sigma g(p')
$$

$$
+ f(p)\rho_1\sigma]v(p')e^{\frac{i}{\hbar}\left[(E+h\nu-E')t-\left(p+n\frac{h\nu}{c}-p'\right)r\right]}
$$

$$
+ \text{complex conjugate part} \Big\},
\tag{30}
$$

where J_0 denotes the current density belonging to the unperturbed eigenfunctions φ_0, ψ_0.

Now let $\mathcal{U}'$ be the vector potential belonging to $\mathcal{J}$ at a point whose distance r is very large compared to the dimensions of the region available for the electron, which in turn are to be regarded as large in comparison with the wavelength of the light and the de Broglie waves. These assumptions should just correspond to the observations of the Compton effect. Further, let n' be a unit vector which gives the direction of observation, i.e. the direction of a radius vector drawn from a point in the region of the electron to the point of observation. In a well-known way one then gets[7]

$$
\mathcal{U}' = \frac{e}{r} \iint dp\,dp' \left\{ e^{\frac{i}{\hbar}(E+h\nu-E')(t-\frac{r}{c})} \int dr\, u(p)[\rho_1 \sigma g(p') \right.
$$

$$
- f(p)\rho_1 \sigma] v(p') e^{-\frac{i}{\hbar}\left[p-p'+n\frac{h\nu}{c}-n'\frac{E+h\nu-E'}{c} \right] r}
$$

$$
\left. + \text{complex conjugate part} \right\}, \tag{31}
$$

where $\int dV$ indicates the integration $\int dx_1 \int dx_2 \int dx_3$ over the whole region at the electron's disposal. Following Gordon[8] one can evaluate this integral with the help of the Fourier theorem by introducing, instead of p and p', certain new vectors β and β' with components P_1, P_2, P_3, P'_1, P'_2, P'_3 through the relations

$$
\mathcal{B} = p + n\frac{h\nu}{c} - n'\frac{E+h\nu}{c}, \qquad \mathcal{B}' = p' - \frac{E'}{c}n'. \tag{32}
$$

It follows that

$$
\mathcal{U}' = \frac{(2\pi h)^3}{r} \int \frac{d\mathcal{B}}{\Delta\Delta'} \left\{ e^{i\nu'(t-\frac{r}{c})} u(p)[\rho_1 \sigma g(p') \right.
$$

$$
\left. + f(p)\rho_1 \sigma] v(p') + \text{complex conjugate part} \right\}, \tag{33}
$$

where Δ and Δ', respectively, mean the functional determinants of P_k with respect to p_k and P'_k with respect to p'_k, while p' is the special value of this quantity which for a given p follows from the relation $P = P'$, i.e. from

$$
p + n\frac{h\nu}{c} = p' + n'\frac{h\nu'}{c}, \tag{34}
$$

where v' is given by the relation

$$E + hv = E' + hv'. \tag{35}$$

These are just the well-known Compton relations which assign a definite final state to the state p and the incident light quantum for a given direction of observation.

From (33) there now follows, according to Gordon,[9] the following expression for the radiation potential $\mathcal{U}(p, p')$ belonging to a transition from the state p to the state p':

$$\mathcal{U}(p, p') = \frac{(2\pi h)^3}{r} \frac{1}{\sqrt{\Delta\Delta'}} \left\{ e^{iv'\left(t - \frac{r}{c}\right)} u(p)[\rho_1 \sigma g(p') \right.$$

$$\left. + f(p)\rho_1\sigma]v(p') + \text{complex conjugate part} \right\}, \tag{36}$$

where $u(p)$, $v(p)$ and $u(p')$, $v(p')$ from now on are to be regarded as normalized.

The radiation field belonging to a definite Compton effect is completely determined by the quantity $\mathcal{U}(p, p')$. From it the magnetic field strength $\mathcal{H}(p, p')$ follows through the relation

$$\mathcal{H}(p, p') = \text{rot}\,\mathcal{U}(p, p'), \tag{37}$$

and from this in turn the electric field strength $\mathcal{E}(p, p')$ through the relation

$$\mathcal{E} = [\mathcal{H}(p, p'), n']. \tag{38}$$

We now must obtain the quantity $\mathcal{U}(p, p')$ or, better, $\mathcal{H}(p, p')$ in a form suitable for a numerical calculation of the intensity and polarization of the scattered radiation. To this end, we have to insert in (36) the expressions for $f, \bar{f}, g, \bar{g}$ given by (25) and then use the calculational rules for the matrices ρ and σ given by Dirac, in order to obtain the result in terms of ordinary numbers. In this connection we will assume for simplicity that the initial state denoted by p corresponds to an electron at rest. We denote the corresponding values of $\mathcal{U}(p, p')$ and $\mathcal{H}(p, p')$ by $\mathcal{U}_0$ and $\mathcal{H}_0$. Thus,

$$p = 0, \quad E = mc^2 \tag{39}$$

should be valid.

With the help of Eqs. (5) for u and v we can reduce all matrices appearing in the expression for $\mathcal{U}_0$ to the unit matrix and the matrix vector σ. Based on the relations (2) it is then possible to give the result in a expression linear in the components of σ. Here the following calculational rule drawn from Dirac,[10] which is intimately connected with quaternion calculus, is of use, and easily proven with the help of (2):

$$(\mathcal{B}\sigma)(\mathcal{C}\sigma) = (\mathcal{B}\mathcal{C}) + i(\sigma, [\mathcal{B}\mathcal{C}]), \tag{40}$$

as well as the one following from it,

$$\sigma(\sigma\mathcal{B}) = \mathcal{B} + i[\sigma\mathcal{B}],$$

where $\mathcal{B}$ and $\mathcal{C}$ stand for two arbitrary vectors, commuting with σ.

As follows from (5), in the quantities f and g there appears, besides the matrix σ, the matrix $\rho_1\sigma$. In (36) these matrices are multiplied by $\rho_1\sigma$. The resulting matrix is, according to (2) and (40), respectively, a linear combination of the unit matrix, ρ_1, $\rho_1\sigma$ and σ. This matrix then comes to stand between $u(p)$ and $v(p')$ and we can show that both $u(p)\rho_1 v(p')$ and $u(p)\rho_1\sigma v(p')$ can be reduced to the quantities $u(p)v(p')$ and $u(p)\sigma v(p')$. Indeed, it follows from

$$u(p)(1 + \rho_3) = 0,$$
$$\{E'/c + \rho_1(\sigma p') + \rho_3 mc\}v(p'),$$

through multiplication by $\rho_1 v(p')$ and $u(p)\rho_1$, respectively, that

$$u(p)(\rho_1 + i\rho_2)v(p') = 0,$$
$$u(p)\{E'/c\rho_1 + (\sigma p') - i\rho_2 mc\}v(p') = 0,$$

or through eliminating $v(p)\rho_2 u(p')$,

$$u(p)\rho_1 v(p') = -\frac{u(p)(p'\sigma)v(p')}{E'/c + mc}. \tag{41}$$

In a similar way one finds that

$$u(p)\rho_1\sigma u(p') = -\frac{u(p)\sigma(\sigma p')v(p')}{E'/c + mc}, \tag{42}$$

a relation which according to (40) can be reduced to an expression linear in σ.

We shall introduce the abbreviations

$$\left.\begin{array}{ll} u(p)\sigma v(p') = s, & u(p')\sigma v(p) = \bar{s}, \\[2mm] u(p)v(p') = d, & u(p')v(p) = \bar{d}, \end{array}\right\} \tag{43}$$

where s and $\bar{s}$ denote two vectors, complex conjugate to each other, while d and $\bar{d}$ are two complex conjugate scalars. With the help of the relations (41) and (42) we now obtain for the potential $\mathcal{U}_0$ an expression linear in s, $\bar{s}$ and $d, \bar{d}$. We shall not write this down here; rather, we content ourselves with the magnetic field strength to be calculated from (27), which is decisive for both the polarization and the intensity of the scattered radiation. It follows after some calculation, where the relations (34) and (35) are employed,

$$\mathcal{H}_0 = \frac{(2\pi h)^3 e^2 v'}{2mc^2 r\left(v - v' + \frac{2mc^2}{h}\right)} \sqrt{\frac{E'v'}{mc^2 v}} \left\{ d\left(\frac{1}{v}(n'\varepsilon)(v' - v)[n'n]\right.\right.$$

$$\left. - v'\left(\frac{1}{v} + \frac{1}{v'}\right)^2 \frac{mc^2}{h}[n'\varepsilon]\right) - i\left[\left(\frac{1}{v'} - \frac{1}{v}\right)\left((s, nv - n'v')((n'\varepsilon)n\right.\right.$$

$$- (nn')\varepsilon) + \left(v - v' + \frac{2mc^2}{h}\right)((sn')\varepsilon - (n'\varepsilon)s)\right)$$

$$+ \frac{2}{v}(\varepsilon n')((n's)(nv - n'v') + (v' - (nn')v)s)$$

$$\left.\left. - \left(\frac{1}{v} + \frac{1}{v'}\right)((n[\varepsilon s])v[n'n] + v'(n'[n\varepsilon])[n's])\right]\right\} e^{iv'\left(t - \frac{r}{c}\right)}$$

$$+ \text{complex conjugate part,} \tag{44}$$

where we have used[11] the values for Δ and Δ' following from (32) with the help of (34), namely

$$\Delta = 1 - \frac{c}{E}(n'p) = 1, \quad \Delta' = 1 - \frac{c}{E'}(n'p') = \frac{mc^2 v}{E'v'}. \tag{45}$$

We will not here enter into a closer discussion of this expression, but rather content ourselves with investigating the intensity of the scattered radiation in various directions.[12] To this end we form the quantity $\mathcal{H}_0^2$. In this context we want to choose as initial state one of the eigenfunctions belonging to $p = 0$ mentioned on page 4, i.e. for this state either only u_1, v_1 or only u_2, v_2 are different from zero and normalized according to (20). Thus, we look at the scattered radiation from an electron which has been magnetized by a field in the direction of the x_3-axis. For the final state we have to assume here that both u_1^*, v_1^* and u_2^*, v_2^* are different from zero and normalized. According to the usual quantum-mechanical methods for calculating transition probabilities, this corresponds to the possibility of such transitions where the magnetic moment of the electron is unchanged as well as such where its direction is changed.

By forming $\mathcal{H}_0^2$ we now obtain an expression bilinear in the quantities d, s and $\bar{d}, \bar{s}$, which we have to average over the phases δ_1 and δ_2 of the initial and final states. Thus we are dealing with a sum of terms of the form $u(p')\alpha v(p) \cdot u(p)\beta v(p')$ which we have to average with respect to the phases, where α and β are two matrices. If we first assume that $u_1(p)$ and $v_1(p)$ are different from zero, we write this as

$$\sum_{i,k} u_i(p')\alpha_{i1}v_1(p)u_1(p)\beta_{1k}v_k(p').$$

This, however, is equal to $u_1(p)v_1(p) \cdot u(p')\alpha\mu\beta v(p')$, where μ represents the matrix

$$\begin{pmatrix} 1 & 0 & 0 & 0 \\ 0 & 0 & 0 & 0 \\ 0 & 0 & 0 & 0 \\ 0 & 0 & 0 & 0 \end{pmatrix}.$$

We can, according to (8) and (9), represent μ in the following way with the help of ρ_3 and σ_3:

$$\mu = \left(\frac{1+\sigma_3}{2}\right)\left(\frac{1-\rho_3}{2}\right), \tag{46}$$

and it thus holds that

$$u(p')\alpha v(p) \cdot u(p)\beta v(p') = (2\pi h)^{-3} u(p')\alpha\mu\beta v(p'), \tag{47}$$

where we, according to (20), have put $u_1v_1 = (2\pi h)^{-3}$. Had we chosen as initial state the state belonging to u_2, v_2, the quantity to take for μ would have been $(\frac{1-\sigma_3}{2})(\frac{1-\rho_3}{2})$.

We now assume that $\alpha = (C\sigma)$ and $\beta = (B\sigma)$, where B and C are vectors commuting with σ. Then $u(p')\alpha v(p) = (C\bar{s})$ and $u(p)\beta v(p') = (Bs)$, and we have

$$(Bs)(C\bar{s}) = (2\pi h)^{-3} u(p')(C\sigma)\mu(B\sigma)v(p').$$

Now the relation

$$(C\sigma)\left(\frac{1+\sigma_3}{2}\right) = \left(\frac{1-\sigma_3}{2}\right)(C\sigma) + C_3$$

holds, so that

$$(Bs)(C\bar{s}) = (2\pi h)^{-3} u(p')\left(\frac{1-\rho_3}{2}\right)$$

$$\times \left\{\left(\frac{1-\sigma_3}{2}\right)(C\sigma)(B\sigma) + C_s(B\sigma)\right\}v(p'). \quad (48)$$

On the right-hand side of this equation there stand between $u(p')$ and $v(p')$ matrices, which are either of the form $\frac{1-\rho_3}{2}$ or of the form $\frac{1-\rho_3}{2}\sigma$. We can easily transform these in the following way. From the equations

$$u(p')\{E'/c + \rho_1(\sigma p') + \rho_3 mc\} = 0, \quad \{E'/c + \rho_1(\sigma p') + \rho_3 mc\}v(p') = 0$$

it follows, in the same way as on page 11, that

$$u(p')\left(\frac{1-\rho_3}{2}\right)v(p') = \frac{1}{2}\left(1 + \frac{mc^2}{E'}\right)u(p')v(p'). \quad (49)$$

Further, it follows that

$$u(p')\left(\frac{1-\rho_3}{2}\right)\sigma v(p') = \frac{1}{2}\left(1 + \frac{E'}{mc^2}\right)u(p')\sigma v(p')$$

$$- \frac{p'}{2mE'}u(p')(\sigma p')v(p'), \quad (50)$$

so that the quantities $u(p')(\frac{1-\rho_3}{2})v(p')$ and $u(p')(\frac{1-\rho_3}{2})\sigma v(p')$ can be expressed through the quantities $u(p')v(p')$ and $u(p')\sigma v(p')$. As now the final state characterized by p' should contain the two independent solutions in equal strength, it is clear from the outset that the average of $u(p')\sigma v(p')$ over the phases must vanish. In fact, this also follows from the representations (18) and (21) of the quantities $u(p')$, $v(p')$. Thus, we assume that

$$\overline{u(p')\sigma v(p')} = 0, \tag{51}$$

where the bar indicates the averaging over the phases $\delta_1(p')$ and $\delta_2(p')$. This assumption is now enough to enable the calculation of all other averages appearing. First we must have the value of $u(p')v(p')$. It follows from (18) and (20) that

$$u(p')v(p') = u^*(p')v^*(p') = 2(2\pi h)^{-3}. \tag{52}$$

Further, it follows from (48) that

$$\overline{(\mathcal{B}s)(\mathcal{C}\bar{s})} = \frac{1}{2}(2\pi h)^{-3}u(p')\left(\frac{1-\rho_3}{2}\right)v(p')\{(\mathcal{B}\mathcal{C}) + i[\mathcal{B}\mathcal{C}]_3\}, \tag{53}$$

or, according to (49) and (52), if we put

$$\gamma = \frac{1}{2}(2\pi h)^{-6}\left(1 + \frac{mc^2}{E'}\right), \tag{54}$$

$$\overline{(\mathcal{B}s)(\mathcal{C}\bar{s})} = \gamma\{(\mathcal{B}\mathcal{C}) + i[\mathcal{B}\mathcal{C}]_3\}. \tag{55}$$

We can easily liberate ourselves from the special choice of coordinates. If $\mathcal{I}$ is a unit vector in the direction of the magnetic moment of the magnetized electron in the initial state, we obviously have in general

$$\overline{(\mathcal{B}s)(\mathcal{C}\bar{s})} = \gamma\{(\mathcal{B}\mathcal{C}) + i(\mathcal{I}[\mathcal{B}\mathcal{C}])\},$$

and similarly $\qquad \overline{[\mathcal{B}s][\mathcal{C}\bar{s}]} = \gamma\{2(\mathcal{B}\mathcal{C}) + i(\mathcal{I}[\mathcal{B}\mathcal{C}])\},\qquad\Big\} \tag{56}$

also $\qquad\qquad \overline{(\mathcal{B}s)\bar{d}} = \overline{d(\mathcal{B}\bar{s})} = \gamma(\mathcal{I}\mathcal{B}),$

and finally

$$d\bar{d} = \gamma. \tag{57}$$

When calculating $\overline{\mathcal{H}_0^2}$ we now assume for simplicity that the incident light linearly polarized so that $\epsilon = \bar{\epsilon}$. It then follows after some calculation that

$$\overline{\mathcal{H}_0^2} = \frac{e^4}{m^2 c^4 r^2} \left(\frac{v'}{v}\right)^3 \left\{\left(\frac{v}{v'} + \frac{v'}{v}\right) \varepsilon^2 - 2(n'\varepsilon)^2\right\} \tag{58}$$

or, if we denote the angle between the direction of observation and the wave normal of the incident light by Θ and the angle between the direction of observation and the electric force of the incident wave by ϑ,

$$I = I_0 \frac{e^4}{m^2 c^4 r^2} \frac{\sin^2 \vartheta}{(1 + \alpha(1 - \cos\Theta))^3}$$

$$\times \left(1 + \alpha^2 \frac{(1 - \cos\Theta)^2}{2\sin^2\vartheta(1 + \alpha(1 - \cos\Theta))}\right), \tag{59}$$

where $\alpha = \frac{hv}{mc^2}$, while I_0 denotes the intensity of the incident radiation, and I the intensity of the scattered radiation, of which the former stands in the same relation to $2\epsilon^2$ as the latter to $\overline{\mathcal{H}_0^2}$. If the incident radiation is unpolarized we must further average over all directions of ϵ for a given direction of the incident radiation. The average of $\sin^2 \vartheta$ here becomes $\frac{1}{2}(1 + \cos^2 \Theta)$, so that

$$\bar{I} = I_0 \frac{e^4}{2m^2 c^4 r^2} \frac{1 + \cos^2 \Theta}{(1 + \alpha(1 - \cos\Theta))^3}$$

$$\times \left(1 + \alpha^2 \frac{(1 - \cos\Theta)^2}{(1 + \cos^2 \Theta)(1 + \alpha(1 - \cos\Theta))}\right). \tag{60}$$

If we look at the last formulae, (58)–(60), we see that the vector $\mathcal{I}$ does not appear therein. The results are thus independent of whether the scattering electrons have first become magnetized or not. As is gathered from a closer look at the expression (44), this is, however, valid in general only if the primary radiation is linearly polarized. The scattering for elliptically polarized incoming light is hence not equal to the sum of the scatterings belonging to the individual polarized components, a circumstance which is intimately connected with the curious polarization of the scattered radiation.[13]

Further, we see that the expressions (59) and (60) only differ through the factors

$$\left(1 + \alpha^2 \frac{(1 - \cos \Theta)^2}{2 \sin^2 \vartheta (1 + \alpha(1 - \cos \Theta))}\right)$$

and

$$\left(1 + \alpha^2 \frac{(1 - \cos \Theta)^2}{(1 + \cos^2 \Theta)(1 + \alpha(1 - \cos \Theta))}\right),$$

respectively, from the corresponding formulas of the Dirac–Gordon theory, i.e. the deviations of the two theories are of the order of magnitude of $(\frac{h\nu}{mc^2})^2$, while the deviations from the earlier expressions of the classical theory of scattered radiation, developed by J. J. Thomson, are of the order of magnitude of[14] $\frac{h\nu}{mc^2}$.

From (60) we can easily derive an expression for the scattering coefficient of a substance which contains N electrons per unit volume. Namely, if we multiply the expression (60) by $\frac{d\Omega}{h\nu'}$, we obtain the number of quanta scattered by an electron in a solid angle $d\Omega$ in the direction Θ. To each scattered light quantum belongs, however, an incoming light quantum of the magnitude $h\nu$, so that the energy loss of the incident radiation through scattering in this direction on an electron is equal to $\frac{\nu}{\nu'} d\Omega$ times the expression (60). By integrating over all directions and multiplying by the number of electrons per unit volume, we thus obtain for the scattering coefficient

$$S = \frac{2\pi N e^4}{m^2 c^4} \left\{ \frac{1 + \alpha}{\alpha^2} \left[\frac{2(1 + \alpha)}{1 + 2\alpha} - \frac{1}{\alpha} \log(1 + 2\alpha) \right] \right.$$
$$\left. + \frac{1}{2\alpha} \log(1 + 2\alpha) - \frac{1 + 3\alpha}{(1 + 2\alpha)^2} \right\}, \tag{61}$$

which again differs from the corresponding expression given by Dirac by quantities of order α^2.

Copenhagen, Universitetets Institut for Teoretisk Fysik, October 1928

References

1. P. A. M. Dirac, *Proc. R. Soc.* **A111**, 405 (1926); will be quoted as A in the following.
2. W. Gordon, *Z. Phys.* **40**, 117 (1927).
3. P. A. M. Dirac, *Proc. R. Soc.* **A117**, 610 (1928); will be quoted as B in the following.
4. Y. Nishina, *Z. Phys.* **52**, 869 (1929).
5. P. A. M. Dirac, *loc. cit.* B.
6. See: O. Klein, *Z. Phys.* **41**, 407, 1927.
7. See e.g.: O. Klein, therein, *loc. cit.*, p. 422.
8. W. Gordon, *loc. cit.*, p. 129.
9. W. Gordon, *loc. cit.*, p. 130; see also: I. Waller, *Phil. Mag.* **4**, 1228, 1927.
10. P. A.M. Dirac, *loc. cit.* B., p. 618.
11. W. Gordon, *loc. cit.*, p. 130.
12. Cf. the accompanying work by Y. Nishina, where the question of the polarization of the scattered radiation is discussed.
13. Cf. Y. Nishina, *loc. cit.*, where this point has been more closely investigated.
14. Cf. *Nature* **122**, 398 (1928); where the relationship between our results and experiment is briefly treated.

THE REFLECTION OF ELECTRONS AT A
POTENTIAL JUMP ACCORDING
TO DIRAC'S RELATIVISTIC DYNAMICS*

O. Klein

in Copenhagen
Received December 24, 1928

The reflection of electrons at a potential jump is investigated according to the new dynamical theory of Dirac. At very large values of the potential jump, electrons may according to the theory penetrate against the electric force acting on them through the surface of the jump and arrive at the other side with a negative kinetic energy. This could be regarded as a particularly sharp example of the difficulty emphasized by Dirac of the relativistic dynamics.

1. Introduction: As Dirac[1] has emphasized, a serious difficulty of the relativistic quantum theory consists in the circumstance that an electron in a field of force may, according to the theory, acquire negative values of the energy, which are in general connected with the physically meaningful positive energy values through the possibility of transitions. Also in his new, in other respects so successful treatment of relativistic quantum dynamics, he has not succeeded in overcoming this difficulty. In the following lines an elementary example will be pointed out where this difficulty appears in a particularly sharp way. This has to do with the reflection and refraction of electron waves at a boundary surface, where the electrostatic potential has a jump.

---

*Original in Z. Phys. **53**, 157 (1929); reproduced here with permission from Springer-Verlag. Translated from the German by Dr Lars Bergström.

2. Let E be the total energy of an electron moving in a force-free part of space, while p_1, p_2, p_3 may denote the components of its momentum with respect to the axes of an orthogonal coordinate system, where the electron has the coordinates x_1, x_2, x_3. We will assume that the electrostatic potential is different from zero in this part of space; in fact the electron should have the constant potential energy P. This specification of course only has a meaning if we compare this part of space with another part of space, where the potential has another value. According to the ordinary relativistic mechanics the following relationship holds between the energy $E - P$, which we will call the kinetic energy of the electron (although for an electron at rest it is not zero, but rather $m_0 c^2$), and the momentum

$$\left(\frac{E - P}{c}\right)^2 = p_1^2 + p_2^2 + p_3^2 + m_0^2 c^2, \tag{1}$$

where m_0 stands for the rest mass of the electron and c the velocity of light. The difficulty in question is connected to the fact that the kinetic energy can take both positive and negative values, as a result of which, besides the physically meaningful solutions, further solutions are present, which cannot be assigned a physical meaning. In the usual relativistic mechanics there is no difficulty with this as the square of the momentum can never become negative so that according to (1) the kinetic energy can never become zero; since in this theory only continuous transitions occur, this means that the negative values of energy can never be reached. In the quantum theory, however, the solutions in question cannot in general be separated, since on the one hand, discontinuous transitions are possible and, on the other hand, the electron waves here can penetrate through regions where the electron, classically speaking, has an imaginary momentum.

3. For an electron in an electrostatic force field, where the potential is V, we can according to Dirac reduce the problem to the following wave equation:

$$\left\{\frac{E + eV}{c} + \beta mc\right\} \psi - ih \sum_1^3 \alpha_k \frac{\partial \psi}{\partial x_k} = 0 \tag{2}$$

with the adjoint equation

$$\varphi \left\{\frac{E + eV}{c} + \beta mc\right\} + ih \sum_1^3 \frac{\partial \varphi}{\partial x_k} \alpha_k = 0, \tag{2a}$$

where again E means the total energy of the electron, which we view as being given, while $-e$ denotes its charge and h is the Planck constant divided by 2π. The quantities α_1, α_2, α_3 and β are matrices with four rows and columns, fulfilling the relations

$$\alpha_i \alpha_k + \alpha_k \alpha_i = 0, \quad i \neq k, \quad \alpha_i \beta + \beta \alpha_i = 0,$$

$$\alpha_1^2 = \alpha_2^2 = \alpha_3^2 = \beta^2 = 1. \tag{3}$$

Correspondingly, the functions φ and ψ consist of four components each, $\varphi_1, \varphi_2, \varphi_3, \varphi_4$ and $\psi_1, \psi_2, \psi_3, \psi_4$, respectively. With γ denoting a matrix with four rows and columns, $\gamma\psi$ shall here stand as an abbreviation for the four quantities $\sum_{k=1}^4 \gamma_{ik} \psi_k (i = 1, 2, 3, 4)$, where γ_{ik} denotes the matrix elements of γ. In the same way, $\varphi\gamma$ should be interpreted as $\sum_{k=1}^4 \varphi_k \gamma_{ki} (i = 1, 2, 3, 4)$. One sees that it is then allowed to multiply (2) from the left and (2a) from the right by any matrix without spoiling their validity. This just means a linear transformation of the system of equations.

We now want to assume that to the left of the plane $x_1 = 0$ the potential V equals zero, whereas right of this plane $eV = -P$ is valid, where P means a positive quantity. One would thus expect that the electrons lose a part P of their kinetic energy when passing this plane. To be able to investigate the reflection and refraction of electron waves at this surface of discontinuity, it is necessary to find the matching conditions of the Dirac wave equations at such surfaces. One can, as is customary,[2] derive this by regarding the surface of discontinuity as a limiting case of a region of finite thickness in which the discontinuous quantity, in this case the potential, changes rapidly. Since Eqs. (2) can be solved with respect to the derivatives of the components of ψ perpendicular to the surface of discontinuity, in this case $\frac{\partial \psi_1}{\partial x_1}, \frac{\partial \psi_2}{\partial x_1}, \frac{\partial \psi_3}{\partial x_1}, \frac{\partial \psi_4}{\partial x_1}$, it follows immediately, if only the potential remains finite in the transition region, that the four quantities $\psi_1, \psi_2, \psi_3, \psi_4$ and of course also $\varphi_1, \varphi_2, \varphi_3, \varphi_4$ remain continuous at the passage of the surface of discontinuity.[a]

---

[a]Through the solubility of Eq. (2) as regards the derivatives of the four components of ψ from an arbitrary direction in space, it also follows that for a continuous solution not all four components can equal zero at a surface without vanishing completely. The boundary condition $\psi = 0$ used for the Schrödinger equation at a wall is thus senseless in the theory of Dirac and must be replaced by conditions which are to be searched for in a more complete specification of the physical properties of the wall.

4. Without any important restriction we can now look at a purely harmonic incident wave which hits the plane $x_1 = 0$ perpendicularly. We thus set for the incident wave, where we simply write x for x_1,

$$\psi_e = v_e e^{\frac{i}{\hbar}(px - Et)}, \tag{4}$$

where t denotes the time and p the momentum of the electron. By inserting this expression in (2) we obtain a system of linear algebraic equations for the four components of the amplitude v_e, which can be written in the following way:

$$\{E/c + \alpha p + \beta m_0 c\} v_e = 0, \tag{5}$$

where we have put α for α_1. If v_e should not vanish identically, the condition below follows:

$$E^2/c^2 = p^2 + m_0^2 c^2, \tag{6}$$

which is a special case of (1). We will here choose for E the positive value. Now one can freely choose two of the components of ψ, which just corresponds to the two possibilities of orientation of the electron in a magnetic field.[3] From (6) it follows that the momentum of the reflected wave has to be $-p$, while for the refracted wave a value of momentum $\bar{p}$ follows, given by the relationship

$$\left(\frac{E - P}{c}\right)^2 = \bar{p}^2 + m_0^2 c^2. \tag{7}$$

We first want to assume P to be so small that a positive value of $\bar{p}^2$ follows from (7). We can then put

$$\psi_r = v_r e^{\frac{i}{\hbar}(-px - Et)}, \qquad \psi_g = v_g e^{\frac{i}{\hbar}(\bar{p}x - Et)}, \tag{8}$$

where ψ_r and ψ_g are associated with the reflected and refracted wave, respectively. From (2) follows

$$\left\{\frac{E}{c} - \alpha p + \beta m_0 c\right\} v_r = 0, \qquad \left\{\frac{E - P}{c} + \alpha \bar{p} + \beta m_0 c\right\} v_g = 0. \tag{9}$$

The boundary condition now simply reads

$$v_e + v_r = v_g. \qquad (10)$$

If we regard the four components of the incident wave as given, we have eight unknowns, namely the four components of v_r and the four components of v_g. However, because of (9) only four of these are independent so that (10) gives just the sufficient number of equations for their calculation. We can easily obtain the solution of the equations with respect to v_r in the following way. From (5) and the first equation of (9) follows

$$(E/c + \beta m_0 c)(v_e + v_r) = -\alpha p(v_e - v_r).$$

From (9) it follows, based on (10), that

$$(E/c + \beta m_0 c)(v_e + v_r) = (P/c - \alpha \bar{p})(v_e + v_r)$$

and thus

$$(P/c - \alpha \bar{p})(v_e + v_r) = -\alpha p(v_e - v_r)$$

or

$$\{P/c - \alpha(p + \bar{p})\}v_r = -\{P/c + \alpha(p - \bar{p})\}v_e.$$

Through multiplication of both sides of this relation by $\frac{p}{c} + \alpha(p + \bar{p})$, it follows that, by taking into account $\alpha^2 = 1$ with the help of (6) and (7),

$$v_r = \frac{2P/c(E/c + \alpha p)}{P^2/c^2 - (p + \bar{p})^2} v_e. \qquad (11)$$

To be able to utilize this result physically we must seek the corresponding solution to the adjoint wave equation (2a), since according to Dirac $\varphi \psi dv = \sum_1^4 \varphi_k \psi_k dv$ gives the probability that we find the electron in the volume element dv. From this follows[4] for the probability that the electron traverses a surface element df perpendicular to the x-axis in the time dt, $-c\varphi \alpha \psi df dt$, where $\varphi \alpha \psi$ is an abbreviation for $\sum_{i,k=1}^4 \varphi_i \alpha_{ik} \psi_k$. As Dirac has shown, it is now possible to choose Hermitian matrices for α and β. If ψ stands for a solution of (2), then $\varphi = $ complex conjugate of ψ will be a solution of (2a). As we choose with Hermitian matrices complex conjugate values for φ and ψ, we obviously get real expressions for $\varphi\psi$ and $\varphi\alpha\psi$. We put accordingly

$$\varphi_e = u_e e^{-\frac{i}{\hbar}(px-Et)}, \quad \varphi_r = u_r e^{-\frac{i}{\hbar}(-px-Et)},$$

$$\varphi_g = u_g e^{-\frac{i}{\hbar}(\bar{p}x-Et)}, \tag{12}$$

where, if α and β are Hermitian, u_e, u_r and u_g are the complex conjugates of v_e, v_r and v_g. From (2a) and (12) follows

$$u_e\{E/c + \beta m_0 c - \alpha p\} = 0, \quad u_r\{E/c + \beta m_0 c + \alpha p\} = 0,$$

$$u_g \left\{ \frac{E-P}{c} + \beta m_0 c + \alpha\bar{p} \right\} = 0. \tag{13}$$

We now derive from (5) and (13) a useful identity as we multiply (5) from the left by $u_e\alpha$ and the first equation (13) from the right by αv_e. Since α and β anticommute and $\alpha^2 = 1$, we get by addition

$$E/c u_e \alpha v_e + p u_e v_e = 0$$

or

$$-c u_e \alpha v_e = \frac{p c^2}{E} u_e v_e. \tag{14}$$

In the particle interpretation $\frac{pc^2}{E}$ means the velocity of the electron (group velocity), so that this equation gives a connection between current density and density in correspondence with the usual hydrodynamics. A similar result is of course obtained for the reflected and refracted waves (for the latter E is to be replaced by the kinetic energy $E - P$). To calculate the fraction of electrons that is reflected and refracted, respectively, it is thus enough to express the quantities $u_r v_r$ and $u_g v_g$ in terms of the components of the incoming wave.

Through a similar calculation to that which led to the expression (11) we now find

$$u_r = -u_e \frac{2P/c(E/c + \alpha p)}{P^2/c^2 - (p + \bar{p})^2}. \tag{15}$$

From (10) and (15) then follows

$$u_r v_r \left(\frac{2P/c}{P^2/c^2 - (p + \bar{p})^2}\right)^2 u_e (E/c + \alpha p)^2 v_e$$

$$= \left(\frac{2P/c}{P^2/c^2 - (p + \bar{p})^2}\right)^2 \left\{(E^2/c^2 + p^2)u_e v_e + \frac{2Ep}{c}u_e \alpha v_e\right\}$$

or according to (14) and (6)

$$u_r v_r = \left(\frac{2Pm_0}{P^2/c^2 - (p + \bar{p})^2}\right)^2 u_e v_e. \tag{16}$$

The quantity $(\frac{2Pm_0}{P^2/c^2 - (p+\bar{p})^2})^2$ thus gives the fraction of electrons that becomes reflected. As one can easily prove with the help of (6) and (7), the reflection coefficient grows with increasing P from zero for $P = 0$ and reaches the value of one for $P = E - m_0 c^2$. Just here $\bar{p}^2$ now equals zero, and by further increase of P we enter into the region of imaginary $\bar{p}$, which we shall now investigate.

In the classical theory, the fact that $\bar{p}$ becomes imaginary means that the electron penetrates so deep into the field that its velocity becomes zero and it is then thrown back. In the wave theory the wave functions will have finite values also to the right of the boundary surface; as we shall see, however, the conditions primarily correspond to total reflection in optics.

If $\bar{p}$ is imaginary, we can put

$$\psi_g = v_g e^{-\mu x - i\frac{E}{\hbar}t}, \quad \varphi_g = u_g e^{-\mu x + i\frac{E}{\hbar}t}, \tag{17}$$

where μ means a real quantity which must obviously be positive as otherwise the density to the right of the boundary surface would grow to infinity with x. Just because in this case μ is real, the exponent proportional to x in ψ_g and φ_g must according to the general prescriptions of the theory have the same sign. This means, however, that we put for ψ_g the quantity $\bar{p}$ equal to $ih\mu$, for φ_g on the other hand equal to $-ih\mu$. With this prescription we obtain from (11) and (15)

$$v_r = -\frac{2P/c(E/c + \alpha p)v_e}{P^2/c^2 - (p + ih\mu)^2}, \quad \mu_r = -\mu_e \frac{2P/c(E/c + \alpha p)}{P^2/c^2 - (p - ih\mu)^2} \tag{18}$$

and thus

$$u_r v_r = \frac{(2P/c)^2(E^2/c^2 - p^2)}{[(P/c + p)^2 + \mu^2 h^2][(P/c - p)^2 + \mu^2 h^2]} u_e v_e.$$

We can simplify this expression with the help of (6) and (7). First follows

$$\bar{p}^2 = p^2 - \frac{P(2E - P)}{c^2} \qquad (19)$$

and thus with $\bar{p}^2 = -\mu^2 h^2$

$$(P/c \pm p)^2 + \mu^2 h^2 = 2P/c(E/c \pm p).$$

Therefore it simply follows that

$$u_r v_r = u_e v_e. \qquad (20)$$

Thus the reflected current equals the incident current, while an exponentially decreasing wave solution exists behind the boundary surface. The condition for this case is according to (19) $p^2 < \frac{P(2E-P)}{c^2}$, a condition which with increasing P is first fulfilled when P exceeds the value $E - c\sqrt{E^2/c^2 - p^2} = E - m_0 c^2$. If P grows even further the quantity μ will first increase; due to the quadratic term in P in (19) it reaches however a maximum, which occurs for $P = E$. From there on μ becomes smaller and is again zero for $P = E + c\sqrt{E^2/c^2 - p^2} = E + m_0 c^2$. For even larger P, $\bar{p}$ again takes real values so that Eqs. (11), (15) and (16) again constitute the solution of the problem. In this region the kinetic energy $E - P$ is instead negative, so that we really have reached the mechanically forbidden region. This has the consequence that the group velocity, which is given by $\frac{c^2}{E-P}\bar{p}$, is directed opposite to the momentum and we must take[b] a negative value for $\bar{p}$. This one can easily see if one takes as initial state a wave packet which moves towards the boundary surface from the left.

We have thus arrived at the curious result that for values of P larger than $E + m_0 c^2$, a fraction of the electrons crosses the potential threshold, while their kinetic energy is transformed from the original positive to a negative

---

[b]I had originally not taken this circumstance into account; rather, it arose in a conversation with Mr W. Pauli, whom I here would like to cordially thank.

value. It is of interest to calculate the group velocity of these through-going electrons. For these it follows from (7) that

$$\frac{c^2}{P - E}|\bar{p}| = c\sqrt{1 - \left(\frac{m_0 c^2}{P - E}\right)^2}. \tag{21}$$

For $P = E + m_0 c^2$ this velocity is, as could be expected, just equal to zero. It then grows with growing P to reach the velocity of light for $P = \infty$.

As one can see from the expression (16), the coefficient of reflection, which for $P = E + m_0 c^2$ equals unity, decreases gradually for increasing P, to the value $\frac{E/c - p}{E/c + p}$ for $P = \infty$. The corresponding limiting value for the fraction of electrons that penetrates the boundary surface is thus $\frac{2p}{E/c + p}$, i.e. of the same order of magnitude as the ratio between the velocity of the incident electrons and the velocity of light, and can, for large values of p, attain considerable values. For $p = m_0 c$, corresponding to a velocity of the incident electrons of around 70% of the velocity of light, we get for instance the value $2(\sqrt{2} - 1)$, i.e. around 83%. It is of course not important that we have here assumed $P = \infty$; evidently one would get numbers of the same order of magnitude as soon as P is several times larger than the rest mass $m_0 c^2$ of the electrons. We will not here enter into the question whether it is possible to realize such potential jumps experimentally. It should only be emphasized that the difficulty in question is not connected to the assumption of a discontinuity, which was chosen only for mathematical simplicity. Also if the jump surface is replaced by a small region where the potential grows rapidly but continuously, electrons will, as follows from this whole procedure of calculation, penetrate into the forbidden region where they possess negative kinetic energy. This is closely connected to the fact that in the case discussed above the total reflection of the wave solution behind the boundary surface does not disappear even though the electron here would have an imaginary momentum according to classical mechanics. It should also be mentioned that following from the expression (16) the reflection coefficient of electrons that fall towards the boundary surface with momentum $\bar{p}$ from a part of space where the potential energy is P, has the same value of the reflection coefficient for the reversed process.

As a result of our investigations we can thus establish that the difficulty of the relativistic quantum mechanics emphasized by Dirac can appear already in purely mechanical problems where no radiation processes are involved.

At the end of this note I would like to cordially thank Professor Niels Bohr for many conversations, which have contributed significantly towards clarifying the considerations above.

Copenhagen, Universitetets Institut for Teoretisk Fysik, December 1928

References

1. P. A. M. Dirac, *Proc. R. Soc.* **117**, 612 (1928).
2. Cf. H. Faxén and J. Holtsmark, *Z. Phys.* **45**, 311 (1927), where a similar study was performed for the wave equation of Schrödinger.
3. Cf. C. G. Darwin, *Proc. R. Soc.* **118**, 654 (1928).
4. P. A. M. Dirac, *Proc. R. Soc.* **118**, 351 (1928).

EXCERPTS FROM SOME
AUTOBIOGRAPHICAL NOTES*

O. Klein

in Stockholm

Not until the summer of 1935 did I make new attempts concerning the five-dimensional theory, in connection with a conversation with Pauli, who had in the meantime himself published two articles containing a modified form thereof, which had been proposed by Veblen. However, it was not until 1937 that I once again took hold of it — this time in connection with Yukawa's idea concerning the interaction of the nucleons through an intermediate particle, which gradually appeared to be essentially correct and utterly fertile for all of nuclear physics.

As concerns this work, about which I gave a talk in Warsaw in the summer of 1938 at a meeting on the new theories of physics (published in *Institut International de Coopération Intellectuelle, Collection Scientifique*, 1939),[1] I have to say that this — although it pointed towards aspects that later, in another context, have become of importance — was suffering from the fact that I, like on some other occasions, tried to apply a mathematical formalism to the too incomplete knowledge one had at the time of what is now called elementary particle physics. However, for me it became of use as a background for much later work.

These later works, like the one mentioned, have as a starting point Dirac's electron theory in a general relativistic form and at the same time my old five-dimensional studies, which I therefore have expanded upon

---

*Original in *Svensk Naturvetenskop*, 1973, reproduced here with permission of the Swedish Science Research Council. Translated from the Swedish by Dr Lars Bergström.

very thoroughly in the preceding work.[2] In their present form we are first of all dealing with an extension of the principle of equivalence of Einstein, fundamental for gravitation. My further occupation with this principle was, however, connected to another area of problems, named relativistic cosmology.

During earlier years I had written a small book on Einstein's theory of relativity (*Natur och Kultur*, 1933), in which I — due to more general doubts — had excluded Einstein's cosmological attempts, which constituted the introduction to the more recent cosmology. On the other hand, I had thoroughly accounted for Mach's idea about the inertial forces as a kind of gravitational action from the assembly of distant celestial bodies in the Universe, which was the main background for this attempt.

That later I was to occupy myself in detail with these questions was due to the following: around the mid-1940's my collaborators (Göran Beskow and Lars Treffenberg) and I had further developed an attempt to explain the relative abundance of the elements as a result of an early thermal equilibrium at very high temperature in a kind of primordial stars. The result looked promising, but simultaneously difficulties appeared. At the same time Gamow and his collaborators Alpher and Herman had tried to explain the same thing in a completely different way: through a series of nuclear reactions during a very early epoch of the evolution of the Universe according to the expanding relativistic cosmology, usually called the Big Bang. After an encounter with them in America in the beginning of 1950, it was clear to me that their attempt was carried out very well, that the results were about as promising as ours, but that they too had encountered difficulties.

This equivalence, as it appeared to be for better or worse, between two so different attempts, made me start becoming more interested in the question of the state of the world at the time of formation of the elements. Thus, I had started to become interested in cosmology, but was still sceptical due to the great difference between it and ordinary physics ever since the days of Galileo — where the concern was about finding ever more general laws of nature and not the structure of the whole world. In other words, cosmology appeared unnatural to me.

Already somewhat earlier I had convinced myself that a system of stars, formed in a natural way through the contraction of a cloud of small density, cannot be compressed indefinitely. The average density for a given mass has a particular limit which is inversely proportional to the square of the mass. For a system of the size of the Milky Way the average density can thus not be greater than about one part in 100 000 of the density of water. Now it seems that the part of the Universe reached by telescopes contains around ten billion Milky Ways, which would mean that its average density before the start of the expansion can hardly have been greater than 10^{-25} g/cm^3, a remarkably good vacuum! The most important conclusions from the Big Bang cosmology, an eventual formation of helium and above all the so-called fireball radiation, which following a suggestion by Alpher and Herman should now have cooled to a temperature a few degrees above absolute zero — and which after the observation of such a radiation constitutes the main argument of cosmologists — is, according to the theory, connected to a state, the density of which is several orders of magnitude higher than that of water.

In the meantime I had, however, found still another reason to replace the infinite Universe of cosmology with a model for the expanding system of Milky Ways — a meta-galaxy — which certainly is very large yet finite, such that it like all other systems of stars has originated in a natural way. Probably there exist also other such Milky Way systems, although they have not been observed so far. My starting point for this consisted of some relations of orders of magnitude found by the great astronomer Eddington, according to which there would be a relation between the mass and dimensions of a closed Universe on the one hand, and similar atomic quantities on the other, that were regarded by him as a deep connection between micro-cosmos and macro-cosmos. In both these relations there appears a very large number, the ratio of the electrostatic and gravitational attraction between an electron and a proton. Later this discovery led Eddington to highly curious speculations — totally strange to real physics — concerning the exact value of this and other numbers.

I had long known about these relations, but my real interest in them only arose when I read Bondi's excellent book on cosmology, where he devotes a chapter to them. Since I was looking for a natural explanation for them, I came to think about something that Gamow had told me at our encounter in 1950, namely that the Milky Way, if it were dissolved into protons and electrons, would be at the limit of transparency for ordinary light. This made me investigate what dimensions and what mass a model for a large bounded system of this kind would have, if it simultaneously were at the limit of transparency and close to the above-mentioned density limit. It appeared that this just corresponded to the Eddington relations, but that this density was somewhat larger than the present estimates of the density of the Milky Way system. This result made me tentatively make the following hypothesis concerning the approximate evolution of the system: that it had started as a very thin and large cloud of hydrogen, that subsequently radiation and ionization appeared through atomic collisions, when the density increased, that the radiation by scattering on the electrons slowed down the expansion, which gradually proceeded to the present expansion, all this because the opacity at some stage of the process became large enough. Naturally, this is a most schematic account of the content of my first published, very cautious version; after a talk at an astrophysics conference in Liège in September 1953 (published 1954 in the conference proceedings).

Further investigation of this was slow, partly due to the difficulty of the problem, partly because I was simultaneously heavily occupied with other work. However, in the meantime I got some opportunities to present my views, among others in a talk at a Solvay conference in 1958, published that year.

Only gradually did I get enough confidence in these ideas to draw the in fact obvious conclusion that such a large system as the one in question had to have started out with practically as much anti-hydrogen as ordinary hydrogen — any difference to speak of would be utterly improbable. Thus, some natural process of separation must have taken place, as at least our solar system does not contain more than extremely small amounts of antimatter. The fact that we cannot see whether other stars consist of ordinary matter or antimatter was certainly already realized by many physicists shortly after the discovery of the anti-electron and Dirac's theory of electrons and anti-electrons, although it would take a long time before real antimatter was discovered. I had written about this myself in the book *Cause and Effect* (*Natur och Kultur*, 1935).

I now asked Alfvén — because of his great ingenuity and knowledge of electromagnetic processes in a plasma — whether he could imagine any such separation processes. This was in the spring of 1961, and already in the summer the same year, he could report some promising suggestions to me.

He was completely taken by the thought that the annihilation of matter and antimatter not only would be able to contribute to the understanding of the development of the meta-galaxy but also constitute a hitherto overlooked energy source for cosmic processes that were not understood. Through his enthusiasm he contributed considerably to the interest in these problems. A valuable result of that was that Bertel Laurent of Stockholm University, one of my best former students, came to devote himself with fervour to a more thorough study of the development of the meta-galaxy based on the views mentioned, in particular a more realistic treatment of the importance of the annihilation than that of Alfvén and myself, published in connection with his separation attempt (in *Arkiv för fysik*, 1962).

In connection with the problems of cosmology, I have first of all occupied myself with two things after the introductory works: one attempt to show that the so-called fireball radiation — the *pièce de résistance* of cosmologists — gets its natural explanation in the framework of the evolution of the metagalaxy. And further a careful analysis of Einstein's road to his general theory of relativity, especially his relationship to Mach's idea concerning the forces of inertia.

Einstein's earlier contributions to what he later called the general theory of relativity consisted mainly of some thought experiments, as interesting as they were simple, where he started from the long-known but isolated fact that all bodies get the same acceleration in a given gravitational field. In the most important of the experiments he imagined what would happen in a freely falling room, namely that all the usual effects of gravity disappear. Nowadays, this indeed belongs to the less pleasant experiences of our space travellers. However, already Newton, had he had any reason to reflect upon it, would immediately have known that it is like that. But the novelty was that Einstein traced therein a then unknown principle, which immediately led him to new predictions concerning the behavior of light in a gravitational field, which were demonstrated later. In addition, he realised, as was shown by another of his thought experiments, that the inertial force which appears if the room is instead accelerated in a region without gravitation, is not in any essential way different from what we otherwise call gravitation or force of gravity. Now it is easy for us to see how this consistently led to the entire foundation of the general theory of gravitation of Einstein.

But this is wisdom after the event. In reality this caused him several years of effort, and in the meantime the matter was complicated to him through musings about why nature seemed to prefer uniform motion, why relativity would not be valid for arbitrary motion. Of course it is, but the question is: with respect to what? To Mach, whom Einstein followed regarding this question, it was simple; to him vacuum was empty and the opposite of matter. Therefore he, unjustly, criticized the absolute space of Newton and had the opinion that the effects of accelerated motion — the forces of inertia — were a kind of gravitation unknown until then, whose source in this case consisted of the collection of celestial bodies in the universe, but which always arose when a body moved nonuniformly with respect to another body, although it takes a very large mass before it is noticeable.

It is understandable that Einstein caught this idea with fervour, which seemed to give him *the general relativity* and thus the main source of inspiration for his cosmology (which in turn inspired the later somewhat more realistic cosmology attempts). However, I cannot see otherwise than that we are here dealing with a common mistake — the power of word over thought — in this case the world vacuum = absolute empty space. In contrast to this the starting point of the principle of equivalence, as we have seen, is a very general *experience* — that all bodies obtain the same acceleration in a gravitational field. But experiences are always limited. Therefore it takes an unusual sense of tact — and in addition certainly luck — to build on such a foundation a new theory, correct in all essentiality. And here the need for explanation is likely to cause trouble, as the explanation, when a really new principle is concerned, has to work with perceptions that lack the generality one had the reason to think. This was true for Newton's theory of gravitation as well as Einstein's theory of relativity, where he himself had been forced to refrain from the kind of explanations that he now demanded concerning his own principle of equivalence.

Now a few words about this. There the concept of inertial frame plays a fundamental role. It is a frame of reference, where gravity is eliminated in the same way as in Einstein's experiment with the freely falling room. But to achieve such weightlessness during a finite time obviously a constant field of gravitation is needed. In a gravitational field of a general kind, i.e.

one where the strength and direction of gravitation is changing from point to point in both space and time, one defines a local inertial frame at each point and instant by a transition to the infinitely small, i.e. to let the region surrounding the point and the duration of the observation become smaller and smaller, thus a similar procedure to that by which a tangent surface is defined at every point on a curved surface. In the same way as the geometry of the curved surface is defined by the plane geometry, namely by lengths and directions within an infinitely small region, the principle of equivalence implies that the known physical laws, valid in an inertial frame, first of all the theory of relativity, can be translated to the corresponding laws in a given gravitational field. Thus, for example, the equations of motion of a particle in an arbitrary gravitational field follow from the law of inertia, which is indeed valid in an inertial frame.

An important consequence of this is that space and time in the gravitational field in principle are based on measurements in the local inertial frames, something that Einstein himself has emphasized, in that he at the same time pointed out the analogy with Gauss's theory for the curved surfaces. From this it now also follows that this is valid for the properties of all kinds of isolated systems, be it ordinary bodies or atoms or systems of stars, and in particular for the mass of a given body. It was therefore curious that Einstein overlooked this, in that he, in connection with Mach's idea, thought himself to be able to prove that the mass of a given body increases, when other bodies approach it.

However, there exists something that probably made him doubt somewhat the possibility of unifying the general relativity with the principle of equivalence, namely that his field equations for gravitation, i.e. the analog of Newton's law of gravity, cannot be derived from this principle although the concepts contained in the equations are defined through it. And these equations are independent of coordinates in which they are expressed, which could appear as a verification of the general relativity. However, that this is not the case is clear from the fact that their physical content comes from the inertial frames. That Einstein, with time, got some doubts about the principle of equivalence, but not the general theory of relativity, is indicated by the considerable effort he made to derive the equations of particle motion from the field equations mentioned, while these earlier, also for him, were an obvious and particularly simple result of the equivalence principle.

After this somewhat sketchy account of the various points of view of Einstein towards the theory that he himself had created, certainly something of the most important in the history of physics, I will say something in brief about my attempts to enlarge the principle of equivalence with the help of quantum field theory.

This work, which is far from finished, is based on my ever stronger conviction that gravity is the simplest, the most universal and at the same time the most understood of all forces of nature — a conviction that at present is hardly shared by many physicists. A main reason for this is just the principle of equivalence. But already earlier I found another reason for this, namely that the difficulties of divergencies of quantum electrodynamics probably would disappear if one took gravity into account. That it was possible despite the infinities to enlarge the region of applicability of this theory through the procedure of renormalization — a rational subtraction of the infinite terms — naturally impressed me to a high degree. But I never thought that this was a real solution to the problem.

Before,[2] I mentioned the extremely small length that seemed to play a key role in the five-dimensional theory, which besides the gravitational constant and the velocity of light is formed by the smallest electric charge. But when I got acquainted with the procedure of renormalization, which also affects the latter, I had the opinion that it should be replaced by Planck's quantum of action. By introducing this even somewhat smaller length as one of the three units in the equations of quantum electrodynamics with gravity these become particularly simple. Some time later I found the following expression for the importance of this length for the problem of divergencies. If, namely, any particle whatever could be confined to a sphere of the same order of magnitude as this length, its own gravitation would reach the aforementioned limit for density relative to mass. From this I drew the conclusion that two particles would never be able to come closer together than that, by which the infinities would cease to exist. A remark by Landau in the *Festschrift* on the occasion of Bohr's 70th birthday indicates that he had similar thoughts.

The starting point for the following works, as for the work of 1938 mentioned, is the Dirac equation with gravitation, which had already been developed by several researchers in the beginning of the 1930's — during the period when I had ceased to believe in the importance for quantum field theory of gravitation and its five-dimensional generalization.

Something relatively easy was finding a transformation that disposes of the gravitational field in accordance with the equivalence principle, i.e. at one point with respect to space and time. A considerably more difficult problem, to which I so far have been unable to find a solution, although I think that one should exist, is to derive the field equations for the gravitational field without any additional term for it, by the same method by which the Dirac equation is derived, i.e. through one and the same variational principle. One reason that this should be successful is that the concepts of vacuum and matter have come in a new situation through quantum theory. Instead of being pure opposites they here relate to one another as the normal state of an atom to its excited states. While the vacuum expectation value of a field of force thus vanishes according to quantum theory, this is not valid for the average of the squares, which gives rise to the so-called vacuum fluctuations. Another reason is that here — contrary to what is the case for the electromagnetic field in the usual treatment — one has both the gravitational potentials and their derivatives in the Dirac equation and therefore also in the expression for the variational principle. Thereby one obtains, so to speak, normal field equations also for gravitation. And just via the vacuum fluctuations of the Dirac field these obtain a form which in any case resembles the field equations of Einstein.

Before I end this brief account I will mention one thing which has reinforced my belief that a rational enlargement of the equivalence principle, in connection with the five-dimensional formulation of gravitation and electromagnetism, constitutes a hopeful beginning of a more extensive quantum field theory. It has to do with the unexpected lack of mirror symmetry discovered after a thorough analysis by Lee and Yang, for which they were awarded the Nobel Prize of 1957. The continued investigations led to a new law of nature, which, as Landau showed shortly after the discovery, can be expressed in such a way that the antiparticles are the mirror images of particles. I was namely able to show that this gets a natural interpretation in the framework of the five-dimensional theory (*Nuclear Physics*, 1957).

Interested readers are referred to a review article on this programme (*Nuclear Physics* **B21** (1970)), and a simple derivation of the equivalence principle for the Dirac equation, dedicated to Harald Wergeland on his 60th birthday (*Physica Norvegica* 5, 1971).

References

1. This article is contained in *The Oskar Klein Memorial Lectures*, Vol. 1, ed. G. Ekspong (World Scientific, 1991), p. 85.
2. The first part of these autobiographical notes can be found in *The Oskar Klein Memorial Lectures*, Vol. 1, ed. G. Ekspong (World Scentific, 1991), p. 103.

PART III

PREFACE

The Oskar Klein Memorial Lecture series has been running for thirteen consecutive years since its start in 1988. In 1994, to commemorate the 100th anniversary of Klein's birth, the lectures were extended to a symposium.

The first four lectures, delivered by C.N. Yang, S. Weinberg, A. Bethe and A. Guth have been published in Volumes I and II of "The Oskar Klein Memorial Lectures" (ed. G. Ekspong) and the symposium is published as "The Oskar Klein Centenary" (ed. U. Lindström).

The present volume is thus the fourth in a series and has some expectations to live up to. Hopefully they are well met by the contributions by T.D. Lee, N. Seiberg, A. Polyakov, P.J.E. Peebles, E. Witten and G. 't Hooft.

The typical format of the Klein Lectures is that the speaker gives one general colloquium type lecture and one specialized seminar. We have included most of the colloquia and also some of the seminars. In the first two volumes, it was still possible to also include some material from Klein's own research that had not been generally available. Here, we have instead included more lectures. For this reason we have had to wait a while until there was enough material to fill a volume. However, the character of the lectures is such that they remain topical, and the outstanding quality of the contributors help to make these written versions timeless. In fact, one of the instructions we usually give to the speakers is to cover the general aspects of their fields as well as the latest results. An example of this more timeless aspect is provided by the contribution by A. Polyakov where he has collected ideas on the parallels between different subjects that he has clearly pondered over some years.

The lecture series has contained some very exciting moments during the time covered in the present volume. We recall, e.g., the interesting discussion of symmetry by T.D. Lee, the presentation by N. Seiberg of wonderful new types of duality in supersymmetric theories, and by E. Witten of the then new ADS/CFT correspondence and its application to the long-standing problem of quark confinement, or the lecture in 1999 by G. 't Hooft, who went on to receive the Nobel Prize later on the same year.

Theoretical high-energy physics dominates the subjects of the lectures, mirroring one of Klein's own main interests. He had other interests as well, cosmology being one of them (particularly in his later work). It is therefore very befitting that the contibution by Peebles in a nice way summarizes the phenomenal improvement in the understanding of the Universe which has taken place since Klein's days. However, we also get sobering reminders that much still remains to be understood, and that it may be dangerous to overinterpret current results — a well-placed remark by one of the founders of physical cosmology.

All in all the yearly Klein Lecture has become a very sucessful tradition in Swedish physics and it has inspired several similar series. We are very grateful to the Swedish Royal Academy of Sciences whose support through its Nobel Institute for Physics has made these lectures possible over the years. We also take this opportunity to express our gratitude to FYSIKUM, Stockholm University, where the lectures and the traditional dinner are held.

Stockholm, April 2001
Lars Bergström and Ulf Lindström, *editors*.

THE WEAK INTERACTION: ITS HISTORY AND IMPACT ON PHYSICS

T. D. Lee

Columbia University, New York, N.Y. 10027, USA

It is a pleasure and an honor for me to give this lecture in honor of Oscar Klein who made major contributions to field theory, quantum electrodynamics and particle physics, including weak interactions. He was the first one to observe that the μ decay and the β decay could be described by the same interaction with the same coupling constant; this led to the discovery of the Universal Fermi Interaction.

Perhaps I should begin my discussion of the history of weak interactions by separating it into three periods:

1. Classical Period, 1898–1949
2. Transition Period, 1949–1956
3. Modern Period, 1956–

1. Classical Period (β Decay)

In 1898 Lord Rutherford[1] discovered that the so-called Becquerel ray actually consisted of two distinct components: one that is readily absorbed, which he called alpha radiation, and another of a more penetrating character, which he called beta radiation. With that began the history of the weak interaction. Then, in 1900,[2] the Curies measured the electric charge of the β particle and found it to be negative.

Sometimes when we think of physics in those old days, we have the impression that life was more leisurely and physicists worked under less pressure. Actually, from the very start the road of discovery was tortuous and the competition intense. A letter written in 1902 by Rutherford (then 32) to his mother expressed the spirit of research at that time[3,4]:

> "I have to keep going, as there are always people on my track. I have to publish my present work as rapidly as possible in order to keep in the race. The best sprinters in this road of investigation are Becquerel and the Curies. . . ."

Most of the people in this room can appreciate these words. Rutherford's predicament is still very much shared by us to this day. Soon many fast runners came: Hahn, Meitner, Wilson, Von Baeyer, Chadwick, Ellis, Bohr, Pauli, Fermi and many others.

In preparing this lecture, I was reminded once more of how relatively recent these early developments are. We know that to reach where we are today took more than 90 years and a large cast of illustrious physicists. I recall that when Lise Meitner came to New York in the mid '60s, I had lunch with her at a restaurant near Columbia. Later K.K. Darrow joined us. Meitner said, "It's wonderful to see young people." To appreciate this comment, you must realize that Darrow was one of the earliest members of the American Physical Society and at that lunch he was over 70. But Lise Meitner was near 90.

I was quite surprised when Meitner told me that she started her first postdoc job in theory with Ludwig Boltzmann. Now, Boltzmann was a contemporary of Maxwell. That shows us how recent even the "ancient" period of our profession is.

After Boltzmann's unfortunate death in 1906, Meitner had to find another job. She said she was grateful that Planck invited her to Berlin. However, upon arrival she found that because she was a woman she could only work at Planck's institute in the basement, and only through the servant's entrance. At that time, Otto Hahn had just set up his laboratory in an old carpenter shop nearby. Lise Meitner decided to join him and to become an experimentalist. For the next thirty years, their joint work shaped the course of modern physics.

In 1906, Hahn and Meitner published a paper[5] stating that the β ray carries a unique energy. Their evidence was that the absorption curve of a β ray shows an exponential decrease along its path when passing through matter, like the α ray.

Then W. Wilson,[6] in 1909, said "no", the β ray does not have a unique energy. By observing the absorption curve through matter of an electron of unique energy, Wilson found electrons to exhibit totally different behavior from the α particle; the absorption curve of a unique energy electron is not exponential. Consequently, Wilson deduced that the apparent exponential behavior of the absorption curve of β decay implies that the β does not have a unique energy, the same experimental observation on β but with a totally opposite conclusion.

In 1910, Von Baeyer and Hahn[7] applied a magnetic field to the β ray; they found the β to have several discrete energies. In this way, they also reconciled the conclusion of W. Wilson. Then, in 1914, Chadwick[8] said "no". The β energy spans a continuous spectrum, instead of discrete values. The discrete energy observed by Von Baeyer and Hahn was due to the secondary electron from a nuclear γ transition, with the γ energy absorbed by the atomic electron. In this process, the discrete energy refers to the nuclear γ emission.

Then came World War I and scientific progress was arrested. In 1922, Lise Meitner[9] again argued that the β energy should be discrete, like α and γ. The apparent continuum manifestation is due to the subsequent electrostatic interaction between β and the nucleus. From 1922 to 1927, through a series of careful measurements, Ellis[10] again said "no" to Meitner's hypothesis. The β energy is indeed continuous. Furthermore, Ellis proved that the maximum β energy equals the difference of the initial and final nuclear energy.

There would then appear a missing energy. This was incorporated by Niels Bohr,[11] who proposed the hypothesis of non-conservation of energy.

Very soon, Pauli said "no" to Bohr's proposition. Pauli[12] suggested that in the β decay energy is conserved, but accompanying the β particle there is always emission of a neutral particle of extremely small mass and with almost no interaction with matter. Since such a weakly interacting neutral particle is not detected, there appears to be an apparent nonconservation of energy.

Fermi[13] then followed with his celebrated theory of β decay. This in turn stimulated further investigation of the spectrum shape of the β decay, which did not agree with Fermi's theoretical prediction. This led Konopinski and Uhlenbeck[14] to introduce the derivative coupling. The confusion was only cleared up completely after World War II, in 1949, by Wu and Albert,[15] signalling the end of one era and the beginning of a new one.

2. Classical Period (Other Weak Interactions)

When I began my graduate study of physics at the University of Chicago, in 1946, the pion was not known. Fermi and Teller[16] had just completed their theoretical analysis of the important experiment of Conversi, Pancini and Piccioni.[17] I attended a seminar by Fermi on this work. He cut right through the complex slowing-down process of the mesotron, the capture rate versus the decay rate, and arrived at the conclusion that the mesotron could not possibly be the carrier of strong forces hypothesized by Yukawa. Fermi's lectures were always superb, but that one to me, a young man not yet twenty and fresh from China, was absolutely electrifying. I left the lecture with the impression that, instead of Yukawa's idea, perhaps one should accept Heisenberg's suggestion[18] that the origin of strong forces could be due to higher-order processes of β interaction. As was known, these were highly singular.

At that time, the β interaction was thought to be reasonably well understood. Fermi's original vector-coupling form,

$$G\left(\psi_n^\dagger \gamma_4 \gamma_\lambda \psi_p\right)\left(\psi_4^\dagger \gamma_4 \gamma_\lambda \gamma_5 \psi_\nu\right)$$

was, after all, too simple; to conform to reality, it should be extended to include a Gamow-Teller term. Fermi told me that his interaction was modelled after the electromagnetic forces between charged particles, and his coupling G was inspired by Newton's constant. His paper was, however, rejected by *Nature* for being unrealistic. It was published later in Italy, and then in *Zeitschrift für Physik*.[13] Fermi wrote his γ matrices explicitly in terms of their matrix elements. His lepton current differs from his hadron current by a γ_5 factor; of course the presence of this γ_5 factor has no physical significance. Nevertheless, it is curious why Fermi should choose this particular expression, which resembles the V-A interaction, but with parity conservation. Unfortunately, by 1956, when I noticed this, it was too late to ask Fermi.

A year later, the discovery of the pion through its decay sequence $\pi \to \mu \to e$ by Lattes, Muirhead, Occhialini and Powell[19] dramatically confirmed the original idea of Yukawa. The fact that the higher-order β interaction is singular is not a good argument that it should simply become the strong force.

This was then followed by Professor Klein's important discovery that μ decay and β decay can be described by the same four-fermion interaction. An excerpt from his writing is reproduced below.

No. 4101 June 5. 1948   N A T U R E   897

MESONS AND NUCLEONS

By Prof. O. KLEIN

Institut för Mekanik och Matematisk Fysik,
Stockholms Högskolas

Since, according to the above assumptions, the decay of the ordinary meson is, so to speak, the prototype of all β-processes, it is important that the value of the life-time, $\tau = 2 \times 10^{-6}$ sec., and the energy available in the process $\sim 100\ m_e c^2$, fit in very well with the value to be expected from our knowledge of the β-decay. Let ε be the energy available divided by $m_e c^2$, and $F(\varepsilon)$ the well-known Fermi function, which for large ε-values may be taken as $\varepsilon^5/50$. We then obtain for the product $\tau.F(\varepsilon)$— putting $\tau = 2 \times 10^{-6}$ sec. and $\varepsilon = 100$—the value 667, which is of about the expected magnitude.

· · · · ·   It was then found that the interaction constant corresponding to the meson life-time is of the same order of magnitude as that to be expected from Fermi's original theory of β-disintegration, a result which is, of course, identical with that of the above calculation

At Chicago we were not aware of Professor Klein's work. In January 1949 my fellow student, Jack Steinberger, submitted a paper[21] to *The Physical Review* in which he established that the μ meson disintegrates into three light particles, one electron and two neutrinos. This made it look very much like any other β decay, and stimulated Rosenbluth, Yang and myself to launch a systematic investigation. Are there other interactions, besides β decay, that could be described by Fermi's theory?

We found that if μ decay and μ capture were described by a four-fermion interaction similar to β decay, all their coupling constants appeared to be of the same magnitude. This was the beginning of the *Universal Fermi Interaction*. We then went on to speculate that, in analogy with electromagnetic forces, the basic weak interaction could be carried by a universal coupling through an *intermediate heavy boson*,[22] which I later called $W^{\pm}$ for weak. Naturally I went to my thesis adviser, Enrico Fermi, and told him of our discoveries. Fermi was extremely encouraging. With his usual deep insight, he immediately recognized the further implications beyond our results. He put forward the problem that if this is to be the universal interaction, then there must be reasons why some pairs of fermions should have such interactions, and some pairs should not. For example, why does

$$p \not\rightarrow e^+ + \gamma$$

and

$$p \not\rightarrow e^+ + 2\nu?$$

A few days later, he told us that he had found the answer; he then proceeded to assign various sets of numbers, $+1$, -1 and 0, to each of these particles. This was the first time to my knowledge that both the laws of baryon-number conservation and of lepton-number conservation were formulated together to give selection rules. However, at that time (1948), my own reaction to such a scheme was to be quite unimpressed: surely, I thought, it is not necessary to explain why $p \not\rightarrow e + \gamma$, since everyone knows that the identity of a particle is never changed through the emission and absorption of a photon; as for the weak interaction, why should one bother to introduce a long list of mysterious numbers, when all one needs is to say that only three combinations $(\bar{\eta}p)$, $(\bar{e}\nu)$ and $(\bar{\mu}\nu)$ can have interactions with the intermediate boson. (Little did I expect that soon there would be many other pairs joining these three.)

Most discoveries in physics are made because the time is ripe. If one person does not make it, then almost inevitably another person will do it at about the same time. In looking back, what we did in establishing the Universal Fermi Interaction was a discovery of exactly this nature. This is clear, since the same universal Fermi coupling observations were made independently by at least three other groups, Klein,[20] Puppi,[23] and Tiomno and Wheeler,[24] all at about the same time. Yet Fermi's thinking was of a more profound nature. Unfortunately for physics, his proposal was never published. The full significance of these conservation laws was not realized until years later. While this might be the first time that I failed to recognize a great idea in physics when it was presented to me, unfortunately it did not turn out to be the last.

Thus, in 1949, there existed a simple theoretical frame based on the Fermi theory, describing the three weak interaction processes:

$$n \rightarrow p + e + \nu,$$

$$\mu + p \rightarrow n + \nu$$

and

$$\mu \rightarrow e + 2\nu.$$

So, at the end of the classical period, we moved from the observation of β decay to the discovery of the Universal Fermi Interaction.

3. Transition Period (1949–1956)

Beginning in 1949, extensive work was done on the shape of the electron spectrum from μ decay. From the analysis of L. Michel,[25] it was found that this distribution is given by

$$N(x) = x^2 \left\{ \left(2 - \frac{4}{3}\rho\right) - \left(2 - \frac{16}{9}\rho\right) x \right\}$$

where

$$x = (\text{momentum of } e)/(\text{maximum } e - \text{momentum}),$$

and ρ is the well-known Michel parameter, which can be any real number between 0 and 1, and measures the height of the end point at $x = 1$, as shown in Fig. 1.

It is instructive to plot the experimental value of ρ against the year when the measurement was made. As shown in Fig. 2, historically it began

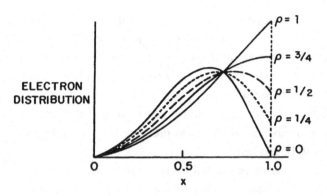

Fig. 1.   The ρ parameter in μ decay.

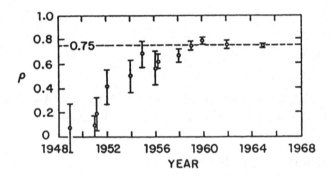

Fig. 2.   Variation of the ρ parameter over time.

with $\rho = 0$ in 1949, at the beginning of the transition period. Then it slowly drifted upwards; only after the end of the transition period with the theoretical prediction in 1957 did it gradually become $\rho = \frac{3}{4}$. Yet, it is remarkable that at no time did the 'new' experimental value lie outside the error bars of the preceding one.

In the same period (1949–56), a large amount of effort was also made on β decay experiments. By then, the Konopinski-Uhlenbeck interaction was definitely ruled out. The absence of the Fierz interference term[26] in the spectrum shows that the β interaction must be either V, A or S, T. These two possibilities were further resolved by a series of $\beta - \nu$ angular correlation experiments. In an allowed transition, the distribution for the angle θ between β and ν is given by (neglecting the Fierz term)

$$[1 + \lambda(P/E)_e \cos \theta] d \cos \theta,$$

where the subscript e refers to the momentum P and energy E of the electron. For a $\Delta J = 1$ transition,

$$\lambda = \begin{cases} +\dfrac{1}{3}, & \text{for } T \\ -\dfrac{1}{3}, & \text{for } A. \end{cases}$$

The experiment on ^{6}He decay by Rustad and Ruby gave[27]

$$\lambda = +0.34 \pm 0.09,$$

which seemed to establish unquestionably that the β decay interaction should be S, T with perhaps some unknown admixture of an additional pseudoscalar interaction.

I was quite depressed at that time because, with this new result, the theoretical idea of the intermediate boson seemed to be definitely ruled out. It is bad enough to assume the possiblity of two kinds of intermediate bosons of different spin-parity, one for the Fermi coupling and the other for the Gamow-Teller coupling. However, a tensor interaction with no derivative coupling simply cannot be transmitted by a spin-2 boson, since the former is described by an antisymmetric tensor and the latter by a symmetric one.

4. New Horizon in the Transition Period

We now come to the $\theta - \tau$ puzzle.

During a recent physics graduate qualifying examination in a well-known American university, one of the questions was on the $\theta - \tau$ problem. Most of the students were puzzled over what θ was; of course they all knew that τ is the heavy lepton, the charged member of the third generation. So much for the history of physics.

In the early 1950s, θ referred to the meson which decays into 2π, whereas τ referred to the one decaying into 3π:

$$\theta \to 2\pi$$

and

$$\tau \to 3\pi.$$

The spin-parity of θ is clearly 0^+, 1^-, 2^+, etc. As early as 1953, Dalitz[28] had already pointed out that the spin-parity of τ can be analyzed through his Dalitz plot and, by 1954, the then-existing data were more consistent with the assignment 0^- than 1^-. Although both mesons were known to have comparable masses (within $\sim 20\,\mathrm{MeV}$), there was, at that time, nothing too extraordinary about this situation. The masses of θ and τ are very near three times the pion mass, the phase space available for the θ decay is much bigger than that for τ decay; therefore one expects the θ decay rate to be much faster. However, when accurate lifetime measurements were made in 1955, it turned out that θ and τ have the same lifetime (within a few percent, which was the experimental accuracy). This, together with a statistically much more significant Dalitz plot of τ decay, presented a very puzzling picture indeed. The spin-parity of τ was determined to be 0^-; therefore it appeared to be definitely a different particle from θ. Yet, these two particles seemed to have the same lifetime, and also the same mass. This was the $\theta - \tau$ puzzle.

My first efforts were all on the wrong track. In the summer of 1955, Jay Orear and I proposed[29] a scheme to explain the $\theta - \tau$ puzzle within the bounds of conventional theory. We suggested a cascade mechanism, which turned out to be incorrect.

The idea that parity is perhaps not conserved in the decay of $\theta - \tau$ flickered through my mind. After all, strange particles are by definition strange, so why should they respect parity? The problem was that, after you say parity is not conserved in $\theta - \tau$ decay, then what do you do? Because if parity nonconservation exists only in $\theta - \tau$, *then we already have all the observable facts*, namely the same particle can decay into either 2π or 3π with different parity. I discussed this possibility with Yang, but we were not able to make any progress.[30] So we instead wrote papers on parity doublets, which was another wrong try.[31]

5. The Breakthrough (1956)

The Rochester meeting on high energy physics was held from April 3 to 7, 1956.

At that time, Steinberger and others were conducting extensive experiments on the production and decay of the hyperons Λ^0 and Σ^-:

$$\pi^- + p \rightarrow \begin{cases} \Lambda^0 + K^0 \\ \Sigma^- + K^+ \end{cases} \tag{1}$$

and

$$\begin{cases} \Lambda^0 \\ \Sigma^- \end{cases} \rightarrow \pi + N. \tag{2}$$

The dihedral angle ϕ between the production plane and the decay plane (which will be defined below) is of importance for the determination of the hyperon spin.

Let $\vec{\pi}$, $\vec{\Lambda}$ and $\vec{N}$ be the momenta of π, Λ in process (1) and N in (2), all, say, in the respective center-of-mass systems of the reactions. The normal to the production plane is parallel to $\vec{\pi} \times \vec{\Lambda}$, and that to the decay plane to $\vec{\Lambda} \times \vec{N}$. Hence the dihedral angle ϕ is defined through its cosine:

$$\cos \phi \propto (\vec{\pi} \times \vec{\Lambda}) \cdot (\vec{\Lambda} \times \vec{N}). \tag{3}$$

Its distribution is

$$D(\phi) = \begin{cases} 1 & \text{if the hyperon-spin is } \frac{1}{2}, \\ 1 + \alpha \cos^2 \phi & \text{if the hyperon-spin is } \frac{3}{2}, \end{cases} \tag{4}$$

etc. By this definition, ϕ varies from 0 to π. Furthermore, $D(\phi)$ is identical to $D(\pi - \phi)$. At the Rochester Conference, Jack Steinberger gave a talk and plotted his data on $D(\phi)$ with ϕ varying from 0 to π. However other physicists, W.D. Walker and R.P. Shutt, plotted $D(\phi) + D(\pi - \phi)$: in this way ϕ can only vary from 0 to $\frac{\pi}{2}$. After the conference, Jack came to my office to discuss a letter which he had just received from R. Karplus. In this letter Karplus questioned why Jack did not join the others, since the total number of events was (at that time) quite limited, and a folding of $D(\pi - \phi)$ onto $D(\phi)$ would increase the experimental sensitivity of the

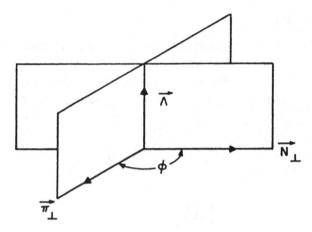

Fig. 3.   The dihedral angle ϕ between the production plane and the decay plane.

spin determination. Jack wanted to know how certain was the relationship that $D(\phi)$ is an even function of $\cos \phi$.

The dihedral angle, as defined by expression (3), has nothing to do with parity, since it is a scalar. In the course of explaining to Jack the $\cos^2 \phi$ dependence of $D(\phi)$, I suddenly realized that if one changes the definition of ϕ to be the angle of rotation around the Λ momentum-vector, which is the intersection of these two planes, then the range of ϕ can be extended from 0 to 2π; that is, in place of (3), one defines ϕ through the pseudoscalar

$$\sin \phi \propto (\vec{\pi}_\perp \times \vec{N}_\perp) \cdot \vec{\Lambda}, \tag{5}$$

where $\vec{\pi}_\perp$ and $\vec{N}_\perp$ refer to the components of $\vec{\pi}$ and $\vec{N}$ perpendicular to $\vec{\Lambda}$, as shown in Fig. 3. In this case, ϕ can vary from 0 to 2π.

If parity is not conserved in strange particle decays, there could be an asymmetry between events with ϕ from 0 to π and those with ϕ from π to 2π. *This is the missing key!* I was quite excited, and urged Jack to re-analyse his data immediately and test the idea experimentally. This led to the very first experiment on parity nonconservation. Very soon, within a week, Jack and his collaborators (Budde, Chrétien, Leitner, Samios and Schwartz) had their results, and the data were published[32] even before the theoretical paper[33] on parity nonconservation. The odds turned out to be 13 to 3 in Σ^- decay and 7 to 15 in Λ^0 decay (see Fig. 4).

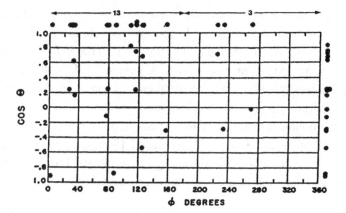

Fig. 4a.   Angular correlation plot of the cosine of the polar angle against the azimuthal angle ϕ for the reaction $\pi^- + p \rightarrow \Sigma^- + K^+$, $\Sigma^- \rightarrow \pi^- + n$. The point histograms on the edges of the figure represent the data after integration over the other coordinate. (Taken from Ref. 32)

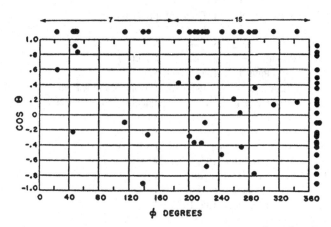

Fig. 4b.   Angular correlation plot for the reaction $\pi^- + p \rightarrow \Lambda^0 + \theta^0$, $\Lambda^0 \rightarrow \pi^- + p$. (Taken from Ref. 32)

Of course, because of the limited statistics, no definitive conclusion on parity violation could be drawn. Nevertheless, had the statistics been ten times more, then with the same kind of ratio one could have made a decisive statement on parity conservation. This showed clearly that parity violation could be tested experimentally provided one measured a pseudoscalar, such as (5).

However, on the theoretical side there was still the question of parity conservation in ordinary β decay. In this connection, at the beginning of May, C.N. Yang came to see me and wished to join me in the examination of β decay. This led to our discovery that, in spite of the extensive use of parity in nuclear physics and β decay, there existed no evidence at all of parity conservation in any weak interaction.

Several months later followed the decisive experiments by Wu, Ambler, Hayward, Hoppes and Hudson, at the end of 1956, on β decay,[34] and by Garwin, Lederman and Weinrich[35] and by Friedman and Telegdi[36] on $\pi - \mu$ decay. Table 1 lists the important papers on symmetry violation in weak interactions.

Table 1.

T.D.L. and C.N. Yang,
"Question of Parity Nonconservation in Weak Interactions,"
Phys. Rev. **104**, 254 (1956).
"A Two-component Theory of the Neutrino,"
Phys. Rev. **105**, 1671 (1957).

T.D.L., R. Oehme and C.N. Yang,
"Possible Noninvariance of T, C and CP,"
Phys. Rev. **106**, 340 (1957).

C.S. Wu, E. Ambler, R.W. Hayward, D. Hoppes and R.P. Hudson,
"Experimental Test of Parity Nonconservation in Beta Decay,"
Phys. Rev. **105**, 1413 (1957).

R.L. Garwin, L.M. Lederman and M. Weinrich,
"Nonconservation of P and C in Meson Decays,"
Phys. Rev. **105**, 1415 (1957).

J.I. Friedman and V.L. Telegdi,
"Parity Nonconservation in $\pi^+ - \mu^+ - e^+$,"
Phys. Rev. **105**, 1681 (1957).

M. Goldhaber, L. Grodzins and A.W. Sunyar,
"Helicity of Neutrinos,"
Phys. Rev. **109**, 1015 (1958).

R.E. Marshak and E.C.G. Sudarshan,
"Chirality and the Universal Fermi Interaction,"
Phys. Rev. **109**, 1860 (1958).

R. Feynman and M. Gell-Mann,
"Theory of the Fermi Interaction,"
Phys. Rev. **109**, 193 (1958).

R. Christenson, J. Cronin, V.L. Fitch and R. Turlay,
"2π Decay of the K_2° Meson,"
Phys. Rev. Lett. **13**, 138 (1964).

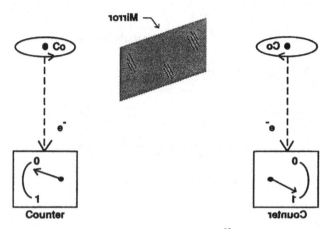

Fig. 5.   The initial setups of these two experiments on Co^{60} decay are exact mirror images, but the final electron distributions are not, as indicated by the different readings on the counters.

6. Symmetry Violations

Wu *et al.* investigated the decay of polarized cobalt nuclei, Co^{60}, into electrons. Because these nuclei are polarized they rotate parallel to each other. The experiment consisted of two setups, identical except that the directions of rotation of the initial nuclei were opposite; that is, each was a looking-glass image of the other. The experimenters found, however, that the patterns of the final electron distributions in these two setups are not mirror images of each other. In short, the initial states are mirror images, but the final configurations are not (see Fig. 5). This established parity nonconservation, i.e., right-left asymmetry.

The β decay experiment by Wu *et al.* and the $\pi - \mu$ decay experiment by Garwin, Lederman and Weinrich and by Friedman and Telegdi proved not only parity P violation, but also asymmetry under particle-antiparticle conjugation C. This is illustrated in Fig. 6. Both the neutrino and the antineutrino possess a spin (angular momentum). For a neutrino, if one aligns one's left thumb parallel to its momentum, then the curling of one's four fingers would always be in the direction of its spin, indicating P violation. Therefore, the spin and momentum direction of a neutrino defines a perfect left-hand screw, whereas the spin and momentum direction of an antineutrino defines a perfect right-hand screw, showing C violation in

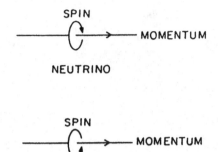

SPIN

NEUTRINO

SPIN

ANTINEUTRINO

Fig. 6.   The spin momentum of a neutrino defines a left-hand screw; that of an antineutrino defines a right-hand screw.

addition to P violation. This property holds for neutrinos and antineutrinos everywhere, independently of how they are produced.

In 1964, the experiment by Christenson *et al.*[37] established CP violation. An example of how such a violation can be established experimentally is to examine the decay[38] of K^0_L into $e^+\pi^-\nu$ and $e^-\pi^+\bar{\nu}$. The long-lived neutral kaon K^0_L is a spherically symmetric particle with zero spin; it carries no electric charge and no electromagnetic form factor of any kind. Yet these two decay modes have different rates:

$$\frac{\text{rate}\left(K^0_L \to e^+ + \pi^- + \nu\right)}{\text{rate}\left(K^0_L \to e^- + \pi^+\bar{\nu}\right)} = \mathbf{1.00648} \pm 0.00035.$$

Consequently, by using a time-clocking device it is possible to differentiate the positive sign of electricity vs. the negative sign of electricity, showing C violation. Because this rate difference remains true under mirror reflection, CP is also violated. The present status of these discrete symmetries and asymmetries is given in Table 2.

At present, only the combined symmetries CPT remain valid. In other words, only when we interchange

particle $\leftrightarrow$ antiparticle,

right $\leftrightarrow$ left,

past $\leftrightarrow$ future,

Table 2.

Present Status	
CPT	Good
P	X
C	X
CP	X
T	X ← without assuming CPT
CT	X
PT	X

do all physical laws appear to be invariant. From *CPT* symmetry, it follows that CP violation also implies time reversal *T* asymmetry.

7. Time Reversal

Time reversal symmetry *T* means that the time-reversed sequence of any motion is also a possible motion. Some of you may think this absurd, since we are all getting older, never younger. So why should we even contemplate that the laws of nature should be time-reversal symmetrical?

In this sense we must distinguish between the evolution of a small system and a large system. Let me give an example. In Fig. 7a each circle represents an airport, and a line indicates an air corridor. We assume that between any two of these airports the number of flights going both ways along any route is the same (this property will be referred to as microscopic reversability). Thus a person in "X" can travel to Stockholm (for this discussion we assume the only air connection from "X" is to Stockholm), then through Stockholm to Paris or New York. At any point in his travel, he can return to "X" with the same ease. But suppose that in every airport we were to remove all the signs and flight information, while maintaining exactly the same number of flights, as shown in Figure 7b. A person starting from "X" would still arrive in Stockholm, since that is the only airport connected to "X". However, without the signs to guide him, it would be very difficult for him to pick out the return flight to "X" from the many gates in the Stockholm airport. The plane he gets on may be headed for New York. If, in

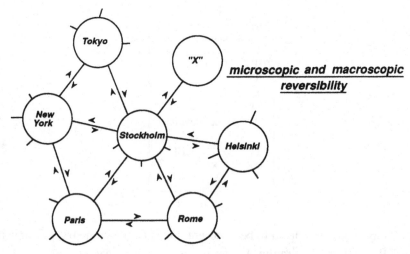

microscopic and macroscopic reversibility

Fig. 7a.   In this example, microscopic reversibility means an equal number of flights in either direction on every air route. When the names of the airports, the numbers of the gates, and all flight information are known, there is also macroscopic reversibility.

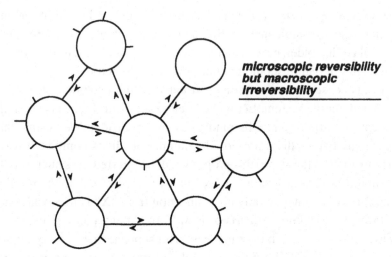

microscopic reversibility but macroscopic irreversibility

Fig. 7b.   If we maintain the same number of flights in each direction on any route (i.e., microscopic reversibility), but remove destination signs, gate numbers, and all other information, then it is nearly impossible to find our way back (i.e., macroscopic irreversibility).

New York, he then tries another plane again without any guidance, he could perhaps arrive in Tokyo. If he keeps on going this way, his chance of getting back to "X" is very slim indeed. In this example, we see that microscopic reversibility is strictly maintained. When all the airport destination signs and other flight information are given clearly, then macroscopically we also have reversibility. On the other hand, if all such information is withheld, then the whole macroscopic process appears irreversible. Thus, macroscopic irreversibility is not in conflict with microscopic reversibility.

Time reversal symmetry in physics refers to *microscopic reversiblity* between all molecular, atomic, nuclear and subnuclear reactions. Since none of these molecules, atoms, nuclei or subnuclear particles can be easily marked, any microscopic system in nature would exhibit irreversibility. This result is independent of microscopic reversbility. In any macroscopic process, we have to average over an immense number of unmarked microscopic units of atoms, molecules, and so forth (as in the example of unmarked airports and air routes), and that gives rise to *macroscopic irreversibility*. It is in this statistical sense of ever increasing disorder (entropy) that we define the direction of our macroscopic time flow. We may recall the words from *H.M.S. Pinafore*, by Gilbert and Sullivan,

> What never? No, never!
> What never? Well, hardly ever!

The existence of the macroscopic time direction then leaves open the important question whether time-reversal symmetry, or microscopic reversibility, is true or not. Since 1964, after the discovery of *CP* violation by Christenson *et al.*, through a series of remarkable experiments involving the kaons, it was found that the microscopic reversibility is indeed violated. Nature does not seem to respect time-reversal symmetry!

8. Present Status

In 1967, the paper of S. Weinberg[39] unified the weak interaction with the electromagnetic force, called the electroweak interaction, which is mediated by the photon γ, the charged intermediate bosons $W^{\pm}$ and their neutral partner Z^0 (these particles were discovered at CERN in 1983[40]).

At present, our theoretical structure of electroweak, strong and gravitational forces can be summarized as follows:

QCD (strong interaction)
$SU(2) \times U(1)$ **Theory** (electro-weak)
General Relativity (gravitation).

However, in order to apply these theories to the real world, we need a set of about 17 parameters, all of unknown origins. Thus, this theoretical edifice cannot be considered complete.

The two outstanding puzzles that confront us today are:

(i) **Missing symmetries** — All present theories are based on symmetry, but most symmetry quantum numbers are *not* conserved.
(ii) **Unseen quarks** — All hadrons are made of quarks; yet, no individual quark can be seen.

As we shall see, the resolutions of these two puzzles probably are tied to the structure of our physical vacuum.

The puzzle of missing symmetries implies the existence of an entirely new class of fundamental forces, the one that is responsible for symmetry breaking. Of this new force, we know only of its existence, and very little else. Since the masses of all known particles break these symmetries, an understanding of the symmetry-breaking forces will lead to a comprehension of the origin of the masses of all known particles. One of the promising directions is the spontaneous symmetry-breaking mechanism in which one assumes that the physical laws remain symmetric, but the physical vacuum is not. If so, then the solution of this puzzle is closely connected to the structure of the physical vacuum; the excitations of the physical vacuum may lead to the discovery of Higgs-type mesons.

In some textbooks, the second puzzle is often "explained" by using the analogy of the magnet. A magnet has two poles, north and south. Yet, if one breaks a bar magnet in two, each half becomes a complete magnet with two poles. By splitting a magnet open one will never find a single pole (magnetic monopole). However, in our usual description, a magnetic monopole can be considered as either a fictitious object (and therefore unseeable) or a real object but with exceedingly heavy mass beyond our

Table 3.

Stable particles poles on the "physical" sheet	Unstable particles poles on the "second" sheet	Particles that cannot be seen individually
GRAVITON	μ	GLUONS
	π	
γ	K	QUARKS
	⋮	⋮
ν	n	
	J/Ψ	
e	ϒ	
	⋮	
p	W	
	Z	
⋮	⋮	

Table 4.

How to detect particles that cannot be seen individually?
1. Jets (KLN Theorem on mass singularities (1964))
2. Entropy density change (Vacuum excitations T.D.L. and G.C. Wick (1974))

present energy range (and therefore not yet seen). In the case of quarks, we believe them to be real physical objects and of relatively low masses (except the top quark); furthermore, their interaction becomes extremely weak at high energy. If so, why don't we ever see free quarks? This is, then, the real puzzle.

The fact that quarks and gluons cannot be seen individually suggests an entirely new direction in our understanding of particles. Traditionally, any particle is either stable or unstable; the former is represented by a pole on the physical sheet and the latter by a pole on the "second" sheet. But now, we have a new third class: particles that cannot be seen individually. This is illustrated in Table 3, and its solution in Table 4. Following the KLN

Theorem,[41] gluons and quarks are indeed discovered by observing high energy jets. The second method, of using entropy change, will be discussed shortly. Before doing that, we may recall what has happened in the past.

At the end of the last century, there were also two physics puzzles:

1. No absolute inertial frame (Michelson-Morley Experiment 1887),
2. Wave-particle duality (Planck's formula 1900).

These two seemingly esoteric problems struck classical physics at its very foundation. The first became the basis for Einstein's theory of relativity and the second laid the foundation for us to construct quantum mechanics. In this century, all the modern scientific and technological developments — *nuclear energy, atomic physics, molecular structure, lasers, x-ray technology, semiconductors, superconductors, supercomputers — only* exist because we have relativity and quantum mechanics. To humanity and to our understanding of nature, these are all-encompassing.

Now, near the end of the twentieth century, we must ask what will be the legacy we give to the next generation in the next century? At present, like the physicists at the end of the 1890's, we are also faced with two profound puzzles. It seems likely that the present two puzzles may bring us as important a change in the development of science and technology in the twenty-first century.

9. Physical Vacuum

The current explanation of the quark confinement puzzle is again to invoke the vacuum. We assume the QCD vacuum to be a condensate of gluon pairs and quark-antiquark pairs so that it is a perfect color dia-electric[42] (i.e., color dielectric constant $\kappa = 0$). This is in analogy to the description of a superconductor as a condensate of electron pairs in BCS theory, which results in making the superconductor a perfect dia-magnet (with magnetic susceptibility $\mu = 0$). When we switch from QED to QCD we replace the magnetic field $\vec{H}$ by the color electric field $\vec{E}_{\text{color}}$, the superconductor by the QCD vacuum, and the QED vacuum by the interior of the hadron. Just as the magnetic field is expelled outward from the superconductor, the color electric field is pushed into the hadron by the QCD vacuum, and that leads to color confinement, or the formation of bags.[43] This situation is summarized below.

QED superconductivity as a perfect dia-magnet		QCD vacuum as a perfect color dia-electric
$\vec{H}$	$\longleftrightarrow$	$\vec{E}_{color}$
$\mu_{inside} = 0$	$\longleftrightarrow$	$\kappa_{vacuum} = 0$
$\mu_{vacuum} = 1$	$\longleftrightarrow$	$\kappa_{inside} = 1$
inside	$\longleftrightarrow$	outside
outside	$\longleftrightarrow$	inside

In the resolution of both puzzles, missing symmetry and quark confinement, the system of elementary particles no longer forms a self-contained unit. The microscopic particle physics depends on the coherent properties of the macroscopic world, represented by the appropriate operator averages in the physical vacuum state.

If we pause and think about it, this represents a rather startling conclusion, contrary to the traditional view of particle physics which holds that the microscopic world can be regarded as an isolated system. To a very good approximation it is separate and uninfluenced by the macroscopic world at large. Now, however, we need these vacuum averages; they are due to some long-range ordering in the state vector. At present our theoretical technique for handling such coherent effects is far from being developed. Each of these vacuum averages appears as an independent parameter, and that accounts for the large number of constants needed in the present theoretical formulation.

On the experimental side, there has hardly been any direct investigation of these coherent phenomena. This is because hitherto in most high-energy experiments, the higher the energy the smaller has been the spatial region we are able to examine. In order to explore physics in this fundamental area, relativistic heavy ion collisions offer an important new direction.[44] The basic idea is to collide heavy ions, say gold on gold, at an ultra-relativistic region. Before the collision, the vacuum between the ions is the usual physical vacuum; at a sufficiently high energy, after the collision almost all of the baryon numbers are in the forward and backward regions (in the center-of-mass system). The central region is essentially free of

T. D. Lee

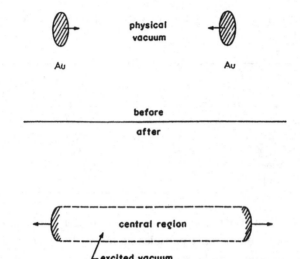

Fig. 8.   Vacuum excitation through relativistic heavy ion collisions.

baryons and, for a short duration, it is of a much higher energy density than the physical vacuum. Therefore, the central region represents the excited vacuum (Fig. 8).

As we shall see, we need RHIC, the 100 GeV × 100 GeV (per nucleon) relativistic heavy ion collider at the Brookhaven National Laboratory, to explore the QCD vacuum.

10. Phase Diagram of the QCD Vacuum

A normal nucleus of baryon number A has an average radius $r_A \approx 1.2A^{\frac{1}{3}}$ *fm* and an average energy density

$$\mathcal{E}_A \approx \frac{m_A}{(4\pi/3)r_A^3} \approx 130\,\text{MeV}/fm^3.$$

Each of the A nucleons inside the nucleus can be viewed as a smaller bag which contains three relativistic quarks inside; the nucleon radius is $r_N \approx 0.8$ *fm* and its average energy density is

$$\mathcal{E}_N \approx \frac{m_N}{(4\pi/3)r_N^3} \approx 440\,\text{MeV}/fm^3.$$

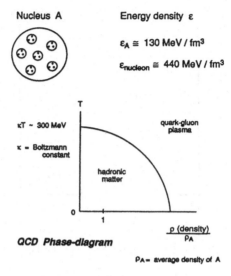

Fig. 9.

Consequently, even without any sophisticated theoretical analysis we expect the QCD phase diagram to be of the form given by Fig. 9.

In Fig. 9, the ordinate is κT (κ = Boltzmann constant, T = temperature), the abscissa is ρ/ρ_A (ρ = nucleon density, ρ_A = average nucleon density in a normal nucleus A). A typical nucleus A is when $\rho = \rho_A$. The scale can be estimated by noting that the critical $\kappa T \sim 300\,\text{MeV}$ is about the difference of 1 fm^3 times $\mathcal{E}_N - \mathcal{E}_A$ and the critical $\rho/\rho_A \sim 4$ is just the nearest integer larger than $(1.2/0.8)^3$.

Accurate theoretical calculation exists only for pure lattice QCD (i.e., without dynamical quarks). The result is shown in Fig. 10.

If one assumes scaling, then the phase transition in pure QCD (zero baryon number, $\rho = 0$) occurs at $\kappa T \sim 340\,\text{MeV}$ with the energy density of the gluon plasma

$$\mathcal{E}_P \sim 3\,\text{GeV}/fm^3.$$

To explore this phase transition in a relativistic heavy ion collision, we must examine the central region. Since only a small fraction of the total energy is retained in the central region, it is necessary to have a beam energy (per nucleon) at least an order of magnitude larger than $\mathcal{E}_N \times (1.2\,fm)^3 \sim 5\,\text{GeV}$;

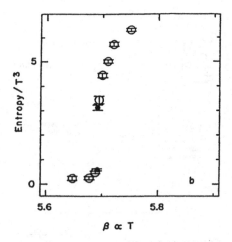

Fig. 10.   Phase transition[45] (pure QCD).

this makes it necessary to have an ion collider of $100\,\mathrm{GeV} \times 100\,\mathrm{GeV}$ (per nucleon) for the study of the QCD vacuum.

Another reason is that at $100\,\mathrm{GeV} \times 100\,\mathrm{GeV}$ the heavy nuclei are almost transparent, leaving the central region (in Fig. 10) to be one almost without any baryon number. As remarked before, this makes it an ideal situation for the study of the excited vacuum. Suppose that the central region does become a quark-gluon plasma. How can we detect it? This will be discussed in the following.

11. $\pi\pi$-Interferometry

As shown in Fig. 11, the emission amplitude of two pions of the same charge with momenta $\vec{k}_1$ and $\vec{k}_2$ from points $\vec{r}_1$ and $\vec{r}_2$ is proportional to

$$A \equiv e^{i\vec{k}_1 \cdot \vec{r}_1 + i\vec{k}_2 \cdot \vec{r}_2} + e^{i\vec{k}_2 \cdot \vec{r}_1 + i\vec{k}_1 \cdot \vec{r}_2},$$

because of Bose statistics. Let

$$\vec{q} \equiv \vec{k}_1 - \vec{k}_2 \quad \text{and} \quad \vec{r} \equiv \vec{r}_1 - \vec{r}_2.$$

Since

$$|A|^2 \equiv 1 + \cos \vec{q} \cdot \vec{r}$$

Hanbury - Brown / Twiss - type Experiment

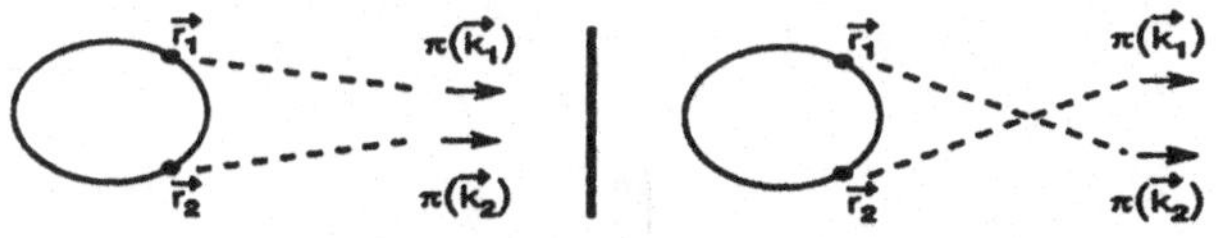

$$|\text{Amp}|^2 \propto |e^{i\vec{K}_1 \cdot \vec{r}_1 + i\vec{K}_2 \cdot \vec{r}_2} + e^{i\vec{K}_2 \cdot \vec{r}_1 + i\vec{K}_1 \cdot \vec{r}_2}|^2$$

$$= 1 + \cos \vec{q} \cdot \vec{r}. \quad \text{(varies from 1 to 2)}$$

$$\vec{q} = \vec{K}_1 - \vec{K}_2 \qquad \& \qquad \vec{r} = \vec{r}_1 - \vec{r}_2$$

If the central rapidity region is a plasma
of entropy density S_{plasma}, which later
hadronizes (of entropy density S_{Had}),

$$(\text{vol} \cdot S)_{plasma} \leq (\text{vol} \cdot S)_{Had}$$

$$S_{Had} \ll S_{plasma} \quad \therefore \quad (\text{vol})_{Had} \gg (\text{vol})_{plasma}$$

Fig. 11.

changes from $|A|^2 = 2$ as $\vec{q} \to 0$, to $|A|^2 = 1$ as $\vec{q} = \infty$, a measurement of the $\pi\pi$ correlation gives a determination of the geometrical size R of the region that emits these pions, like the Hanbury-Brown/Twiss determination of the stellar radius.

Now, if the central region is a plasma of entropy density S_P occupying a volume V_P, which later hadronizes to ordinary hadronic matter (of entropy density S_H and volume V_H), the total final entropy $S_H V_H$ must be larger than the total initial entropy $S_P V_P$. Since $S_P > S_H$, we have

$$V_H > V_P.$$

The experimental configurations and results[46] are given in Fig. 12 and Table 5. One sees that the hadronization radius in the central region is indeed much larger than that in the fragmentation region. This is, at best, only indicative of the quark-gluon plasma. Much work and higher energy are needed for a more definitive proof. Nevertheless, it does show that relativistic heavy ion can be an effective means of exploring the structure of the vacuum.

T. D. Lee

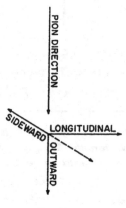

Fig. 12.

Table 5. $\pi\pi$-interference result[46] from the collision of an O beam (200 GeV/nucleon) on a stationary Au target.

	Rapidity interval	R_T(fm)	Gaussian R_L(fm)	Λ
	$1 < y < 2$	4.3 ± 0.6	2.6 ± 0.6	$0.34^{+0.09}_{-0.06}$
		$R_T^{\text{side}} = 4.0 \pm 1.0\,\text{fm}$	2.6 ± 0.6	$0.34^{+0.09}_{-0.06}$
		$R_T^{\text{out}} = 4.4 \pm 1.0\,\text{fm}$	2.6 ± 0.6	$0.34^{+0.09}_{-0.06}$
Central region (mid-rapidity)	$2 < y < 3$	8.1 ± 1.6	$5.6^{+1.2}_{-0.8}$	0.77 ± 0.19
		$R_T^{\text{side}} = 6.6 \pm 1.8\,\text{fm}$	$5.6^{+1.2}_{-0.8}$	0.77 ± 0.19
		$R_T^{\text{out}} = 11.2 \pm 2.3\,\text{fm}$	$5.6^{+1.2}_{-0.8}$	0.77 ± 0.19

R (Oxygen) $\cong$ 3 fm

12. Concluding Remarks

To conclude we emphasize, once again, that the most challenging problems in physics are

(1) the symmetry-breaking force, and
(2) the structure of the vacuum.

It is quite likely that the answer to these two problems lies in the same direction. They can only be solved when we learn how to excite the vacuum. In the traditional way of thinking, our world is the world of particles. Larger units are made of small ones, which in turn are made of even smaller elements. The search for the smallest building block that everything is made of drives us to explore physical phenomena within smaller and smaller distances; that necessitates energies higher and higher in inverse proportion to the distance in question. On the other hand, the puzzles of missing symmetry and quark confinement have forced us to face the profound possibility that the vacuum could be a physical medium.

As we look into the future, the completion of RHIC in 1997 offers an unprecedented opportunity for physicists to explore the possibility of exciting the vacuum and to examine whether it is indeed a physical medium.

If the vacuum is the underlying cause for the strange phenomena in the microscopic world of particle physics, it must also have been actively responsive to the macroscopic distribution of matter and energy in the universe. Because the vacuum is everywhere and forever, these two, the micro- and the macro-, have to be linked together; neither can be considered a separate entity. Future history books will record that ours was a time when humankind was able to forge this bond on a scientific basis.

Acknowledgment

This research was supported in part by the U.S. Department of Energy

References

1. E. Rutherford, *Philos. Mag.* **42**, 392 (1898).
2. M. and P. Curie, *C.R. Acad. Sci.* **130**, 647 (1900).
3. A.S. Eve, *Rutherford* (Cambridge, Cambridge University Press, 1939), p. 80.
4. A. Pais, *Inward Bound* (Oxford, The Clarendon Press, 1985).
5. O. Hahn and L. Meitner, *Phys. Z.* **9**, 321, 697 (1908).
6. W. Wilson, *Proc. Roy. Soc.* **A82**, 612 (1909).
7. O. von Baeyer and O. Hahn, *Phys. Z.* **11**, 488 (1910).
8. J. Chadwick, *Verh. Dtsch. Phys. Ges.* **16**, 383 (1914).
9. L. Meitner, *Z. Phys.* **9**, 131, 145 (1922).
10. E.D. Ellis, *Proc. Cambridge Philos. Soc.* **21**, 121 (1922).

11. N. Bohr, *J. Chem. Soc.* **135**, 349 (1932).
12. W. Pauli, American Physical Society Meeting in Pasadena, June 1931.
13. E. Fermi, *Ric. Scient.* **4**, 491 (1934); *Nuovo Cimento* **11**, 1 (1934); *Z. Phys.* **88**, 161 (1934).
14. E.J. Konopinski and G.E. Uhlenbeck, *Phys. Rev.* **48**, 7 (1935).
15. C.S. Wu and R.D. Albert, *Phys. Rev.* **75**, 315 (1949).
16. E. Fermi and E. Teller, *Phys. Rev.* **72**, 399 (1947).
17. M. Conversi, E. Pancini and O. Piccioni, *Phys. Rev.* **68**, 232 (1945).
18. W. Heisenberg, *Z. Phys.* **101**, 533 (1936).
19. C.M.G. Lattes, H. Muirhead, G.P.S. Occhialini and C.F. Powell, *Nature* **159**, 694 (1947).
20. O. Klein, *Nature* **161**, 897 (1948).
21. J. Steinberger, *Phys. Rev.* **75**, 1136 (1949).
22. T.D. Lee, M. Rosenbluth and C.N. Yang, *Phys. Rev.* **75**, 905 (1949).
23. G. Puppi, *Nuovo Cimento* **6**, 194 (1949).
24. J. Tiomno and J.A. Wheeler, *Rev. Mod. Phys.* **21**, 153 (1949).
25. L. Michel, *Nuovo Cimento* **10**, 319 (1953).
26. M. Fierz, *Z. Phys.* **104**, 553 (1937).
27. B.M. Rustad and S.L. Ruby, *Phys. Rev.* **89**, 880 (1953).
28. R.H. Dalitz, *Philos. Mag.* **44**, 1068 (1953); *Phys. Rev.* **94**, 1046 (1954).
29. T.D. Lee and J. Orear, *Phys. Rev.* **100**, 932 (1955).
30. *High Energy Nuclear Physics*, Proc. 6th Annual Rochester Conference, April 3–7 1956 (New York, Interscience, 1956), p. VI-20.
31. T.D. Lee and C.N. Yang, *Phys. Rev.* **102**, 290 (1956).
32. R. Budde, M. Chrétien, J. Leitner, N.P. Samios, M. Schwartz and J. Steinberger, *Phys. Rev.* **103**, 1827 (1956).
33. T.D. Lee and C.N. Yang, *Phys. Rev.* **104**, 254 (1956).
34. C.S. Wu, E. Ambler, R.W. Hayward, D.D. Hoppes and R.P. Hudson, *Phys. Rev.* **105**, 1413 (1957).
35. R.L. Garwin, L.M. Lederman and M. Weinrich, *Phys. Rev.* **105**, 1415 (1957).
36. J.I. Friedman and V.L. Telegdi, *Phys. Rev.* **105**, 1681 (1957).
37. J.H. Christenson, J.W. Cronin, V.L. Fitch and R. Turlay, *Phys. Rev. Lett.* **13**, 138 (1964).
38. C. Alff-Steinberger *et al.*, *Phys. Lett.* **20**, 207 (1966). See also *Particles and Detectors, Festschrift for Jack Steinberger*, eds. K. Kleinknecht and T.D. Lee (Berlin, Springer Verlag, 1986), p. 285ff for other references.
39. S. Weinberg, *Phys. Rev. Lett.* **19**, 1264 (1967). See also S.L. Glashow, *Nucl. Phys.* **22**, 579 (1961); A. Salam and J.C. Ward, *Nuovo Cimento* **11**, 568

(1959). *Phys. Lett.* **13**, 168 (1964); A. Salam, in *Proceedings of the 8th Nobel Symposium*, ed. N. Svartholm (Stockholm, Almqvist and Wiksell, 1968), p. 367.

40. C. Rubbia, in *Proceedings of the International Conference on High Energy Physics*, eds. J. Guy and C. Costain (Didcot, Rutherford Appleton Laboratory, 1983), p. 880.

41. T. Kinoshita, *J. Math. Phys.* **3**, 650 (1950). T.D. Lee and M. Nauenberg, *Phys. Rev.* **133**, B1549 (1964).

42. R. Friedberg and T.D. Lee, *Phys. Rev.* **D18**, 2623 (1978).

43. A. Chodos, R.J. Jaffe, K. Johnson, C.B. Thorn and V.F. Weisskopf, *Phys. Rev.* **D9**, 3471 (1974).

44. T.D. Lee and G.C. Wick, *Phys. Rev.* **D9**, 2291 (1974); T.D. Lee, "Relativistic Heavy Ion Collisions and Future Physics," in *Symmetries in Particle Physics*, ed. I. Bars, A. Chodos and C.H. Tze (New York, Plenum Press, 1984), p. 93.

45. F.R. Brown, N.H. Christ, Y.F. Oeng, M.S. Gao and T.J. Woch, *Phys. Rev. Lett.* **61**, 2058 (1988).

46. W. Willis, "Experimental Studies on States of the Vacuum," in *Relativistic Heavy-Ion Collisions*, ed. R.C. Hwa, C.S. Gao and M.H. Ye (New York, Gordon and Breach Science Publishers, 1990). p. 39.

ELECTRON ORBITS AND SUPERCONDUCTIVITY
OF CARBON 60

T. D. Lee

Columbia University, New York, N.Y. 10027, USA

1. Introduction

The discovery of C_{60} and the superconductivity of its alloy compounds[1-4] provides a rich field for new theoretical and experimental investigations.[5-8] In this paper, we show that the near degeneracy of the low-lying vector (T_{1u}) and axial-vector (T_{1g}) levels of an isolated C_{60}^- may provide a simple pairing mechanism in C_{60}^{--}, C_{60}^{---} and the K_nC_{60} crystal.

We will begin with a review of the single-electron energy levels of a C_{60} molecule. Above the 60-closed shell lie two triplets possessing the same icosahedral symmetry label T_1 but of opposite parity. We shall derive their wave functions in the tight-binding limit of the molecular orbit approximation, and exhibit these functions in a simple and useful form which prompts us to refer to them as "vector" and "axial vector". Our analysis is then extended to C_{60}^{--}. By taking into account the strong Coulomb energy between electrons, we find that the spectrum of C_{60}^{--} contains two low-lying states invariant under the proper icosahedral symmetry transformations, but again of opposite parity and hence naturally called "scalar" and "pseudoscalar", with an energy degeneracy within 1% of their excitation energy (in our approximate calculation), even closer than the vector axial-vector doublets in C_{60}^-.

Since parity doublets, two states of opposite parity with nearly the same energy, are not a common occurrence in physics, it is worthwhile to examine the mechanism that gives rise to them in C_{60}. For a microscopic system, such as mesons and baryons, composite particles of opposite parities usually have quite different internal structures which make them unlikely to be degenerate. In most macroscopic systems, as in the case of left-handed or right-handed sugar molecules, it is conceptually simple to contstruct coherent mixtures of opposite parities that are nearly degenerate; the difficulty lies in the practical realization of such coherent mixtures.

2. Icosahedral Group

In order to have the necessary tools for our subsequent analysis, we give a brief discussion of the molecular orbits of C_{60}. As will be shown below, these can be expressed in terms of the regular representation of the proper

$$\text{icosahedral group } \mathcal{G} = \{g\} \tag{1}$$

where g denotes its sixty group elements. Corresponding to the identity element e, we may take any point $\hat{e}$ on a unit sphere with center 0. The application of the finite rotation that associates with g transforms $\hat{e}$ to $\hat{g}$. For $\hat{e}$ not lying on any of the symmetry axes, the resulting sixty $\hat{g}$ are all different. (A convenient notation is to use the same $\hat{e}$ and $\hat{g}$ to denote the corresponding unit radial vectors of the sphere, as well as their end points on the sphere.) For definiteness, let the unit sphere be the *circumsphere* of a regular icosahedron of vertices $N, A_1, \ldots, A_5', S$. As in Fig. 1, choose the point e (corresponding to the identity element e in $\mathcal{G}$) on the edge $\overline{A_1 N}$, with $\overline{eN} = \frac{1}{3}\overline{A_1 N}$ and $\vec{0e} \| \hat{e}$. Each unit radial vector $\hat{g}$ intersects a point g on one of the edges of the icosahedron. The location of each point g, thus generated, gives the position of a carbon nucleus in C_{60}, provided the unit scale of the radius is $R \cong 4.03 \text{Å}$, so that the circumspherical radius of C_{60} is its actual value $R_0 \cong 3.5 \text{Å}$, with the ratio

$$\frac{R}{R_0} = 3\left(5 + \frac{4}{\sqrt{5}}\right)^{-\frac{1}{2}} \cong 1.15. \tag{2}$$

(The same notation g denotes both the group element as well as the position of the carbon nucleus.) Any function $f(g)$ of these sixty positions g can be expressed in terms of the irreducible representations of the icosahedral group $\mathcal{G}$, which consist of a singlet **1**, two triplets **3** and $\tilde{\mathbf{3}}$, a quartet **4** and

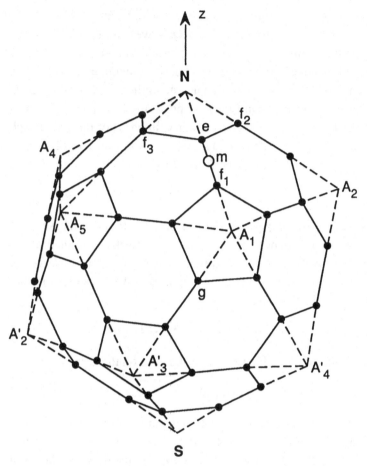

Fig. 1. Set on a unit sphere the 12 vertices of an icosahedron: N, S, A_i and $A_i'(i = 1, 2, \ldots, 5)$ with $\overline{NS}$, $\overline{A_i A_i'}$ the diameters of the sphere whose center is 0. Each of the sixty dots e, f_1, f_2, f_3, g, $\ldots$ denote the positions of C atoms in C_{60}. The three nearest neighbors of e are f_1, f_2 and f_3. The distance $\overline{ef_2} = \overline{ef_3}$ is slightly different from $\overline{ef_1}$.

a quintet **5**. This gives

$$60 = 1^2 + 3^2 + 3^2 + 4^2 + 5^2. \tag{3}$$

In the literature,[9,10] these representations are often referred to as A, T_1, T_2, G and H respectively. Our notations follow those used in particle physics, with the dimensionality of the irreducible representation shown explicitly.

The irreducible representations **1**, **3** and **5** can be readily identified with the usual $\ell = 0$ (s-wave), $\ell = 1$ (p-wave) and $\ell = 2$ (d-wave) of the spherical harmonics $Y_{l,m}(\theta, \phi)$, where $m = -\ell, -\ell + 1, \ldots, \ell$, and θ, ϕ are the polar and azimuthal angles. To derive the remaining two irreducible representations $\tilde{\mathbf{3}}$ and **4**, we designate one of the icosahedral vertices as the north pole $N(\theta = 0)$, as in Fig. 1. Express any function F of θ, ϕ defined on the twelve vertices of the icosahedron in terms of the sixteen spherical harmonics: $\ell = 0, 1, 2$ and 3 of $Y_{\ell,m}(\theta, \phi)$. (Note that $16 = 1 + 3 + 5 + 7$.) On the other hand, the function F has only

$$12 = 1 + 3 + 5 + 3 \tag{4}$$

values. Hence, we expect four linear combinations of the seven $\ell = 3$ $Y_{\ell,m}(\theta, \phi)$ to be identically zero on these twelve icosahedral vertices, leaving a triplet $\tilde{\mathbf{3}}$. For an explicit construction, we observe that the geodesic arc between two nearest neighboring vertices of an icosahedron is $\cos^{-1}(1/\sqrt{5})$; therefore,

$$Y_{3,\pm 1}(\theta, \phi) \propto (5\cos^2\theta - 1)\sin\theta e^{\pm i\phi} \tag{5}$$

is zero on all these twelve vertices. Likewise, on these sites

$$Y_{3,\pm 3}(\theta, \phi) \propto Y_{3,\mp 2}(\theta, \phi), \tag{6}$$

from which we can readily form two linear combinations of these $\ell = 3$ spherical harmonics that are also zero. In (6), both sides have the same azimuthal variation because of the five-fold symmetry along the north pole-south pole diameter; hence, (6) holds in the northern hemisphere, since excluding the pole, all the other five icosahedral vertices are at the same latitude. Its validity in the southern hemisphere then follows from inversion symmetry. In this way, we see that under the icosahedral rotations, the seven $Y_{3,m}(\theta, \phi)$ functions decompose into a quartet **4** and another triplet $\tilde{\mathbf{3}}$, different from **3**.

3. C_{60}

Returning to the C_{60} structure on the unit sphere, we denote by $|g\rangle$ a vector (*ket*-vector in Dirac's notation) with sixty components. Each element g corresponds to a definite finite rotation of the icosahedral group.

The radial position vector $\hat{g}$ of each C atom on the sphere corresponds to an element g of the proper icosahedral group. Thus the 60-dimensional Hilbert space of the tight-binding limit supports the regular representation of this group. For a typical "hopping" Hamiltonian of icosahedral symmetry, the lowest single-particle state is that having constant wave function, reminiscent of the s-wave (i.e., the orbital angular momentum quantum number $\ell = 0$) on a sphere. In the icosahedral group this singlet representation $\mathbf{1}^+$ is usually labeled A, or more fully A_g (the subscript $g = gerade$ meaning even parity).

The next-lowest state has a threefold degeneracy (excluding spin) which may be described as follows. For each C atom it turns out[11] that there is a certain unit vector $\hat{e}_0(g)$, very close to $\hat{g}$, within an angle of about $1°$; the wave function at site $\hat{g}$ can be chosen to be proportional to $\hat{e}_0(g) \cdot \hat{n}$ where $\hat{n}$ is some fixed vector. Clearly one obtains in this way three independent wave functions of the same energy, corresponding to the three independent choices of $\hat{n}$. If we approximate $\hat{e}_0(g) \cong \hat{g}$ (as we shall do hereafter in this discussion), then these three states correspond to the p-wave triplet ($\ell = 1$) on the sphere. Under the icosahedral group this triplet representation $\mathbf{3}^-$ is usually labeled T_1, or more fully T_{1u} ($u = ungerade$ meaning odd parity, since $\hat{g}$ changes sign for oppositely placed C atoms).

It is natural to call the singlet and triplet states described above "scalar" and "vector" referring to the way they transform when the laboratory coordinate system is rotated or inverted. Thus, the designation "vector" refers to the vector $\hat{n}$ which selects a T_1 state. Since multiplicity and parity play a prominent role in our analysis, we shall write $\mathbf{1}^+$ (scalar) for A_g, $\mathbf{3}^-$ (vector) for T_{1u}, etc., with $\mathbf{1}^+$ denoting a singlet of even parity, $\mathbf{3}^-$ a triplet of odd parity, etc. (All these multiplicities are doubled by spin.)

In the regular representation, each irreducible representation of dimension d appears d times; thus, the triplet T_1 must appear three times. Our interest will be in the *other* two triplets T_1, whose energy lies far above that of the "p-wave" state described above. Their wave functions can be expressed simply as follows:

Because of the detailed structure of the molecule, each C atom defines not only a radial vector $\hat{g}$ but a local coordinate system, with its three orthonormal basis vectors denoted by

$$\hat{g}, \quad \hat{e}_-(g) \quad \text{and} \quad \hat{e}_+(g), \tag{7}$$

where

$$\hat{e}_+(g) = \hat{g} \times \hat{e}_-(g). \tag{8}$$

Under an inversion, $\hat{g}$ changes sign; consequently, $\hat{e}_-(g)$ and $\hat{e}_+(g)$ are of opposite parities. Let $\hat{e}_-(g)$ be of odd parity, then $\hat{e}_+(g)$ is of even parity. By taking the wave function proportional to $\hat{e}_-(g) \cdot \hat{n}$ (where, as before, $\hat{n}$ is some fixed vector) one obtains a triplet $\mathbf{3}^-$ or T_{1u}, the LUMO (lowest unoccupied molecular orbital) state in neutral C_{60}. By taking $\hat{e}_+(g) \cdot \hat{n}$, one obtains a $\mathbf{3}^+$ or T_{1g}, the second LUMO state in neutral C_{60}. It is natural to call these triplets "vector" and "axial-vector" respectively. Together, (7) determines the set of three triplets T_1 contained in the regular representation.

To transform a C atom to one of its nearest neighbors by an element of the icosahedral group, one must rotate through a large angle ($180°$ for one of the nearest neighbors, from e to f_1 in Fig. 1), about an axis rather close to $\hat{g}$. Hence $\hat{g}$ varies slowly between neighbors but, as we shall see, $\hat{e}_-(g)$ and $\hat{e}_+(g)$ vary rapidly and, because of (8), by about the same amount. This is why the $\hat{e}_- \cdot \hat{n}$ and $\hat{e}_+ \cdot \hat{n}$ triplets have energy far above that of the $\hat{g} \cdot \hat{n}$ triplet, but close to each other. (Although the high-lying $\mathbf{3}^-$ and $\mathbf{3}^+$ states have the same formal structure under the *icosahedral group* as the low-lying $\mathbf{3}^-$, if one attempts to fill in a smooth wave function between the C atoms, one will need mostly spherical harmonics of $\ell = 5$ and $\ell = 6$ for $\hat{e}_-$ and $\hat{e}_+$ respectively, instead of $\ell = 1$ as for $\hat{g}$.)

Since $\hat{e}_- \cdot \hat{n}$ and $\hat{e}_+ \cdot \hat{n}$ are the two low-lying levels above the 60-closed shell, in C_{60}^- the extra electron may be in either of these triplets. The nearness of these two levels of opposite parity, combined with their compatibility (both T_1) under the icosahedral group, render C_{60}^- highly polarizable; the polarizability can be readily calculated.

4. Pairing Mechanism

In C_{60}^{--}, the Coulomb energy between the two extra electrons plays a dominant role. Without this term, the lowest-energy state would be of the form $T_{1u}T_{1u}$ or $\mathbf{3}^- \times \mathbf{3}^-$. But we find that two other states have considerably lower Coulomb energy. One is obtained by combining T_{1u} and T_{1g} to make a two-particle state $\mathbf{1}^-$:

$T_{1u}T_{1g} + T_{1g}T_{1u}$ with wave function

$$[\hat{e}_-(g) \cdot \hat{e}_+(g') + \hat{e}_+(g) \cdot \hat{e}_-(g')](\uparrow\downarrow' - \downarrow\uparrow'). \tag{9}$$

The other is made by mixing the 1^+ combination of $T_{1u}T_{1u}$ with that of $T_{1g}T_{1g}$:

$T_{1u}T_{1u} + T_{1g}T_{1g}$ with wave function

$$[\hat{e}_-(g) \cdot \hat{e}_-(g') - \hat{e}_+(g) \cdot \hat{e}_+(g')](\uparrow\downarrow' - \downarrow\uparrow'), \tag{10}$$

where $\hat{g}$, $\uparrow$ or $\downarrow$ and $\hat{g}'$, $\uparrow'$ or $\downarrow'$ are the position and spin variables of the two electrons. Because both wave functions vanish when the two electrons coalesce, $\hat{g} = \hat{g}'$, their mutual Coulomb energy is greatly reduced; this leads to the spin-0 parity doublets in C_{60}^{--}, labeled 1^- and 1^+.

The wave functions (9) and (10) represent singlets in both position and spin; no external vector $\hat{n}$ appears. It is natural to call these paired states pseudoscalar and scalar. It is evident that if the energy difference $\Delta\epsilon$ between 3^- (T_{1u}) and 3^+ (T_{1g}) is small, then it will contribute only to order $(\Delta\epsilon)^2$ to the splitting between scalar and pseudoscalar paired states. This is why the paired states in C_{60}^{--} form a much tighter parity doublet than the component one-particle states in C_{60}^-.

The closeness of these parity doublets makes it natural to isolate their response to strong interactions, separate from other distant levels. This provides a convenient means to derive analytical expressions for many of the important parts of strong interaction effects. Thus, we can calculate the polarization energy of C_{60}^- in a strong electric field E, showing that it changes from the weak field expression $-\frac{1}{2}\alpha E^2$ to one that depends linearly on E. Another example is to derive the final energy in C_{60}^{--}, within the parity doublet approximation, to all powers of the interaction Hamiltonian.

5. $K_3 C_{60}$

We turn our attention to the K_3C_{60} crystal and calculate the Madelung energy. The result shows that all K are ionized, as is commonly accepted.

Above the 60-closed shell, a typical band calculation reveals a low-lying cluster of three overlapping narrow bands, which is the Bloch wave extension of the three components of the vector wave function $\hat{e}_-(g)$ of an isolated C_{60}^-. These overlapping fermion bands are half-filled in the case of K_3C_{60}. On the other hand, the pseudoscalar pairing wave function (9) in C_{60}^{--} suggests a different Bloch wave extension, one that represents the hopping of such a highly correlated two-particle state in the crystal. The orthogonality of these correlated two-particle Bloch wave functions to any product of two one-particle wave functions in the fermion bands follows from the original orthogonality condition in a single C_{60} molecule:

$$\sum_{g=1}^{60} \hat{e}_-(g)_i \hat{e}_+(g)_j = 0 \tag{11}$$

where i and j denote the vector components. We call the Bloch extension of (9) the "boson band" (or the pseudoscalar band).

The two electrons in the fermion bands have an energy advantage over the boson on account of the excitation energy $\Delta \epsilon$ and the lowering in kinetic energy; however, there is a disadvantage to the fermion bands of having a higher Coulomb energy. The important question concerning the role of bosons versus fermions depends on the delicate balance between these two opposing factors. This is examined in Ref. 11. We start with the Coulomb energy ($\sim 11\,\text{eV}$ per lattice cell) between three electrons in the fermion bands, and compare that with the configuration of placing two electrons in the boson band and the remaining one in the fermion bands. We then take into account the kinetic energy difference and the Van der Waals energy difference; the latter is important because of the large polarizability of the C_{60} negative ion. The final energy balance between these two configurations is estimated to be only $0.22\,\text{eV}$, slightly in favor of the fermion bands. However, considering the highly approximate nature of our calculations and that the final answer is only about 2% of the initial energy that we begin with, it is not possible to make any definitive statement. Nevertheless, a variety of interesting theoretical possibilities emerges. It seems likely that there exists a narrow boson band, which may lie very close to, or overlapping with, the three fermion bands.

The possibility of a close-by low-lying boson band, in addition to the usual fermion bands, has important consequences for superconductivity. The bosons may undergo Bose-Einstein condensation. The zero momentum nature of the Bose condensate necessarily generates charge fluctuations in the coordinate space; thereby it increases the Coulomb energy. It can be shown that this Coulomb energy increase is compensated for, at near distances, by the monopole-dipole interaction between neighboring C_{60} molecules because of their large polarizability. At large distances there is, in addition, the Debye screening of the Coulomb potential generated by these charge fluctuations.

In Ref. 11, through a simplified but explicit field theoretic model, we are able to examine the effect of the Bloch extension of the correlated scalar wave function (10), the parity doublet partner of the pseudoscalar (9). Except in the somewhat unlikely case that the boson band is lower in energy than two times the bottom energy of the fermion bands, the scalar only provides a resonance to the electrons in the fermion bands. Such a resonance can produce an energy gap, as in the BCS theory of superconductivity. While both members of the bosonic parity doublet can be important to superconductivity, their roles are different. The experimentally observed pressure variation of the critical temperature[6] can be shown to be consistent with the model.

Our thesis is that the 3^+ (T_{1g}) level plays a special role because of its coupling and near degeneracy with the 3^- (T_{1u}) at the Fermi level. Therefore the C_{60}^{--} energy should be calculated by starting with the mixed states of (9) and (10), and treating all levels *outside* the parity doublet perturbatively. This makes it possible to handle strong interactions, and leads to high T_c superconductivity.

The parity doublets provide an essential insight into the nature of the paired wave function in C_{60}, as well as the physics associated with polarizability of the C_{60} negative ion. The superconductivity is discussed on the basis of a simplified model Hamiltonian in which there is an effective "attractive and local" four-fermion field interaction term. Its presence is due to the lowering of the strong Coulomb energy in the correlated paired states (9) and (10) in C_{60}^{--} which, in turn, can be justified within the parity doublet approximation. On the other hand, parity doublets are rather special to C_{60}; a natural question is to ask how important are their roles to superconductivity in general. We believe that, for all high T_c superconductors (cupric oxides and C_{60} alloys) with very small coherence length, the essential common feature probably lies in the approximate applicability of an effective "attractive and local" interaction term. For material other than C_{60}, its underlying reason may have nothing to do with parity doublets. The parity doublets of C_{60} simply provide a convenient tool for us to penetrate the maze of strong interactions.

Acknowledgment

This research was supported in part by the U.S. Department of Energy.

References

1. H. W. Kroto *et al.*, *Nature* **318**, 162 (1985); W. Krätschmer *et al.*, *Nature* **347**, 354 (1990); H. Ajie *et al.*, *J. Phys. Chem.* **94**, 8630 (1990).
2. A. F. Hebard *et al.*, *Nature* **350**, 600 (1991).
3. M. J. Rosseinsky *et al.*, *Phys. Rev. Lett.* **66**, 2830 (1991).
4. K. Holczer *et al.*, *Science* **252**, 1154 (1991).
5. K. Holczer *et al.*, *Phys. Rev. Lett.* **67**, 271 (1991).
6. G. Sparn *et al.*, *Science* **252**, 1829 (1991); R. M. Fleming *et al.*, *Nature* **352**, 787 (1991); O. Zhou *et al.*, *Science* **255**, 833 (1992).
7. P.-M. Allemand *et al.*, *Science* **253**, 301 (1991).
8. S. Chakravarty, M.P. Gelfand and S. Kivelson, *Science* **254**, 970 (1991), and UCLA preprint "Electronic Correlation Effects and Superconductivity in Doped Fullerenes".
9. M. Hamermesh, *Group Theory and Its Application to Physical Problems* (Addison-Wesley Publishing Co., 1962).
10. F. Albert Cotton, *Chemical Applications of Group Theory*, 2nd edition (Wiley-Interscience, 1963).
11. R. Friedberg, T. D. Lee and H. C. Ren, *Phys. Rev.* **B46**, 14 150 (1992).

THE POWER OF DUALITY — EXACT RESULTS IN 4D SUSY FIELD THEORY*

N. Seiberg

Department of Physics and Astronomy, Rutgers University
Piscataway, NJ 08855-0849, USA
and
Institute for Advanced Study
Princeton, NJ 08540, USA

Recently the vacuum structure of a large class of four dimensional (supersymmetric) quantum field theories was determined exactly. These theories exhibit a wide range of interesting new physical phenomena. One of the main new insights is the role of "electric-magnetic duality." In its simplest form it describes the long distance behavior of some strongly coupled, and hence complicated, "electric theories" in terms of weakly coupled "magnetic theories." This understanding sheds new light on confinement and the Higgs mechanism and uncovers new phases of four dimensional gauge theories. We review these developments and speculate on the outlook.

1. Introduction

Exact solutions play a crucial role in physics. It is often the case that a simple model exhibits the same phenomena which are also present in more complicated examples. The exact solution of the simple model then teaches us about more generic situations. For example, the exact solutions of the harmonic oscillator and of the hydrogen atom demonstrated many of the crucial aspects of quantum mechanics. Similarly, the Ising model was a useful laboratory in the study of statistical mechanics and quantum field theory. Many other exactly solvable two dimensional field theories like the Schwinger model and others have led to the understanding of many mechanisms in quantum field theory, which are also present in four dimensions.

---

*To appear in the Proc. of PASCOS 95 and in the Proc. of the Oskar Klein Lectures.

The main point of this talk is to show how four dimensional super-symmetric quantum field theories can play a similar role as laboratories and testing grounds for ideas in more generic quantum field theories. This follows from the fact that these theories are more tractable than ordinary, non-supersymmetric theories, and many of their observables can be computed exactly. Nevertheless, it turns out that these theories exhibit explicit examples of various phenomena in quantum field theory. Some of them had been suggested before without an explicit realization and others are completely new. (For a brief review summarizing the understanding as of a year ago, see Ref. 1.)

Before continuing we would like to mention that supersymmetric four dimensional field theories also have two other applications:

1. Many physicists expect supersymmetry to be present in Nature, the reason being that it appears to be the leading candidate for solving the gauge hierarchy problem. If this is indeed the case, it is likely to be discovered experimentally in the next round of accelerators. Independent of the hierarchy problem, supersymmetry plays an important role in string theory as an interesting extension of our ideas about space and time. If Nature is indeed supersymmetric, understanding the dynamics of supersymmetric theories will have direct experimental applications. In particular, we will have to understand how supersymmetry is broken, i.e. why Nature is not exactly supersymmetric.
2. Witten discovered an interesting relation between supersymmetric field theories and four dimensional topology.[2] Using this relation, the exact solutions lead to a simplification of certain topological field theories and with that to advances in topology.[3]

Even though these applications are important, here we will focus on the dynamical issues and will not discuss them.

The main insight that the study of these theories has taught us so far is the role of electric-magnetic duality in strongly coupled non-Abelian gauge theories.[a] Therefore, we will start our discussion in the next section

---

[a] The role of electric-magnetic duality in four dimensional quantum field theory was first suggested by Montonen and Olive.[4] Then, it became clear that the simplest version of their proposal is true only in $N = 4$ supersymmetric field theories[5] and in certain $N =$

by reviewing the duality in Abelian gauge theories, i.e. in electrodynamics. In Section 3 we will present supersymmetric field theories and will outline how they are solved. This discussion will be rather heuristic. In Section 4 we will summarize the results and will present the duality in non-Abelian theories. Finally in Section 5 we will present our conclusions and a speculative outlook.

2. Duality in Electrodynamics

2.1. *The Coulomb phase*

The simplest phase of electrodynamics is the Coulomb phase. It is characterized by massless photons which mediate a long range $\frac{1}{R}$ potential between external sources. In the absence of sources the relevant equations are Maxwell's equations in the vacuum

$$\nabla{\cdot}E = 0$$

$$\nabla \times B - \frac{\partial}{\partial t}E = 0$$

$$\nabla{\cdot}B = 0$$

$$\nabla \times E + \frac{\partial}{\partial t}B = 0$$

(we have set the speed of light $c = 1$). Clearly, they are invariant under the duality transformation

$$E \rightarrow B$$

$$B \rightarrow -E$$

which exchanges electric and magnetic fields.

If charged particles are added to the equations, the duality symmetry will be preserved only if both electric charges and magnetic monopoles are present. However, in Nature we see electric charges but no magnetic monopole has been observed yet. This fact ruins the duality symmetry. We usually also ruin the symmetry by solving two of the equations by introducing the vector potential. Then, these equations are referred to as

---

2 supersymmetric theories.[6] Here we will discuss the extension of these ideas to $N=1$ theories.[7]

the Bianchi identities while the other two equations are the equations of motion. Had there also been magnetic monopoles, this would have been impossible.

Dirac was the first to study the possible existence of magnetic monopoles in the quantum theory. He derived the famous Dirac quantization condition which relates the electric charge e and the magnetic charge g:

$$eg = 2\pi$$

(we have set Planck's constant $\hbar = 1$). This relation has many important consequences. One of them is that since duality exchanges electric and magnetic fields, it also exchanges

$$e \longleftrightarrow g.$$

Since the product $eg = 2\pi$ is fixed, it relates weak coupling ($e \ll 1$) to strong coupling ($g \gg 1$). Therefore, even if we could perform such duality transformations in quantum electrodynamics, they would not be useful. Electrodynamics is weakly coupled because e is small. Expressing it in terms of "magnetic variables" will make it strongly coupled and therefore complicated.

However, one might hope that in other theories like QCD, such a duality transformation exists. If it does, it will map the underlying "electric" degrees of freedom of QCD, which are strongly coupled, to weakly coupled "magnetic" degrees of freedom. We would then have a weakly coupled, and therefore easily understandable, effective description of QCD.

2.2. *The Higgs phase*

When charged matter particles are present, electrodynamics can be in another phase — the superconducting or the Higgs phase. It is characterized by the condensation of a charged field ϕ

$$\langle \phi \rangle \neq 0.$$

This condensation creates a gap in the spectrum by making the photon massive. This phenomenon was first described in the context of superconductivity, where ϕ is the Cooper pair. It has since appeared in different systems including the weak interactions of particle physics where ϕ is the Higgs field.

The condensation of ϕ makes electric currents superconducting. Its effect on magnetic fields is known as the Meissner effect. Magnetic fields cannot penetrate the superconductor except in thin flux tubes. Therefore, when two magnetic monopoles (e.g. the ends of a long magnet) are inserted in a superconductor, the flux lines are not spread. Instead, a thin flux tube is formed between them. The energy stored in the flux tube is linear in its length and therefore the potential between two external magnetic monopoles is linear (as opposed to the $\frac{1}{R}$ potential outside the superconductor). Such a linear potential is known as a *confining* potential.

Mandelstam and 't Hooft considered the dual of this phenomenon: if instead of electric charges, magnetic monopoles condense, then magnetic currents are superconducting while electric charges are confined. Therefore, confinement is the dual of the Higgs mechanism. They suggested that confinement in QCD can be understood in a similar way by the condensation of color magnetic monopoles.

To summarize, we see that duality exchanges weak coupling and strong coupling. Therefore, it exchanges a description of the theory with small quantum fluctuations with a description with large quantum fluctuations. Similarly, it exchanges the weakly coupled phenomenon of the Higgs mechanism with the strong coupling phenomenon of confinement.

Such a transformation between variables which fluctuate rapidly and variables which are almost fixed is similar to a Fourier transform. When a coordinate is localized, its conjugate momentum fluctuates rapidly and vice versa. A duality transformation is like a Fourier transform between electric and magnetic variables.

It should be stressed, however, that except in simple cases (like electrodynamics without charges and some examples in two dimensional field theory) an explicit duality transformation is not known. It is not even known whether such a transformation exists at all. As we will show below, at least in supersymmetric theories such a transformation does exist (even though we do not have an explicit description of it).

3. The Dynamics of Supersymmetric Field Theories

In supersymmetric theories the elementary particles are in representations of supersymmetry. Every gauge boson, a gluon, is accompanied by a fermion, a gluino, and every matter fermion, a quark, is accompanied by a scalar, a squark. The theory is specified by a choice of a gauge group, which determines the coupling between the gluons, and a matter representation, which determines the coupling between the quarks and the gluons. For example, in QCD the gauge group is $SU(N_c)$ (N_c is the number of colors) and the matter representation is $N_f \times (\mathbf{N_c} + \mathbf{\bar{N}_c})$ (N_f is the number of flavors).

An important object in these theories is the superpotential, $W(q)$. It is a holomorphic (independent of $\bar{q}$, the complex conjugate of q) gauge invariant function of the squarks, q. It determines many of the coupling constants and interactions in the theory including the Yukawa couplings of the quarks and the squarks and the scalar potential, $V(q, \bar{q})$. For example, a mass term for the quarks appears as a quadratic term in the superpotential.

The analysis of the classical theory starts by studying the minima of the scalar potential $V(q, \bar{q})$. It is often the case that the potential has many different degenerate minima. Then, the classical theory has many inequivalent ground states. Such a degeneracy between states which are not related by a symmetry is known as an "accidental degeneracy." Typically in field theory such an accidental degeneracy is lifted by quantum corrections. However, in supersymmetric theories, the degeneracy often persists in the quantum theory. Therefore, the quantum theory has a space of inequivalent vacua which can be labeled by the expectation values of the squarks $\langle q \rangle$.

This situation of many different ground states in the quantum theory is reminiscent of the situation when a symmetry is spontaneously broken. However, it should be stressed, that unlike that case, where the different ground states are related by a symmetry, here they are inequivalent. Physical observables vary from one vacuum to the other — they are functions of $\langle q \rangle$.

Our problem is to solve these theories not only as a function of all the parameters (like quark masses) but, for every value of the parameters, also as a function of the ground state $\langle q \rangle$.

As in Landau-Ginzburg theory, the best way to organize the information is in a low energy effective Lagrangian, L_{eff}. It includes only the low lying modes and describes their interactions. This effective theory is also specified by a gauge group and a matter representation. Since the theory is supersymmetric,[b] the interactions of the light particles are characterized by a superpotential, W_{eff}.

The key observation is that this effective superpotential can often be determined exactly by imposing the following constraints[8]:

1. In various limits of the parameter space or the space of ground states the theory is weakly coupled. In these limits W_{eff} can be determined approximately by weak coupling techniques (examples are the instanton calculations of Refs. 9–11).

2. W_{eff} must respect all the symmetries in the problem. It is important to use also the constraints following from symmetries which are explicitly broken by the various coupling constants. Such symmetries lead to selection rules.

3. W_{eff} is holomorphic in the light fields and the parameters. This constraint is the crucial one which makes supersymmetric theories different from non-supersymmetric ones. The use of holomorphy in Ref. 8 generalized previous related ideas.[12–16]

The superpotential determined in this way teaches us about the light particles and their interactions. This information determines the phase structure of the theory and the mechanisms for phase transitions.

By repeating this process for different theories, i.e. different gauge groups and matter representations, a rich spectrum of phenomena has been found. In the next section we will describe some of them.

4. Results — Duality in Non-Abelian Theories

4.1. $SU(N_c)$ with $N_f \times (N_c + \overline{N}_c)$

Here the physical phenomena depend crucially on the number of massless quarks[7,9,15,17,18]:

I. $N_f \geq 3N_c$

In this range the theory is not asymptotically free. This means that, because of screening, the coupling constant becomes smaller at large distances. Therefore, the spectrum of the theory at large distance can be read off from

---

[b]We limit ourselves to theories where supersymmetry is not spontaneously broken.

the Lagrangian — it consists of the elementary quarks and gluons. The long distance behavior of the potential between external electric sources is of the form

$$V \sim \frac{1}{R \log R}.$$

The logarithm in the denominator shows that the interactions between the particles are weaker than in Coulomb theory. Therefore, we refer to this phase of the theory as a free electric phase.

We should add here that, strictly speaking, such a theory is not well defined as an interacting quantum field theory. However, it can be a consistent description of the low energy limit of another theory.

II. $\frac{3}{2}N_c < N_f < 3N_c$

In this range the theory is asymptotically free. This means that at short distance the coupling constant is small and it becomes larger at longer distances. However, in this regime[7,18] rather than growing to infinity, it reaches a finite value — a fixed point of the renormalization group.

Therefore, this is a non-trivial four dimensional conformal field theory. The elementary quarks and gluons are not confined there but appear as interacting massless particles. The potential between external electric sources behaves as

$$V \sim \frac{1}{R}$$

and therefore we refer to this phase of the theory as the non-Abelian Coulomb phase.

It turns out that there is an equivalent, "magnetic," description of the physics at this point [7]. It is based on the gauge group $SU(N_f - N_c)$, with N_f flavors of quarks and some gauge invariant fields. We will refer to this gauge group as the magnetic gauge group and to its quarks as magnetic quarks. This theory is also in a non-Abelian Coulomb phase because $\frac{3}{2}(N_f - N_c) < N_f < 3(N_f - N_c)$. The surprising fact is that its large distance behavior is identical to the large distance behavior of the original, "electric," $SU(N_c)$ theory. Note that the two theories have different gauge groups and different numbers of interacting particles. Nevertheless, they describe the same fixed point. In other words, there is no experimental way to determine whether the $\frac{1}{R}$ potential between external sources is mediated by the interacting electric or the interacting magnetic variables. Such a phenomenon of two different Lagrangians describing the same long distance physics is common in two dimensions and is known there as quantum equivalence. These four dimensional examples generalize the duality[4] in finite $N = 4$ supersymmetric theories[5] and in finite $N = 2$ theories[6] to asymptotically free $N = 1$ theories.

As N_f is reduced (e.g. by giving masses to some quarks and decoupling them) the electric theory becomes stronger — the fixed point of the renormalization group occurs at larger values of the coupling. Correspondingly, the magnetic theory becomes weaker. This can be understood by noting that as N_f is reduced, the magnetic gauge group becomes smaller. This happens by the Higgs mechanism in the magnetic theory.[7]

III. $N_c + 2 \leq N_f \leq \frac{3}{2} N_c$

In this range the electric theory is very strongly coupled. However, since $3(N_f - N_c) \leq N_f$, the equivalent magnetic description based on the gauge group $SU(N_f - N_c)$ is not asymptotically free and it is weakly coupled at large distances. Therefore, the low energy spectrum of the theory consists of the particles in the dual magnetic Lagrangian.[7] These magnetic massless states are composites of the elementary electric degrees of freedom. The massless composite gauge bosons exhibit gauge invariance which is not visible in the underlying electric description. The theory generates new gauge invariance! We will return to this phenomenon in the conclusions.

This understanding allows us to determine the long distance behavior of the potential between magnetic sources:

$$V \sim \frac{1}{R \log R}.$$

Since the magnetic variables are free at long distance, we refer to this phase as a free magnetic phase.

IV. $N_f = N_c + 1, N_c$

As we continue to decouple quarks by giving them masses and thus reducing N_f, the magnetic gauge group is Higgsed more and more. Eventually it is completely broken and there are no massless gauge bosons. This complete

Higgsing of the magnetic theory can be interpreted as complete confinement of the electric variables. We thus see an explicit realization of the ideas of Mandelstam and 'tHooft about confinement.

For $N_f = N_c + 1$ this confinement is not accompanied by chiral symmetry breaking, while for $N_f = N_c$ chiral symmetry is also broken.[18] In the electric language we describe the spectrum in terms of gauge invariant fields. These include massless mesons and baryons. In the magnetic language these are elementary fields. The idea that some of the composites in QCD, in particular the baryons, can be thought of as solitons was suggested in Ref. 19. Here we see an explicit realization of a related idea — the baryons are magnetic monopoles composed of the elementary quarks and gluons.

V. $N_f < N_c$
In this range the theory of massless quarks has no ground state.[9,15]

4.2. $SO(N_c)$ with $N_f \times N_c$

In the $SU(N_c)$ theories there is no invariant distinction between Higgs and confinement.[20] This is not the case in theories based on $SO(N_c)$ with $N_f \times N_c$ and therefore, they lead to a clearer picture of the dynamics. In particular, here the transition from the Higgs phase to the Confining phase occurs with a well defined phase transition.

Many of the results in these $SO(N_c)$ theories[7,9,21−23,24] are similar to the results in $SU(N_c)$ showing that these phenomena are generic. Here the duality map is

$$SO(N_c) \text{ with } N_f \times N_c \longleftrightarrow SO(N_f - N_c + 4) \text{ with }$$
$$N_f \times (N_f - N_c + 4)$$

Let us consider three special cases of the duality:

1. For $N_c = 2$, $N_f = 0$ the map is

$$SO(2) \cong U(1) \longleftrightarrow SO(2) \cong U(1)$$

which is the ordinary duality of electrodynamics. Therefore, our duality is compatible with and generalizes this duality.

2. For $N_c = 3$, $N_f = 1$ the map is:

$$SO(3) \text{ with } \mathbf{3} \longleftrightarrow SO(2) \cong U(1) \text{ with } \mathbf{2}.$$

This theory was first analyzed in Ref. 22. Here it is possible to understand the duality in more detail than in the more general case and in particular to identify the matter fields in the magnetic theory as magnetic monopoles. This theory has $N = 2$ supersymmetry but this did not play a role in our discussion. However, using the extra supersymmetry one can derive more exact results about the massive spectrum of the theory.[22]

3. For $N_f = N_c - 2$ the map is

$$SO(N_c) \text{ with } (N_c - 2) \times \mathbf{N_c} \longleftrightarrow SO(2) \cong U(1) \text{ with } (N_c - 2) \times \mathbf{2}$$

which generalizes the previous example to situations without $N = 2$ supersymmetry. These three examples are simple because the magnetic theory is Abelian. However, in general the magnetic $SO(N_f - N_c + 4)$ gauge group is non-Abelian.

These $SO(N_c)$ theories also exhibit many new phenomena, which are not present in the $SU(N_c)$ examples. The most dramatic of them is oblique confinement,[25,26] driven by the condensation of dyons (particles with both electric and magnetic charges). This phenomenon is best described by another equivalent theory — a dyonic theory. Therefore, these theories exhibit electric-magnetic-dyonic triality.[24]

5. Conclusions

To conclude, supersymmetric field theories are tractable and many of their observables can be computed exactly.[c] The main dynamical lesson we learn is the role of electric-magnetic duality in non-Abelian gauge theories in four dimensions. This duality generalizes the duality in Maxwell's theory and in $N = 4$,[4] and certain $N = 2$ supersymmetric theories.[6]

The magnetic degrees of freedom are related to the underlying electric degrees of freedom in a complicated (non-local) way. They are the effective degrees of freedom useful for describing the long distance behavior of the

---

[c]Although we did not discuss them here, we would like to point out that many other examples were studied[6,7,9,15,21,23,27,28−46] exhibiting many new interesting phenomena.

theory. These variables give a weak coupling description of strong coupling phenomena such as confinement.

Our analysis led us to find new phases of non-Abelian gauge theories, like the non-Abelian Coulomb phase with its quantum equivalence and the free magnetic phase with its massless composite gauge bosons.

Outlook

We would like to end by suggesting some future directions for research and speculate about them.

1. The exploration of supersymmetric models is far from complete. There are many models whose dynamics are not yet understood. It is likely that there are new phenomena in quantum field theory which can be uncovered here.
2. Of particular importance are chiral theories whose matter content is not in a real representation of the gauge group. Only a few of these have been analyzed.[9,15,28] These theories are also interesting as they can lead to dynamical supersymmetry breaking. Finding a nice model of dynamical supersymmetry breaking which is phenomenologically acceptable is an important challenge. For some recent work in this direction see Ref. 47.
3. An important question is to what extent these results are specific to supersymmetric theories. It would be very interesting to extend at least some of these ideas to non-supersymmetric theories and to find the various phases and the mechanisms for the phase transitions without supersymmetry. One way such a study can proceed is by perturbing a supersymmetric theory whose solution is known by soft breaking terms. When these terms are small they do not affect the dynamics significantly and the non-supersymmetric perturbed theory is qualitatively similar to the unperturbed theory (for a recent attempt in this direction see Ref. 48). This makes it clear that the phenomena we found can be present in non-supersymmetric theories as well. Alternatively, one can start with ordinary non-supersymmetric QCD with a large enough number of flavors N_f, where a non-trivial fixed point of the renormalization group exists[49] and the theory is in a non-Abelian Coulomb phase. It is not yet known for which values of N_f this phase exists. It is possible that there is a dual magnetic description of this phase and perhaps even a non-Abelian free magnetic phase exists for some values of N_f. One

might be tempted to speculate that perhaps the confinement of ordinary QCD can be described as a Higgs phenomenon in these variables. Then, perhaps some of the massive particles in the spectrum of QCD (like the ρ meson or the A_1) can be identified as "Higgsed magnetic gluons."

4. The duality points at a big gap in our current understanding of gauge theories. We do not have an explicit transformation relating the underlying electric degrees of freedom to their magnetic counterparts. Finding such an explicit transformation will be extremely interesting. If such a transformation does not exist within the standard framework of local quantum field theory, perhaps a reformulation of quantum field theory will be needed.

5. The duality reflects new gauge invariance, which is not obvious in the fundamental description of the theory. In hindsight this is not surprising because gauge symmetry is not a symmetry. It is merely a redundancy in the description. Perhaps we should conclude that gauge symmetries might not be fundamental! They might only appear as long distance artifacts of our description of the theory. If so, perhaps some of the gauge symmetries of the standard model or even general relativity are similarly long distance artifacts. Then, the corresponding gauge particles are the "magnetic" degrees of freedom of more elementary "electric" variables. In order not to violate the theorem of Ref. 50, the underlying theory cannot have these symmetries as global symmetries. In particular, this can be the case for gravity only if the underlying theory is topological.

6. Recently, there has been enormous interest in dualities and non-perturbative effects in string theory.[51] Some of the phenomena found in string theory are similar to and generalize those found in field theory. It would be nice to use some of the techniques which turned out to be useful in field theory in string theory as well.

Finally we should note that there are obvious relations between these directions. For example, progress in understanding the origin of the duality can help the study of supersymmetric chiral theories and lead to phenomenological theories of supersymmetry breaking.

Acknowledgments

We would like to thank T. Banks, D. Kutasov, R. Leigh, M.R. Plesser, P. Pouliot, S. Shenker, M. Strassler and especially K. Intriligator and E. Witten for many helpful discussions. This work was supported in part by DOE grant #DE-FG05-90ER40559.

References

1. N. Seiberg, The Power of Holomorphy — Exact Results in 4D SUSY Field Theories. To appear in the *Proc. of PASCOS 94.* hep-th/9408013, RU-94-64, IASSNS-HEP-94/57.
2. E. Witten, *Comm. Math. Phys.* **117** (1988) 353.
3. E. Witten, hep-th/9411102, *Math. Res. Lett.* **1** (1994) 769.
4. C. Montonen and D. Olive, *Phys. Lett.* **72B** (1977) 117; P. Goddard, J. Nuyts and D. Olive, *Nucl. Phys.* **B125** (1977) 1.
5. H. Osborn, *Phys. Lett.* **83B** (1979) 321; A. Sen, hep-th/9402032, *Phys. Lett.* **329B** (1994) 217; C. Vafa and E. Witten, hep-th/9408074, *Nucl. Phys.* **B432** (1994) 3.
6. N. Seiberg and E. Witten, hep-th/9408099, *Nucl. Phys.* **B431** (1994) 484.
7. N. Seiberg, hep-th/9411149, *Nucl. Phys.* **B435** (1995) 129.
8. N. Seiberg, hep-ph/9309335, *Phys. Lett.* **318B** (1993) 469.
9. I. Affleck, M. Dine and N. Seiberg, *Nucl. Phys.* **B241** (1984) 493; *Nucl. Phys.* **B256** (1985) 557.
10. N. Seiberg, *Phys. Lett.* **206B** (1988) 75.
11. D. Finnell and P. Pouliot, RU-95-14, SLAC-PUB-95-6768, hep-th/9503115.
12. E. Witten, *Nucl. Phys.* **B268** (1986) 79.
13. M. Dine and N. Seiberg, *Phys. Rev. Lett.* **57** (1986) 2625.
14. M.A. Shifman and A.I. Vainshtein, *Nucl. Phys.* **B277** (1986) 456; *Nucl. Phys.* **B359** (1991) 571.
15. D. Amati, K. Konishi, Y. Meurice, G.C. Rossi and G. Veneziano, *Phys. Rep.* **162** (1988) 169 and references therein.
16. J. Polchinski and N. Seiberg, (1988) unpublished.
17. V.A. Novikov, M.A. Shifman, A. I. Vainshtein and V. I. Zakharov, *Nucl. Phys.* **B223** (1983) 445; *Nucl. Phys.* **B229** (1983) 381; *Nucl. Phys.* **B260** (1985) 157.
18. N. Seiberg, hep-th/9402044, *Phys. Rev.* **D49** (1994) 6857.
19. T.H.R. Skyrme, *Proc. Roy. Soc.* **A260** (1961) 127; E. Witten, *Nucl. Phys.* **B160** (1979) 57; *Nucl. Phys.* **B223** (1983) 422; *Nucl. Phys.* **B223** (1983) 433.
20. T. Banks, E. Rabinovici, *Nucl. Phys.* **B160** (1979) 349; E. Fradkin and S. Shenker, *Phys. Rev.* **D19** (1979) 3682.
21. K. Intriligator, R.G. Leigh and N. Seiberg, hep-th/9403198, *Phys. Rev.* **D50** (1994) 1092; K. Intriligator, hep-th/9407106, *Phys. Lett.* **336B** (1994) 409.

22. N. Seiberg and E. Witten, hep-th/9407087, *Nucl. Phys.* **B426** (1994) 19.

23. K. Intriligator and N. Seiberg, hep-th/9408155, *Nucl. Phys.* **B431** (1994) 551.

24. K. Intriligator and N. Seiberg, RU-95-3, hep-th/9503179; RU-95-40, IASSNS-HEP-95/48, to appear in the Proc. of Strings 95.

25. G. 't Hooft, *Nucl. Phys.* **B190** (1981) 455.

26. J. Cardy and E. Rabinovici, *Nucl. Phys.* **B205** (1982) 1; J. Cardy, *Nucl. Phys.* **B205** (1982) 17.

27. A. Yu. Morozov, M.A. Olshansetsky and M.A. Shifman, *Nucl. Phys.* **B304** (1988) 291.

28. K. Intriligator, N. Seiberg and S. Shenker, hep-ph/9410203, *Phys. Lett.* **342B** (1995) 152.

29. A. Klemm, W. Lerche, S. Theisen and S. Yankielowicz, hep-th/9411048, *Phys. Lett.* **344B** (1995) 169; hep-th/9412158.

30. P. Argyres and A. Faraggi, hep-th/9411057.

31. O. Aharony, hep-th/9502013, TAUP-2232-95.

32. D. Kutasov, hep-th/9503086, EFI-95-11.

33. R. Leigh and M. Strassler, RU-95-2, hep-th/9503121.

34. U. Danielsson and B. Sundborg, USITP-95-06, UUITP-4/95, hep-th/9504102.

35. M.R. Douglas and S.H. Shenker, RU-95-12, RU-95-12, hep-th/9503163.

36. S. Elitzur, A Forge, A. Giveon and E. Rabinovici, RI-4-95 hep-th/9504080.

37. O. Aharony, J. Sonnenschein and S. Yankielowicz, TAUP-2246-95, CERN-TH/95-91, hep-th/9504113.

38. D. Kutasov and A. Schwimmer, EFI-95-20, WIS/4/95, hep-th/9505004.

39. K. Intriligator and P. Pouliot, RU-95-23, hep-th/9505006.

40. K. Intriligator, RU-95-27, hep-th/9505051.

41. P.C. Argyres and M.R. Douglas, RU-95-31, hep-th/9505062.

42. M. Berkooz, RU-95-29, hep-th/9505067.

43. R. Leigh and M. Strassler, hep-th/9505088, RU-95-30.

44. A. Hanany and Y. Oz, TAUP-2248-95, WIS-95/19, hep-th/9505075.

45. P.C. Argyres, M.R. Plesser and A. Shapere, IASSNS-HEP-95/32, UK-REP/95-06, hep-th/9505100.

46. K. Intriligator, R. Leigh and M. Strassler, RU-95-38, to appear.

47. M. Dine and A.E. Nelson and Y. Shirman, hep-ph/9408384, *Phys. Rev.* **D51** (1995) 1362.

48. N. Evans, S.D.H. Hsu, M. Schwetz, hep-th/9503186, YCTP-P8-95.

49. T. Banks and A. Zaks, *Nucl. Phys.* **B196** (1982) 189.

50. S. Weinberg and E. Witten, *Phys. Lett.* **96B** (1980) 59.

51. At the current rate of progress any list of references will be enormous and will immediately become outdated.

STRING THEORY AS A UNIVERSAL LANGUAGE

A. M. Polyakov

Joseph Henry Laboratories
Princeton University
Princeton, New Jersey 08544, USA

This article, based on the Klein lecture, contains some new results and new speculations on various topics. They include discussion of open strings in the AdS space, unusual features of D-branes, conformal gauge theories in higher dimensions. We also comment on the infrared screening of the cosmological constant and on the "brane worlds"

1. Introduction

String theory is a beautiful and dangerous subject. On one hand it is a top achievement of theoretical physics exploiting the most advanced and daring methods. On the other — without a guidance from the experiment it can easily degenerate into a collection of baroque curiosities, some kind of modern alchemy looking for philosopher's stone.

This danger can be somewhat reduced if we try to study string theory in connection with some concrete physical problem and then extrapolate the gained experience to the Planck domain unreachable by experiments. This is a well established strategy in theoretical physics. For example one can learn about the Cherenkov radiation while studying supersonic aerodynamics. And usually there is the "back-reaction": the technical progress at the frontier turns out to be helpful in solving the old problems. Thus, it is conceivable that string theory will provide us with the language for the future theoretical physics.

In this lecture I will examine a number of problems in which the language of string theory is appropriate and effective. We begin with the problem of quark confinement. The task here is to find the string description of the color-electric flux lines emerging in QCD. Recently there has been a considerable progress in this field. Various aspects of it have been reviewed in Refs. 1, 2. I shall not review again these developments and instead concentrate on new results. After that we will discuss some general features of D-branes, conformal gauge theories in higher dimensions, and speculations concerning the cosmological constant.

2. The Image of Gluons and Zigzag Symmetry

As was explained in the above references, a string theory, needed to describe gauge fields in 4 dimensions, must be formulated in the 5d space with the metric

$$ds^2 = d\varphi^2 + a^2(\varphi)d\vec{x}^2 \tag{1}$$

This curved 5d space is a natural habitat for the color-electric flux lines. If apart from the pure gauge fields there are some matter fields in the theory, they must correspond to extra degrees of freedom on the world sheet. In some cases these extra degrees of freedom can be balanced so that the field theory β-function is equal to zero. These cases are the easiest ones, since the conformal symmetry of the field theory requires conformal symmetry of the string background and determines it completely:

$$a(\varphi) \sim e^{\alpha\varphi},$$

where α is some constant. After an obvious change of variables the metric takes the form

$$ds^2 = \sqrt{\lambda}y^{-2}(dy^2 + d\vec{x}^2), \tag{2}$$

where λ is related to the coupling constant of field theory. When $\lambda \gg 1$ the curvature of this 5d space is small and the 2d sigma model describing the string in the above background is weakly coupled. This greatly simplifies the analyses and we will concentrate on this case. Our aim in this section will be to demonstrate that open strings in this background have some very unusual properties, allowing to identify them with gluons.

Before starting let us recall that at present we have two possible approaches to the question of field-string correspondence. None of them is fully justified but both have certain heuristic power. In the first approach one begins with a stack of D-branes describing a gauge theory and then replace the stack by its gravitational background. In the second approach one doesn't introduce D-branes and starts directly with the sigma model action, adjusting the background so that the boundary states of this string describe the gauge theory. The key principle here is the zigzag symmetry. This is a requirement that these boundary states consist of vector gluons (and matter fields, if present) and nothing else.

The situation is very unusual. Normally we have an infinite tower of states in both open and closed string sectors. Here we need a string theory in which the closed string sector contains an infinite number of states, while the open sector has a finite number of the field-theoretic states. Our first task will be to explore how this is possible.

To set the stage, let us remember how open strings are treated in the standard case.[3] One begins with the action

$$S = \frac{1}{2} \int_D (\partial x)^2 + i \int_{\partial D} A_\mu dx_\mu, \tag{3}$$

where D is a unit disc, ∂D — its boundary, and A_μ is a vector condensate of the open string states. The possible fields A_μ are determined from the condition that the functional integral

$$Z[A] = \int Dx e^{-S} \tag{4}$$

is conformally invariant. The explicit form of this condition is derived by splitting

$$x = c + z, \tag{5}$$

where c is a slow variable, while z is fast, and integrating out z. Conformal invariance requires vanishing of the divergent counterterms and that restricts the background fields. It is convenient to integrate first the fields inside the disk with the fields at the boundary being fixed. That gives the standard boundary action

$$S_B = \frac{1}{2} \int \frac{dudv}{(u-v)^2} (x(u) - x(v))^2 + i \int A_\mu dx_\mu. \qquad (6)$$

We see that x_μ are the Gaussian fields with the correlation function (in the momentum space)

$$\langle x_\mu(p) x_\nu(-p) \rangle \sim \delta_{\mu\nu} |p|^{-1}. \qquad (7)$$

Expanding the second term in z we obtain

$$S_B \approx \frac{1}{2} \sum |p| z(p) z(-p) + \int \nabla_\lambda F_{\sigma\mu}(c) \frac{dc_\mu}{ds} z_\lambda(s) z_\sigma(s) ds. \qquad (8)$$

Using the fact that

$$\langle z_\lambda(s) z_\sigma(s) \rangle \sim \delta_{\lambda\sigma} \int^\Lambda \frac{dp}{|p|}, \qquad (9)$$

where Λ is an ultraviolet cut-off, we obtain as condition that the divergence cancel (in this approximation)

$$\nabla_\lambda F_{\lambda\mu} = 0. \qquad (10)$$

This is the on-shell condition for the massless string mode. Qualitatively the same treatment is applicable to the massive states as well. In this case one perturbs S_B with the operator

$$\Delta S_B = \int ds \Psi(x(s)) (x^2(s) h^{-2}(s))^n h(s), \qquad (11)$$

where Ψ is the scalar massive mode at the level $2n$ and $h(s)$ is the boundary metric on the world sheet needed for the general covariance of this expression. Conformal invariance of this perturbation means that the $h(s)$ dependence must cancel. The cancellation occurs between the explicit h-dependence in the above formula and the factors coming from the quantum fluctuations of $x(s)$. These factors appear because in the covariant theory the cut-off is always accompanied by the boundary metric

$$h(s)(\Delta s)^2_{\min} = a^2, \tag{12}$$

$$\Lambda^2 = \frac{1}{(\Delta s)^2_{\min}} = \frac{1}{a^2} h(s), \tag{13}$$

where a is an invariant cut-off.

In the one loop approximation (which is not, strictly speaking, applicable here, but gives a correct qualitative picture) we have

$$\partial^2 \Psi - M_n^2 \Psi = 0, \tag{14}$$

$$M_n^2 \sim n. \tag{15}$$

This is the on-shell condition for the massive string mode and it was obtained, let us stress it again, from the cancellation between the classical and quantum $h(s)$ dependence

Now we are ready to attack the AdS case. Let us consider the string action in this background

$$S = \sqrt{\lambda} \int_D \frac{(\partial x_\mu)^2 + (\partial y)^2}{y^2} + \cdots, \tag{16}$$

where we dropped all fermionic and RR terms. This is legitimate in the WKB limit $\lambda \gg 1$ which we will study in this section. To find the counterterms we must once again calculate the boundary action and $\langle z_\lambda(s) z_\mu(s) \rangle$. It is not as easy as in the previous case, but this well-defined mathematical problem was solved in Refs. 4, 5. The answer has the following form

$$S_{cl} = \sqrt{\lambda} \int ds_1 ds_2 k_{\mu\nu}(s_1, s_2) z_\mu(s_1) z_\nu(s_2). \tag{17}$$

After introducing variables $s = \frac{s_1 + s_2}{2}$ and $\sigma = s_1 - s_2$ and taking the Fourier transform with respect to σ we obtain the following asymptotic for the kernel in the mixed representation

$$\kappa_{\mu\nu}(p, s) \underset{p \to \infty}{\approx} \frac{|p|^3}{(c'(s))^2} \left[3 \frac{c'_\mu c'_\nu}{(c')^2} - \delta_{\mu\nu} \right]. \tag{18}$$

From this it follows that

$$\langle z_\mu(s)z_\nu(s)\rangle \propto \int^\infty \frac{dp}{|p|^3} < \infty. \tag{19}$$

The remarkable feature of this answer is that it implies that *there is no quantum ultraviolet divergences on the world sheet.* Hence, if we add to the action the background fields

$$\Delta S \sim \int A_\mu dx_\mu + \int \Psi(x(s))(x'(s))^2(h(s))^{-1}ds + \cdots \tag{20}$$

and treat it in the one loop approximation, we come to the following conclusions. First of all, as far as the A-term is concerned, it is finite for any $A_\mu(x)$ and thus describes the off-shell gluons. This situation is in the sharp contrast with the standard case in which conformal invariance implied the on-shell condition.

Now let us examine the massive mode (11). The only quantum dependence on the cut-off comes from

$$\langle z'_\mu(s)z'_\nu(s)\rangle \sim \int \frac{dp}{|p|} \tag{21}$$

As a result we obtain the following counterterm

$$\Delta S \sim \int \Psi(x(s))(x'(s))^2 \left(\frac{\log h(s)}{h(s)}\right) ds \tag{22}$$

We see that the only way to keep the theory conformally invariant in this approximation is to set $\Psi = 0$. There is also a possibility that at some fixed value of λ the $h(s)$ dependence will go away. However it is impossible to cancel it by a suitable on-shell condition. All this happens because, due to (19), the kinetic energy for the Ψ-term is not generated.

There is one more massless mode in the AdS string which requires a special treatment. Let us examine

$$\Delta S = \int \Phi(x(s))\partial_\perp y(s)ds, \tag{23}$$

where $\partial_\perp$ is a normal derivative at the boundary of the world sheet (which lies at infinity of the AdS space). In the more general case of $AdS_p \times S_q$

we have also a perturbation

$$\Delta S = \int \Phi^i(x) n^i(s) \partial_\perp y \, ds. \tag{24}$$

Here we must remember that the string action is finite only if[6,5]

$$(\partial_\perp y)^2 = (x')^2. \tag{25}$$

It is easy to see that when we substitute the decomposition (5) into this formula we get the logarithmic divergence (21) once again. We come to the conclusion that the above perturbation is not conformal and must not be present at the boundary. However in the case of (24) there is also a logarithmic term, coming from the fluctuations of $n^i(s)$. In the presence of space-time supersymmetry these two divergences must cancel since the masslessness of the scalar fields is protected by the SUSY. Otherwise, keeping the scalar fields massless requires a special fine tuning of the background. It would be interesting to clarify the corresponding mechanism.

Another interesting problem for the future is the fate of the open string tachyon in the AdS space. So far we assumed that it is excluded by the GSO projection. But in the purely bosonic string it may lead to some interesting effects via Sen mechanism.[7]

We come to the following conclusion concerning the spectrum of the boundary states in the AdS-like background. It consists of a few modes which would have been massless in the flat case. The infinite tower of the massive states can not reach the boundary. The above finite set of states must be associated with the fields of the field theory under consideration.

The full justification of this assertion requires the analyses of the Schwinger–Dyson equations of the Yang–Mills theory. It is still absent, and we give some heuristic arguments instead. The loop equation, expressing the Schwinger–Dyson equations in terms of loops has the form

$$\hat{L}(s) W(C) = W * W \tag{26}$$

where $W(C)$ is the Wilson loop and $\hat{L}$ is the loop laplacian and the right hand side comes from the self-intersecting contours. In recent papers[4,5] we analyzed the action of the loop Laplacian in the AdS space. It was shown that at least in the WKB approximation and in the four-dimensional space-time we have a highly non-trivial relation

$$\hat{L}(s) \sim T_{\perp\parallel}(s) \tag{27}$$

where $T_{\perp\parallel}(s)$ is a component of the world sheet energy-momentum tensor at the boundary. When substituted inside the string functional integral the energy momentum tensor receives contributions from the degenerate metrics only. These metrics describe a pinched disk, that is two discs joined at a point. The corresponding amplitude is saturated by the allowed boundary operators inserted at this point. That gives the equation (26) provided that the boundary operators of the string are the same as the fields of the field theory. Much work is still needed to make this argument completely precise.

3. The D-Brane Picture

An alternative way to understand gauge fields-strings duality is based on the D-brane approach. It is less general than the sigma model approach described above, but in the supersymmetric cases it provides us with an attractive visual picture. The logic of this method is based on the fundamental conjecture that D-branes can be described as some particular solitons in the closed string sector. One of the strongest arguments in favor of this conjecture is that both D-branes and solitons have the same symmetries and are sources of the same RR fields.[8] The gauge fields-strings dualities then follows from the D-branes — solitons duality in the limit $\alpha' \to 0$. On the D-brane side only the massless gauge field modes of the open string survive in this limit. On the string theory side we have a near horizon limit[1] of the soliton metric,[9] given by (1). These two theories must be equivalent, if the basic D-brane conjecture is correct.

The connection with the sigma model approach of the previous section follows from the following argument. First of all the closed string background is the same in both cases. As for the open strings, we placed their ends at the boundary of the AdS space where $a^2(\varphi) \to \infty$. That means that the effective slope of this strings behaves as

$$\alpha'_{\text{open}} \sim a^{-2}(\varphi) \to 0 \tag{28}$$

and thus only the massless modes are present. We said that the D-brane approach is less general, because in the non-supersymmetric cases there could exist solitons with the required boundary behavior, which are not describable by any combination of D-branes in the flat space.

This fact is related to another often overlooked subtlety. The 3-brane soliton has a metric[9]

$$ds^2 = H^{-\frac{1}{2}}(r)(dx)^2 + H^{\frac{1}{2}}(r)(dy)^2 \tag{29}$$

$$r^2 = y^2; \quad H(r) = 1 + \frac{L^4}{r^4} \tag{30}$$

It is often assumed that this metric is an extremum of the action

$$S = S_{\text{bulk}} + S_{BI} \tag{31}$$

where the first term contains the modes of the closed string, while the second is the Born–Infeld action localized on the brane. In the equations of motion the second term will give a delta function of the transverse coordinates.

Would it be the case, where the 3-brane is located? From (30) it is clear that the singularity of the metric is located in the complex domain $r^4 = -L^4$. Let us try to understand the significance of this fact from the string-theoretic point of view. Consider a string diagram describing the D-brane world volume in the arbitrary order in $\lambda = g_s N$. It is represented by a disc with an arbitrary number of holes. At each boundary one imposes the Dirichlet conditions for the transverse coordinates. An important feature of this diagram is that it is finite. This follows from the fact that the only source of divergences in string theory are tadpoles and for 3-branes their contribution is proportional to the integral $\int \frac{d^6k}{k^2}$ where k is the transverse momentum. This expression is infrared finite (which is of course very well known). Thus D-branes in the flat space are described by the well defined string amplitudes. But that contradicts the common wisdom, that one must determine the background from the action (31), because the *flat space is not a solution*, once the Born-Infeld term is added. More over if we try to deform the flat space, the above disc with holes will loose its conformal invariance.

Let us analyze this apparent paradox. It is related to the fact on a sphere conformal invariance is equivalent to the absence of tadpoles since for any $(1, 1)$ vertex operator we have $\langle V \rangle_{\text{sphere}} = 0$. However on a disc this is not true, the conformal symmetry doesn't forbid the expectation values of vertex operators. On a disc conformal symmetry and the absence of tadpoles are two different conditions. Which one should we use?

If we denote the bulk couplings by λ and the boundary couplings by μ we can construct three different objects, the bulk central charge $c(\lambda)$, the "boundary entropy"[10] $b(\lambda, \mu)$ and the effective action generating the S-matrix, $S(\lambda, \mu) = c(\lambda) + b(\lambda, \mu)$. To ensure conformal invariance we must have

$$\frac{\partial c}{\partial \lambda} = 0 \tag{32}$$

$$\frac{\partial b}{\partial \mu} = 0 \tag{33}$$

This does not coincide in general with the "no tadpole condition"

$$\frac{\partial S}{\partial \lambda} = \frac{\partial S}{\partial \mu} = 0 \tag{34}$$

In the case of 3-branes the paradox is resolved in an interesting way. The metric (30) has a horizon at $r = 0$. When we go to the Euclidean signature the horizon, as usual, shrinks to a non-singular point. As a result we have a metric which solves the equation (32) and has no trace of the D-brane singularity in it! The paradox is pushed under the horizon.

The conclusion of this discussion is as following. We have two dual and different descriptions of the D-brane amplitudes. In the first description we calculate the amplitudes of a disc with holes *in the flat space*. In this description it is simply inconsistent to introduce the background fields generated by D-branes.

In the second description we forget about the D-branes and study a non-singular closed string soliton. The D-brane conjecture implies that we must get the same answers in these two cases. The situation is analogous to the one we have in the sine-gordon theory, which admits two dual descriptions, either in terms of solitons or in terms of elementary fermions, but not both.

Let us touch briefly another consequence of these considerations. When minimizing the action (31) one can find a solution which is singular on the 3-brane and is AdS-space outside of it.[11] These solutions are known to "localize" gravitons on the brane and are the basis of the popular "brane-world" scenarios. It is clear that for the string-theoretic branes this is not a physical solution because the world volume does not contain gravity (being described by the open strings). As we argued above, there must be a horizon, not a singularity. Technically this happens because in string theory the Born-Infeld action is corrected with the Einstein term $\sim \int R\sqrt{g}d^{p+1}x$, coming from the finite thickness of the brane. It can be shown that the coefficient in front of this term (which is fully determined by string theory) is tuned so that the localization is destroyed. There are no worlds on D-branes. Of course if one compactifies the ambient transverse space, the 4d graviton reappears by the Kaluza–Klein mechanism.

4. Conformal Gauge Theories in Higher Dimensions

Although our main goal is to find a string theoretic description of the asymptotically free theories, conformal cases are not without interest. They are easier and can be used as a testing ground for the new methods. In this section we briefly discuss conformal bosonic gauge theories in various dimensions.[2] The background in these cases is just the AdS space. We have to perform the non-chiral GSO projection in order to eliminate the boundary tachyon (it would add an instability to the field theory under consideration; we do not consider here an interesting possibility that this instability resolves in some new phase).

The GSO projection in the non-critical string is slightly unusual. Let us consider first $d = 5$ (corresponding to the $d = 4$ gauge theory). In this case we have 4 standard NSR fermions ψ_μ on the world sheet and also a partner of the Liouville field ψ_5. For the former we can use the standard spin fields defined by the OPE

$$\psi_\mu \times \Sigma_A \sim (\gamma_\mu)_{AB}\Sigma_B \tag{35}$$

where we use the usual 4×4 Dirac matrices. The spinor Σ_A can be split into spinors $\Sigma_A^\pm$ with positive and negative chiralities. It is easy to check that the OPE for them have the structure

$$\Sigma^\pm \times \Sigma^\pm \sim (\psi)^{[\text{even}]} \tag{36}$$

$$\Sigma^\pm \times \Sigma^\mp \sim (\psi)^{[\text{odd}]} \tag{37}$$

The symbols on the RHS mean the products of even/odd number of the NSR fermions. Notice that this structure is the opposite to the one in 10 dimension. In order to obtain the spin operators in 5d we have to introduce the Ising order and disorder operators, σ and μ, related to ψ_5. These operators are non-holomorphic and correspond to RR-states. Their OPE have the structure

$$\sigma \times \sigma \sim (\psi_5)^{[\text{even}]} \tag{38}$$

$$\sigma \times \mu \sim (\psi_5)^{[\text{odd}]} \tag{39}$$

Using these relations we obtain the following GSO projected RR spin operator

$$\Sigma = \begin{pmatrix} \sigma \Sigma^+ \overline{\Sigma^+} & \mu \Sigma^+ \overline{\Sigma^-} \\ \mu \Sigma^- \overline{\Sigma^+} & \sigma \Sigma^- \overline{\Sigma^-} \end{pmatrix} \tag{40}$$

It has the property

$$\Sigma \times \Sigma \sim (\psi)^{[\text{even}]} \tag{41}$$

needed for the non-chiral GSO projection, which consists of dropping all operators with the odd number of fermions. Notice also that the non-chiral picture changing operator has even number of fermions. The RR matrix Σ has 16 elements. We can now write down the full string action in the AdS$_5$ space. It has the form

$$S = S_B + S_F + S_{RR} + S_{\text{ghost}} \tag{42}$$

where S_B is given by (16), S_F by

$$S_F = \int d^2\xi \left[\overline{\psi}_M \nabla \psi_M + \frac{1}{\sqrt{\lambda}} (\overline{\psi}_M \gamma_\mu \psi_N)^2 \right] \tag{43}$$

where $M = 1, \ldots 5$, and one have to use the standard spin-connection projected from AdS on the world sheet in the Dirac operator. So far we are describing the usual action of the sigma model with $N = 1$ supersymmetry on the world sheet. The unusual part is the RR term given by

$$S_{RR} = f \int d^2\xi Tr(\gamma_5 \Sigma) e^{-\frac{\phi}{2}} \tag{44}$$

Here $e^{-\frac{\phi}{2}}$ is a spin operator for the bosonic ghost[12] and f is a coupling constant (which in one loop approximation is equal to 1).

I believe that this model can be exactly solved, although it has not been done yet. A promising approach to this solution may be based on the non-abelian bosonization[13] in which the fermions ψ_M are replaced by the orthogonal matrix Ω_{MN} with the WZNW Lagrangian. In this case the RR term is simply a trace of this matrix in the spinor representation. This formalism lies in the middle between the NSR and the Green-Schwartz approaches and hopefully will be useful. Meanwhile we will have to be content with the one loop estimates which are justified in some special cases listed below and help to get a qualitative picture in general. In this approach one begins with the effective action

$$S = \int d^d x \sqrt{G} e^{\Phi} \left[\frac{d-10}{2} - R - (\nabla \Phi)^2 \right] + \int d^d x F_d^2 \sqrt{G} \qquad (45)$$

Here F_d is the RR d-form and $F_d^2 = G^{A_1 A_1'} \cdots G^{A_d A_d'} F_{A_1 \dots A_d} F_{A_1' \dots A_d'}$; the form F_d will be assumed to be proportional to the volume form. The dilaton field Φ is normalized so that $e^{\Phi} = g_s^{-2}$, where g_s is the string coupling constant. Conformal cases involve either constant curvature solutions of the equations of motion or the products of the manifolds with constant curvatures. The dilaton in these cases is also a constant. Such an ansatz is very easy to analyze. Let us begin with the single AdS$_d$ space. Consider the variation of the metric which preserves the constancy of the curvature

$$\delta G_{AB} = \varepsilon G_{AB} \qquad (46)$$

$$\delta R = \delta(G^{AB} R_{AB}) = -\varepsilon R \qquad (47)$$

$$\delta \Phi = \text{const} \qquad (48)$$

That immediately gives the relations

$$\frac{d-10}{2} - R = 0 \qquad (49)$$

$$e^{\Phi} \left(1 - \frac{d}{2} \right) R - \frac{d}{2} F_d^2 = 0 \qquad (50)$$

If the assume that the flux of the RR field is equal to N, we get $F_d^2 = N^2$. If we introduce the coupling constant $\lambda = g_s N = g_{YM}^2 N$, we obtain the background AdS solution[2] with

$$|R| \sim \lambda^2 \sim 10 - d \tag{51}$$

We can trust this solution if $d = 10 - \epsilon$, in which case the curvatures and the RR fields are small and the above one loop approximation is justified. According to the discussion in the preceding sections, this solution must describe the Yang-Mills theory, perhaps with one adjoint scalar, in the space with dimension $9 - \epsilon$. We conclude that this bosonic higher dimensional gauge theory has a conformally invariant fixed point! It may be worth mentioning that this is not a typical for non-renormalizable theories to have such fixed points. For example a nonlinear sigma model in dimension higher than two, where it is non-renormalizable, does have a conformal critical point at which the phase transition to ferromagnetic phase takes place. However it is hard to say up to what values of ϵ we can extrapolate this result.

These considerations allow for several generalizations. First of all we can consider products of spaces with constant curvatures by the same method. Take for example the space $\mathrm{AdS}_p \times S_q$ with curvatures R_1 and R_2 and $d = p + q$. To get the equations of motion in this case it is sufficient to consider the variations

$$\delta G_{ab} = \varepsilon_1 G_{ab} \tag{52}$$

$$\delta G_{ij} = \varepsilon_2 G_{ij} \tag{53}$$

where the first part refers to AdS and the second to the sphere. A simple calculation gives the equations

$$\frac{d - 10}{2} - R_1 - R_2 = 0 \tag{54}$$

$$\left(1 - \frac{2}{p}\right) R_1 + \frac{10 - d}{2} = -e^{-\Phi} F_d^2 \tag{55}$$

$$\left(1 - \frac{2}{q}\right) R_2 + \frac{10 - d}{2} = e^{-\Phi} F_d^2 \tag{56}$$

$$F_d^2 = \lambda^2 R_2^q \tag{57}$$

Here we assume that the RR flux is permeating the AdS component of space only, being given by the volume form. The last equation follows from the normalization condition of this flux and the extra factor proportional to the volume of S_q in the action. Solving this equations we get

$$R_1 = - \left(\frac{10 - p - q}{2} \right) \frac{p(q+2)}{p-q} \tag{58}$$

$$R_2 = \left(\frac{10 - p - q}{2} \right) \frac{q(p+2)}{p-q} \tag{59}$$

This solution describes bosonic gauge theories with $q + 1$ adjoint bosons; again it can be trusted if the curvatures are small. Another generalization is related to the fact that strictly speaking we must include the closed string tachyon in our considerations. It was shown in Ref. 14 that there exists an interesting mechanism for the tachyon condensation, following from its couplings to the RR fields. It is easy to include the constant tachyon field in our action and to show that it doesn't change our results in the small curvature limit. According to Ref. 14, in the critical case $d = 10$ the tachyon leads to the running coupling constants. In the non-critical case there is also a conformal option, described above, in which the tachyon condenses to a constant value.

Finally let us describe the reasons to believe that the conformal solutions can be extrapolated to non-small curvatures and thus the sigma model (42) has a conformal fixed point. The first two terms in (45) are the expansion of the sigma model central charge. When the couplings are not small, we have to replace

$$\frac{d - 10}{2} - R \Rightarrow \frac{c(R) - 10}{2} \tag{60}$$

where $c(R)$ is the central charge of the 2d sigma model with the target space having a curvature R. It decreases, according to Zamolodchikov, along the renormalization group trajectory. We can call this the second law of the renormalization group. Let us conjecture that there is also a *third law*

$$c(R) \to 0 (R \to +\infty) \tag{61}$$

$$c(R) \to \infty (R \to -\infty) \tag{62}$$

The first property follows from the fact that usually the sigma models with positive curvature develop a mass gap and thus there are no degrees of freedom contributing to the central charge. The second equation is harder to justify; we know only that $c(R)$ is increasing in the direction of negative curvature.

With this properties and with some general form of the RR terms it is possible to see that the conformal solution to the equations of motion, obtained by the variations (46), continue to exist when ϵ is not small. Of course this is not a good way to explore these solutions. Instead one must construct the conformal algebra for the sigma model action (42). This has not been done yet.

5. Infrared Screening of the Cosmological Constant and Other Speculations

In this section we will discuss some speculative approaches to the problems of vacuum energy and space-time singularities. I shall try to revive some old ideas,[15] adding some additional thoughts. The motivation to do that comes from the remarkable recent observational findings indicating that the cosmological constant is non-zero, and its scale is defined by the size of the universe (meaning the Hubble constant). These results seem very natural from the point of view advocated in Ref. 15, according to which there is an almost complete screening of the cosmological constant due to the infrared fluctuations of the gravitational field. This phenomenon is analogous to the complete screening of electric charge in quantum electrodynamics found by Landau, Abrikosov and Khalatnikov and nicknamed "Moscow zero". Here we will try to argue in favor for another "zero" of this kind — that of the cosmological constant.

Let us consider at first the Einstein action

$$S = -\int (R - 2\Lambda_0)\sqrt{g}d^4x \qquad (63)$$

Here Λ_0 is a bare cosmological constant which is assumed to be defined by the Planck scale. It is clear from the form of the action that if we consider the infrared fluctuations of the metric (with the wave length much larger then the Planck scale) their interaction will be dominated by the second term in this formula since it doesn't contain derivatives. To get some qualitative understanding of the phenomenon, let us consider conformally flat fluctuations of the metric

$$g_{\mu\nu} = \varphi^2 \delta_{\mu\nu} \tag{64}$$

The action takes the form

$$S = - \int \left[\frac{1}{2}(\partial\varphi)^2 - \Lambda_0\varphi^4 \right] d^4x \tag{65}$$

It has the well known feature of non-positivity. The way to treat it was suggested in Ref. 16 and we will accept it, although it doesn't have good physical justification. To use S in the functional integral we will simply analytically continue $\varphi \Rightarrow i\varphi$; after that the action takes the form

$$S = \int \left[\frac{1}{2}(\partial\varphi)^2 + \Lambda_0\varphi^4 \right] d^4x \tag{66}$$

The infrared fluctuations of φ are relevant and lead to the screening of Λ_0

$$\Lambda \sim \frac{1}{\log(M_{pl}L)} \tag{67}$$

where L is an infrared cut-off. More generally we could represent the metric in the form

$$g_{\mu\nu} = \varphi^2 h_{\mu\nu}; \quad \det(h_{\mu\nu}) = 1 \tag{68}$$

It is not known how to treat the unimodular part of the metric. We can only hope that it will not undo the infrared screening although can change it. Also the screening (67) with $L \sim H^{-1}$ (where H is the Hubble constant) is not strong enough to explain the fact that $\Lambda \sim H^2$. It is not impossible that the two problems cure each other. When we have several relevant degrees of freedom, the renormalization group equations governing the L dependence of various coupling constants including Λ may have a powerlike asymptotic (in contrast with (67)). Such examples exist, starting from the cases with two independent coupling constants. Thus the infrared limit of the Einstein action (perhaps with the dilaton added) may be described by a conformal field theory, giving the cosmological constant defined by the Hubble scale. The renormalization group should take us from Planck to Hubble.

Even if this fantasy is realized, we have to resolve another puzzle. It is certainly unacceptable to have a large cosmological constant in the early Universe, since it will damage the theory of nucleosynthesis. At the first glance it seems to create a serious problem for the screening theory, because when the universe is relatively small, the screening is small also and the cosmological constant is large. The way out of this problem is to conjecture that in the radiation dominated universe the infrared cut-off is provided by the curvature of space-time, while in the matter-dominated era it begins to depend on other quantities characterizing the size of the universe. If this is the case, in the early universe we get the screening law $\Lambda \sim R$ instead of $\Lambda \sim H^2$. Substituting it in the Einstein action we find that at this stage the infrared mechanism simply renormalizes the Newton constant and thus is unobservable. In the matter dominated era the effective Λ begin to depend on other things (like the Friedman warp factor a) and that can easily give the observed acceleration of the universe. Thus the change in the infrared screening must be related to a trace of the energy momentum tensor. We can say that in the correct theory the cosmological constant vanishes without the trace. To be more precise, it is disguised as a Newton constant, until the trace of the energy-momentum tensor reveals its true identity. The above picture have some remote resemblance to the scenario suggested recently in [17].

In spite of the obvious gaps in these arguments, they give a very natural way of relating the cosmological constant to the size of the universe and thus are worth developing. Perhaps the AdS/CFT correspondence will be of some use for this purpose. The main technical problem in testing these ideas is the unusual $\varphi -$ dependent kinetic energy of the h-field.

We can also notice that the above scenario can explain the dimensionality of space-time. Indeed, if this dimensionality is larger than four, the infrared effects are small, the cosmological constant large, and we end up in the universe of the Planck size, which is not much fun.

Acknowledgment

This work was supported in part by the NSF grant PHY9802484.

References

1. O. Aharony, S. Gubser, J. Maldacena, H. Ooguri, *Phys. Rep.* **323** (2000) 183.
2. A. Polyakov, *J. Mod. Phys.* **A14** (1999) 645, hep-th/9809057.
3. C. Callan *et al.*, *Nucl. Phys.* **B308** (1988) 221.
4. A. Polyakov, V. Rychkov, hep-th/0002106.
5. A. Polyakov, V. Rychkov, hep-th/0005173.
6. N. Drukker, D. Gross, H. Ooguri, *Phys. Rev.* **D60** (199) 125, hep-th/9904191.
7. A. Sen, JHEP 9912027, hep-th/9911116.
8. J. Polchinski, *Phys. Rev. Lett.* **75** (1995) 4724, hep-th/9510017.
9. G. Horowitz, A. Strominger, *Nucl. Phys.* **B360** (1991) 197.
10. I. Affleck, A. Ludwig, *Phys. Rev. Lett.* **67** (1991) 161.
11. L. Randall, R. Sundrum, *Phys. Rev. Lett.* **83** (1999) 4690.
12. D. Friedan, E. Martinec, S. Shenker, *Nucl. Phys.* **B271** (1986) 93.
13. E. Witten, *Comm. Math. Phys.* **92** (1984) 455.
14. Ig. Klebanov A. Tseytlin, *Nucl. Phys.* **B547** (1999) hep-th/9812089.
15. A. Polyakov, *Sov. Phys. Uspekhi* **25** (1982) 187.
16. G. Gibbons, S. Hawking, *Phys. Rev.* **D15** (1977) 2725.
17. C. Armendariz-Picon, V. Mukhanov, P. Steinhardt, astro-ph/0004139.

THE COSMOLOGICAL TESTS

P. J. E. Peebles

Joseph Henry Laboratories, Princeton University, Princeton, NJ 08544, USA

Recent observational advances have considerably improved the cosmological tests, adding to the lines of evidence, and showing that some issues under discussion just a few years ago may now be considered resolved or irrelevant. Other issues remain, however, and await resolution before the great program of testing the relativistic Friedmann–Lemaître model, that commenced in the 1930s, may at last be considered complete.

1. Introduction

The search for a well-founded physical cosmology is ambitious, to say the least, and the heavy reliance on philosophy not encouraging. When Einstein (1917) adopted the assumption that the universe is close to homogeneous and isotropic, it was quite contrary to the available astronomical evidence; Klein (1966) was right to seek alternatives. But the observations now strongly support Einstein's homogeneity. Here is a case where philosophy led us to an aspect of physical reality, even though we don't know how to interpret the philosophy. Other aesthetic considerations have been less successful. The steady-state cosmology and the Einstein-de Sitter relativistic cosmology are elegant, but inconsistent with the observational evidence we have now.

   The program of empirical tests of cosmology has been a productive research activity for seven decades. The results have greatly narrowed the options for a viable cosmology, and show that the relativistic Friedmann–Lemaître model passes impressively demanding checks. My survey of the state of the tests four years ago, in a Klein lecture, was organized around

Test	Einstein-de Sitter	$R^{-2} = 0$	$\Lambda = 0$	Issues for Physics	Issues for Astronomy & Astrophysics
		$\Omega = 0.2 \pm 0.15$			
1a. Dynamics on scales $\lesssim 10$ Mpc	X	✓	✓	The inverse square law; nature of the dark mass & its effective Jeans length	What's in the voids? Where and when did galaxies form? What is Ω_{eff} vs. scale?
1b. Dynamics on larger scales	X?	✓	✓		
2a. World time $t(z)$ & ages of stars & elements	X??	✓	✓	Radial displacement-redshift relation; $dl/dz = c\,dt/dz$; angular size distance $r(z)$; Liouville relation; $i \propto (1+z)^{-4}$; gravitational deflection of light	It's the distance scale, stupid. Evolution: galaxy merging & accretion; star populations & SNe; ISM & dust; AGN central engines; ionizing background; Mass structures, dust in lensing objects
2b. Redshift-angular size relation					
2c. Redshift-magnitude relation	✓	?	✓		
2d. Counts: dN/dm, dN/dz, $dN = f(m,z)dm\,dz$	X??	?	✓		
2e. Lensing by galaxies & clusters of galaxies	✓	?	✓		
3a. Baryon mass fractions in clusters of galaxies	X	✓	✓	BBNS; nature of the primeval departure from homogeneity: near scale-invariant adiabatic Gaussian fluctuations? strings? non-Gaussian &/or isocurvature?; models for nonlinear & non-gravitational processes	The structures of clusters of galaxies; evolution of cluster luminosity, baryon & mass functions; angular & redshift surveys; large-scale peculiar motions & lensing How did galaxies form in the first generations?
3b. Cluster structures $\rho(r)$ & number density at $z \ll 1$					
3c. Cluster evolution $N(\sigma, L, M_B, z)$	X	✓?	✓		
3d. $\xi_{gg}(r,z)$, ξ_{cc}, ξ_{cg}, ξ_{gp}, ξ_{pp}, $\Gamma = \Omega h$, superclusters, ...	X??	✓	✓		
3e. Redshifts of assembly of Lα clouds, galaxies, QSOs, ...					
4. Baryon density Ω_B & the origin of light elements				N_ν, m_ν, μ_ν, s/n_B; B-fields; new physics	What is the present baryon density Ω_B?
5. 3 K CBR spectrum, anisotropy, polarization				All of the above	Effects of reionization & interactions at low z

P. J. E. Peebles, June 1997   ✓ = conditional pass;   X = conditional fail

Fig. 1.   The cosmological tests — a scorecard.

Table 1. I discuss here the provenance of this table, and how it would have to be revised to fit the present situation.

The table is crowded, in part because I tried to refer to the main open issues; there were lots of them. Some are resolved, but an updated table would be even more crowded, to reflect the considerable advances in developing new lines of evidence. This greatly enlarges the checks for consistency that are the key to establishing any element of physical science. Abundant checks are particularly important here, because astronomical evidence is limited, by what Nature chooses to show us and by our natural optimism in interpreting it.

2. The Classical Tests

2.1. *The physics of the Friedmann–Lemaître model*

The second to the last column in the table summarizes elements of the physics of the relativistic Friedmann–Lemaître cosmological model we want to test.

The starting assumption follows Einstein in taking the observable universe to be close to homogeneous and isotropic. This agrees with the isotropy of the radiation backgrounds and of counts of sources observed at wavelengths ranging from radio to gamma rays. Isotropy allows a universe that is inhomogeneous but spherically symmetric about a point very close to us. I think it is not overly optimistic to consider this picture unlikely, but in any case it is subject to the cosmological tests, through its effect on the redshift-magnitude relation, for example.

It will be recalled that, if a homogeneous spacetime is described by a single metric tensor, the line element can be written in the Robertson–Walker form,

$$ds^2 = dt^2 - a(t)^2 \left(\frac{dr^2}{1 \pm r^2/R^2} + r^2(d\theta^2 + \sin^2\theta d\phi^2) \right). \qquad (1)$$

This general expression contains one constant, that measures the curvature of sections of constant world time t, and one function of t, the expansion parameter $a(t)$. Under conventional local physics the de Broglie wavelength of a freely moving particle varies as $\lambda \propto a(t)$. In effect, the expansion of the universe stretches the wavelength. The stretching of the wavelengths of observed freely propagating electromagnetic radiation is measured by the redshift, z, in terms of the ratio of the observed wavelength of a spectral feature to the wavelength measured at rest at the source,

$$1 + z = \frac{\lambda_{\text{observed}}}{\lambda_{\text{emitted}}} = \frac{a(t_{\text{observed}})}{a(t_{\text{emitted}})}. \qquad (2)$$

At small redshift the difference δt of world times at emission and detection of the light from a galaxy is relatively small, the physical distance between emitter and observer is close to $r = c\delta t$, and the rate of increase of the proper distance is

$$v = cz = H_o r, \quad H_o = \dot{a}(t_o)/a(t_o), \qquad (3)$$

evaluated at the present epoch, t_o. This is Hubble's law for the general recession of the nebulae.

Under conventional local physics Liouville's theorem applies. It says an object at redshift z with radiation surface brightness i_e, as measured by an observer at rest at the object, has observed surface brightness (integrated over all wavelengths)

$$i_o = i_e(1 + z)^{-4}. \tag{4}$$

Two powers of the expansion factor can be ascribed to aberration, one to the effect of the redshift on the energy of each photon, and one to the effect of time dilation on the rate of reception of photons. In a static "tired light" cosmology one might expect only the decreasing photon energy, which would imply

$$i_o = i_e(1 + z)^{-1}. \tag{5}$$

Measurements of surface brightnesses of galaxies as functions of redshift thus can in principle distinguish these expanding and static models (Hubble & Tolman 1935).

The same surface brightness relations apply to the $3\,\mathrm{K}$ thermal background radiation (the CBR). Under Eq. (4) (and generalized to the surface brightness per frequency interval), a thermal blackbody spectrum remains thermal, the temperature varying as $T \propto a(t)^{-1}$, when the universe is optically thin. The universe now is optically thin at the Hubble length at CBR wavelengths. Thus we can imagine the CBR was thermalized at high redshift, when the universe was hot, dense, and optically thick. Since no one has seen how to account for the thermal CBR spectrum under equation (5), the CBR is strong evidence for the expansion of the universe. But since our imaginations are limited, the check by the application of the Hubble-Tolman test to galaxy surface brightnesses is well motivated (Sandage 1992).

The constant R^{-2} and function $a(t)$ in Eq. (1) are measurable in principle, by the redshift-dependence of counts of objects, their angular sizes, and their ages relative to the present. The metric theory on which these measurements are based is testable from consistency: there are more observable functions than theoretical ones.

The more practical goal of the cosmological tests is to over-constrain the parameters in the relativistic equation for $a(t)$,

$$\left(\frac{\dot{a}}{a}\right)^2 = \frac{8}{3}\pi G\rho_t \pm \frac{1}{a^2 R^2}$$

$$= H_o^2[\Omega_m(1 + z)^3 + \Omega_\Lambda + (1 - \Omega_m - \Omega_\Lambda)(1 + z)^2]. \tag{6}$$

The total energy density, ρ_t, is the time-time part of the stress-energy tensor. The second line assumes ρ_t is a sum of low pressure matter, with energy density that varies as $\rho_m \propto a(t)^{-3} \propto (1 + z)^3$, and a nearly constant component that acts like Einstein's cosmological constant Λ. These terms, and the curvature term, are parametrized by their present contributions to the square of the expansion rate, where Hubble's constant is defined by Eq. (3).

2.2. *Applications of the tests*

When I assembled Table 1 the magnificent program of application of the redshift-magnitude relation to type Ia supernovae was just getting underway, with the somewhat mixed preliminary results entered in line 2c (Perlmutter *et al.* 1997). Now the supernovae measurements clearly point to low Ω_m (Reiss 2000 and references therein), consistent with most other results entered in the table.

The density parameter Ω_m inferred from the application of Eq. (6), as in the redshift-magnitude relation, can be compared to what is derived from the dynamics of peculiar motions of gas and stars, and from the observed growth of mass concentrations with decreasing redshift. The latter is assumed to reflect the theoretical prediction that the expanding universe is gravitationally unstable. Lines 1a and 1b show estimates of the density parameter Ω_m based on dynamical interpretations of measurements of peculiar velocities (relative to the uniform expansion of Hubble's law) on relatively small and large scales. Some of the latter indicated $\Omega_m \sim 1$. A constraint from the evolution of clustering is entered in line 3c. The overall picture was pretty clear then, and now seems well established: within the Friedmann–Lemaître model the mass that clusters with the galaxies almost certainly is well below the Einstein-de Sitter case $\Omega_m = 1$. I discuss the issue of how much mass might be in the voids between the concentrations of observed galaxies and gas clouds in Sec. 4.1.

All these ideas were under discussion, in terms we could recognize, in the 1930s. The entry in line 2e is based on the prediction of gravitational lensing. That was well known in the 1930s, but the recognition that it provides a cosmological test is more recent. The entry refers to the multiple imaging of quasars by foreground galaxies (Fukugita & Turner 1991). The straightforward reading of the evidence from this strong lensing still favors small Λ, but with broad error bars, and it does not yet seriously constrain Ω_m (Helbig 2000). Weak lensing — the distortion of galaxy images by clustered foreground mass — has been detected; the inferred surface mass densities indicate low Ω_m (Mellier *et al.* 2001), again consistent with most of the other constraints.

The other considerations in lines 1 through 4 are tighter, but the situation is not greatly different from what is indicated in the table and reviewed in Lasenby, Jones & Wilkinson (2000).

3. The Paradigm Shift to Low Mass Density

Einstein and de Sitter (1932) argued that the case $\Omega_m = 1$, $\Omega_\Lambda = 0$, is a reasonable working model: the low pressure matter term in Eq. (6) is the only component one could be sure is present, and the Λ and space curvature terms are not logically required in a relativistic expanding universe.

Now everyone agrees that this Einstein-de Sitter model is the elegant case, because it has no characteristic time to compare to the epoch at which we have come on the scene. This, with the perception that the Einstein-de Sitter model offers the most natural fit to the inflation scenario for the very early universe, led to a near consensus in the early 1990s that our universe almost certainly is Einstein-de Sitter.

The arguments make sense, but not the conclusion. For the reasons discussed in Sec. 4.1, it seems to me exceedingly difficult to reconcile $\Omega_m = 1$ with the dynamical evidence.

The paradigm has shifted, to a low density cosmologically flat universe, with

$$\Omega_m = 0.25 \pm 0.10, \quad \Omega_\Lambda = 1 - \Omega_m. \tag{7}$$

There still are useful cautionary discussions (Rowan-Robinson 2000), but the community generally has settled on these numbers. The main driver was not the dynamical evidence, but rather the observational fit to the adiabatic cold dark matter (CDM) model for structure formation (Ostriker & Steinhardt 1995).

I have mixed feelings about this. The low mass density model certainly makes sense from the point of view of dynamics (Bahcall *et al.*, 2000; Peebles, Shaya & Tully 2000; and references therein). It is a beautiful fit to the new evidence from weak lensing and the SNeIa redshift-magnitude relation. But the paradigm shift was driven by a model for structure formation, and the set of assumptions in the model must be added to the list to be checked to complete the cosmological tests. I am driven by aspects of the CDM model that make me feel uneasy, as discussed next.

4. Issues of Structure Formation

A decade ago, at least five models for the origin of galaxies and their clustered spatial distribution were under active discussion (Peebles & Silk 1990). Five years later the community had settled on the adiabatic CDM model. That was in part because simple versions of the competing models were shown to fail, and in part because the CDM model was seen to be successful enough to be worthy of close analysis. But the universe is a complicated place: it would hardly be surprising to learn that several of the processes under discussion in 1990 prove to be significant dynamical actors, maybe along with things we haven't even thought of yet. That motivated my possibly overwrought lists of issues in lines 3a to 3e in the table.

I spent a lot of time devising alternatives to the CDM model. My feeling was that such a simple picture, that we hit on so early in the search for ideas on how structure formed, could easily fail, and it would be prudent to have backups. Each of my alternatives was ruled out by the inexorable advance of the measurements, mainly of the power spectrum of fluctuations in the temperature of the CBR. The details can be traced back through Hu & Peebles (2000). The experience makes me all the more deeply impressed by the dramatic success of the CDM model in relating observationally acceptable cosmological parameters to the measured temperature fluctuation spectrum (Hu *et al.* 2000 and references therein).

This interpretation is not unique. McGaugh (2000) presents a useful though not yet completely developed alternative, that assumes there is no nonbaryonic dark matter — $\Omega_{baryon} = \Omega_m \sim 0.04$ — and assumes a modification to the gravitational inverse square law on large scales — following Milgrom (1983) — drives structure formation. But the broad success of the CDM model makes a strong case that this is a good approximation to what happened as matter and radiation decoupled at redshift $z \sim 1000$. An update of Table 1 would considerably enlarge entry 5.

If the CDM model really is the right interpretation of the CBR anisotropy it leaves considerable room for adjustment of the details. I turn now to two issues buried in the table that might motivate a critical examination of details.

4.1. *Voids*

The issue of what is in the voids defined by the concentrations of observed galaxies and gas clouds is discussed at length in Peebles (2001); here is a summary of the main points.

The familiar textbook, optically selected, galaxies are strongly clustered, leaving large regions — voids — where the number density of galaxies is well below the cosmic mean. Galaxies with low gas content and little evidence of ongoing star formation prefer dense regions; gas-rich galaxies prefer the lower ambient density near the edges of voids. This is the morphology-density correlation.

The CDM model predicts that the morphology-density correlation extends to the voids, where the morphological mix swings to favor dark galaxies. But the observations require this swing to be close to discontinuous. A substantial astronomical literature documents the tendency of galaxies of all known types — dwarfs, irregulars, star forming, low surface brightness, and purely gaseous — to avoid the same void regions. There are some galaxies in voids, but they are not all that unusual, apart from the tendency for greater gas content.

The natural interpretation of these phenomena is that gravity has emptied the voids of most galaxies of all types, and with them drained away most of the low pressure mass. This is is not allowed in the Einstein-de Sitter model. If the mass corresponding to $\Omega_m = 1$ were clustered with the galaxies the gravitational accelerations would be expected to produce peculiar velocities well in excess of what is observed. That is, if $\Omega_m = 1$ most of the mass would have to be in the voids, and the morphology-density correlation would have to include the curious discontinuous swing to a mix dominated by dark galaxies in voids. That is why I put so much weight on the dynamical evidence for low Ω_m.

At $\Omega_m = 0.25 \pm 0.10$ [Eq. (7)] the mass fraction in voids can be as small as the galaxy fraction. That would neatly remove the discontinuity. But gravity does not empty the voids in numerical simulations of the low density CDM model. In the simulations massive dark mass halos that seem to be suitable homes for ordinary optically selected galaxies form in concentrations. This is good. But spreading away from these concentrations are dark mass halos that are too small for ordinary galaxies, but seem to be capable of developing into dwarfs or irregulars. This is contrary to the observations.

The consensus in the theoretical community is that the predicted dark mass clumps in the voids need not be a problem, because we don't know how galaxies form, how to make the connection between dark mass halos in a simulation and galaxies in the real world. The point is valid, but we have some guidance, from what is observed. Here is an example.

The Local Group of galaxies contains two large spirals, our Milky Way and the somewhat more massive Andromeda Nebula. There many smaller galaxies, most tightly clustered around the two spirals. But some half dozen irregular galaxies, similar to the Magellanic clouds, are on the outskirts of the group. These irregulars have small velocities relative to the Local Group. Since they are not near either of the large galaxies they are not likely to have been spawned by tidal tails or other nonlinear process. Since they are at ambient densities close to the cosmic mean their first substantial star populations would have formed under conditions not very different from the voids at the same epoch. In short, these objects seem to prove by their existence that observable galaxies can form under conditions similar to the voids in CDM simulations. Why are such galaxies so rare in the voids?

4.2. *The epoch of galaxy formation*

Numerical simulations of the CDM model indicate galaxies were assembled relatively recently, at redshift $z \sim 1$ (eg. Cen & Ostriker 2000). For definiteness in explaining what bothers me about this I adopt the density parameter in Eq. (7) and Hubble parameter $H_o = 70 \, \mathrm{km \, s^{-1} \, Mpc^{-1}}$.

The mass in the central luminous parts of a spiral galaxy is dominated by stars. The outer parts are thought to be dominated by nonbaryonic dark matter. The circular velocity v_c of a particle gravitationally bound in a circular orbit in the galaxy varies only slowly with the radius of the orbit, and there is not a pronounced change in v_c at the transition between the luminous inner part and the dark outer part. The value of the mean mass density $p(< r)$ averaged within a sphere of radius r centered on the galaxy relative to the cosmic mean mass density, $\bar{\rho}$, is

$$\frac{\rho(<r)}{\bar{\rho}} = \frac{2}{\Omega_m}\left(\frac{v_c}{H_o r}\right)^2 \sim 3 \times 10^5, \quad \text{at } r = 15\,\text{kpc}. \tag{8}$$

At this radius the mass of the typical spiral is thought to be dominated by nonbaryonic dark matter. Why is the dark mass density so large? Options are that

1. at formation the dark mass collapsed by a large factor,
2. massive halos formed by the merging of smaller dense clumps, that formed at high redshift, when the mean mass density was large, or
3. massive halos themselves were assembled at high redshift.

   To avoid confusion let us pause to consider the distinction between interpretations of large density contrasts in the luminous baryonic central regions and in the dark halo of a galaxy. If the baryons and dark matter were well mixed at high redshift, the baryon-dominated central parts of the galaxies would have to have been the result of settling of the baryons relative to the dark matter. Gneddin, Norman, and Ostriker (2000) present a numerical simulation that demonstrates dissipative settling of the baryons to satisfactory stellar bulges. The result is attractive — and hardly surprising since gaseous baryons tend to dissipatively settle — but does not address the issue at hand: how did the dark matter halos that are thought to be made of dissipationless matter get to be so dense?

   We have one guide from the great clusters of galaxies. The cluster mass is thought to be dominated by nonbaryonic matter. A typical line-of-sight velocity dispersion is $\sigma = 750\,\text{km s}^{-1}$. The mean mass density averaged within the Abell radius, $r_A = 2\,\text{Mpc}$, relative to the cosmic mean, is

$$\frac{\rho(<r_A)}{\bar{\rho}} = \frac{4}{\Omega_m}\left(\frac{\sigma}{H_o r_A}\right)^2 \sim 300. \tag{9}$$

Clusters tend to be clumpy at the Abell radius, apparently still relaxing to statistical equilibrium after the last major mergers, but they are thought to be close to dynamic equilibrium, gravity balanced by streaming motions of the galaxies and mass. This argues against the first of the above ideas: here are dark matter concentrations that have relaxed to dynamical support at density contrast well below the dark halo of a galaxy [Eq. (8)].

The second idea to consider is that a dense dark matter halo is assembled at low redshift by the merging of a collection of dense lower mass halos that formed earlier. This is what happens in numerical simulations of the CDM model. Sometimes cited as an example in Nature is the projected merging of the two Local Group spirals, the Milky Way and the Andromeda Nebula. They are 750 kpc apart, and moving together at 100 km s^{-1}. If they moved to a direct hit they would merge in another Hubble time. But that would require either wonderfully close to radial motion or dynamical drag sufficient to eliminate the relative orbital angular momentum. Orbit computations indicate masses that would be contained in $\rho \propto r^{-2}$ dark halos truncated at $r \sim 200$ kpc, which seems small for dissipation of the orbital angular momentum. The computations suggest the transverse relative velocity is comparable to the radial component (Peebles, Shaya & Tully 2000). That would say the next perigalacticon will be at a separation $\sim$300 kpc, not favorable for merging. Thus I suspect the Local Group spirals will remain distinct elements of the galaxy clustering hierarchy well beyond one present Hubble time. If the Local Group is gravitationally bound and remains isolated the two spirals must eventually merge, but not on the time scale of the late galaxy formation picture.

My doubts are reenforced by the failure to observe precursors of galaxies. Galaxy spheroids — elliptical galaxies and the bulges in spirals that look like ellipticals — are dominated by old stars. Thus it is thought that if present-day spheroids were assembled at $z \sim 1$ it would have been by the merging of star clusters. These star clusters might have been observable at redshift $z > 1$, as a strongly clustered population, but they are not.

That leaves the third idea, early galaxy assembly. I am not aware of any conflict with what is observed at $z < 1$. The observations of what happened at higher redshift are rich, growing, and under debate.

5. Concluding Remarks

If the models for cosmology and structure formation on which Table 1 is based could be taken as given, the only uncertainties being the astronomy, the constraints on the cosmological parameters would be clear. The long list of evidence for $\Omega_m = 0.25 \pm 0.10$, in the table and the new results from the SNeIa redshift-magnitude relation and weak lensing, abundantly demonstrates that we live in a low density universe. The CBR demonstrates space sections are flat. Since Ω_m is small there has to be a term in the stress-energy tensor that acts like Einstein's cosmological constant. The SNeIa redshift-magnitude result favors a low density flat universe over low density with open geometry ($\Lambda = 0$), at about three standard deviations. That alone is not compelling, considering the hazards of astronomy, but it is an impressive check of what the CBR anisotropy says.

But we should remember that all this depends on models we are supposed to be testing. The dynamical estimates of Ω_m in lines 1a and 1b assume the inverse square law of gravity. That is appropriate, because it follows from the relativistic cosmology we are testing. We have a check on this aspect of the theory, from consistency with other observations whose theoretical interpretations depend on Ω_m in other ways. The CDM model fitted to observed large-scale structure requires a value of Ω_m that agrees with dynamics. This elegant result was an early driver for the adoption of the low density CDM model. But we can't use it as evidence for both the CDM model and the inverse square law; we must turn to other measures. We have two beautiful new results, from weak lensing and the redshift-magnitude relation, that agree with $\Omega_m \sim 0.25$. The latter does not exclude $\Omega_m = \Omega_{baryons} \sim 0.04$; maybe MOND accounts for flat $v_c(r)$ but does not affect equation (6) (McGaugh 2000). And if we modified local Newtonian dynamics we might want to modify the physics of the gravitational deflection of light.

There are alternative fits to the CBR anisotropy, with new physics (McGaugh 2000), or conventional physics and an arguably desperate model for early structure formation (Peebles, Seager & Hu 2000). They certainly look a lot less elegant than conventional general relativity theory with the CDM model, but we've changed our ideas of elegance before.

In Sec. 4, I reviewed two issues in structure formation that I think challenge the CDM model. They may in fact only illustrate the difficulty of interpreting observations of complex systems. It's just possible that they will lead us to some radical adjustment of the models for structure formation and/or cosmology. I don't give much weight to this, because it would mean the model led us to the right Ω_m for the wrong reason. Relatively fine adjustments are easier to imagine, of course. With them we must be prepared for fine adjustments of the constraints on parameters such as Λ.

This is quite a tangled web. Progress in applying the many tests, including the mapping the CBR temperature and polarization, will be followed with close attention.

We have an impressive case for the Friedmann–Lemaître cosmology, from the successful fit to the CBR anisotropy and the consistency of the evidence for $\Omega_m \sim 0.25$ from a broad range of physics and astronomy. But the cosmological tests certainly are not complete and unambiguous, and since they depend on astronomy the program is not likely to be closed by one critical measurement. Instead, we should expect a continued heavy accumulation of evidence, whose weight will at last unambiguously compel acceptance. We are seeing the accumulation; we all look forward to the outcome.

Acknowledgment

This work was supported in part by the US National Science Foundation.

References

1. Bahcall, N. A., Cen, R., Davé, R., Ostriker, J. P. & Yu, Q. 2000, astroph/0002310.
2. Cen, R. & Ostriker, J. P. 2000, *ApJ* **538**, 83
3. Einstein, A. 1917, S.-B. Preuss. *Akad. Wiss.* **142**
4. Einstein, A. & de Sitter, W. 1932, *Proc NAS* **18**, 213
5. Fukugita, M. & Turner, E. L. 1991, *MNRAS* **253**, 99
6. Gneddin, N. Y., Norman, M. L. & Ostriker, J. P. 2000, astro-ph/9912563
7. Helbig, P. 2000, astro-ph/0011031
8. Hu, W., Fukugita, M., Zaldarriaga, M., & Tegmark, M. 2000, astro-ph/0006436
9. Hu, W. & Peebles, P. J. E. 2000, *ApJ* **528**, L61
10. Hubble, E. & Tolman, R. C. 1935, *ApJ* **82**, 302
11. Klein, O. 1966, *Nature* **211**, 1337
12. Lasenby, A., Jones, A. W. & Wilkinson, A. 2000, New Cosmological Data and the Values of the Cosmological Parameters, IAU Symposium 201
13. McGaugh, S. S. 2000, *ApJ* **541**, L33
14. Mellier *et al.* 2001, astro-ph/0101130
15. Milgrom, M. 1983, *ApJ* **270**, 371
16. Ostriker, J. P. & Steinhardt, P. J. 1995, *Nature* **377**, 600
17. Peebles, P. J. E. 2001, astro-ph/0101127

18. Peebles, P. J. E., Seager, S. & Hu, W. 2000, *ApJ* **539**, L1
19. Peebles, P. J. E., Shaya, E. J. & Tully, R. B. 2000, astro-ph/0010480
20. Peebles, P. J. E. & Silk, J. 1990, *Nature* **346**, 233
21. Perlmutter, S. *et al.* 1997, *ApJ* **483**, 565
22. Reiss, A. G. 2000, *PASP* **112**, 1248
23. Rowan-Robinson, M. 2000, astro-ph/0012026
24. Sandage, A. R. 1992, *Physica Scripta* **T43**, 22

ANTI-DE SITTER SPACE, THERMAL PHASE TRANSITION, AND CONFINEMENT IN GAUGE THEORIES

Edward Witten*

School of Natural Sciences, Institute for Advanced Study
Olden Lane, Princeton, NJ 08540, USA

The correspondence between supergravity (and string theory) on *AdS* space and boundary conformal field theory relates the thermodynamics of $\mathcal{N} = 4$ super Yang-Mills theory in four dimensions to the thermodynamics of Schwarzschild black holes in Anti-de Sitter space. In this description, quantum phenomena such as the spontaneous breaking of the center of the gauge group, magnetic confinement, and the mass gap are coded in classical geometry. The correspondence makes it manifest that the entropy of a very large *AdS* Schwarzschild black hole must scale "holographically" with the volume of its horizon. By similar methods, one can also make a speculative proposal for the description of large N gauge theories in four dimensions without supersymmetry.

1. Introduction

Understanding the large N behavior of gauge theories in four dimensions is a classic and important problem.[1] The structure of the "planar diagrams" that dominate the large N limit gave the first clue that this problem might be solved by interpreting four-dimensional large N gauge theory as a string theory. Attempts in this direction have led to many insights relevant to critical string theory; for an account of the status, see Ref. 2.

Recently, motivated by studies of interactions of branes with external probes,[3-7] and near-extremal brane geometry,[8,10] a concrete proposal in this vein has been made,[11] in the context of certain conformally-invariant

---

*Originally published in *Adv. Theor. Math. Phys.* **2** (1998) 505. Permission for use granted by International Press.

theories such as $\mathcal{N} = 4$ super Yang-Mills theory in four dimensions. The proposal relates supergravity on anti-de Sitter or AdS space (or actually on AdS times a compact manifold) to conformal field theory on the boundary, and thus potentially introduces into the study of conformal field theory the whole vast subject of AdS compactification of supergravity (for a classic review see Ref. 12; see also Ref. 13 for an extensive list of references relevant to current developments). Possible relations of a theory on AdS space to a theory on the boundary have been explored for a long time, both in the abstract (for example, see Ref. 14), and in the context of supergravity and brane theory (for example, see Ref. 15).

More complete references relevant to current developments can be found in papers already cited and in many of the other important recent papers[16-48] in which many aspects of the CF T/AdS correspondence have been extended and better understood.

In,[29,49] a precise recipe was presented for computing CFT observables in terms of AdS space. It will be used in the present paper to study in detail a certain problem in gauge theory dynamics. The problem in question, already discussed in part in Section 3.2 of Ref. 49, is to understand the high temperature behavior of $\mathcal{N} = 4$ super Yang-Mills theory. As we will see, in this theory, the CFT/AdS correspondence implies, in the infinite volume limit, many expected but subtle quantum properties, including a non-zero expectation value for a temporal Wilson loop,[50,51] an area law for spatial Wilson loops, and a mass gap. (The study of Wilson loops is based on a formalism that was introduced recently.)[39,40] These expectations are perhaps more familiar for ordinary four-dimensional Yang-Mills theory without supersymmetry — for a review see Ref. 52. But the incorporation of supersymmetry, even $\mathcal{N} = 4$ supersymmetry, is not expected to affect these particular issues, since non-zero temperature breaks supersymmetry explicitly and makes it possible for the spin zero and spin one-half fields to get mass,[a] very plausibly reducing the high temperature behavior to that of the pure gauge theory. The ability to recover from the CFT/AdS correspondence relatively subtle dynamical properties of high temperature

---

[a]The thermal ensemble on a spatial manifold $\mathbf{R}^3$ can be described by path integrals on $\mathbf{R}^3 \times \mathbf{S}^1$, with a radius for the $\mathbf{S}^1$ equal to $\beta = T^{-1}$, with T the temperature. The fermions obey antiperiodic boundary conditions around the $\mathbf{S}^1$ direction, and so get masses of order $1/T$ at tree level. The spin zero bosons get mass at the one-loop level.

gauge theories, in a situation not governed by supersymmetry or conformal invariance, certainly illustrates the power of this correspondence.

In Section 2, we review the relevant questions about gauge theories and the framework in which we will work, and develop a few necessary properties of the Schwarzschild black hole on *AdS* space. The CFT/*AdS* correspondence implies readily that in the limit of large mass, a Schwarzschild black hole in *AdS* space has an entropy proportional to the volume of the horizon, in agreement with the classic result of Bekenstein[57] and Hawking.[58] (The comparison of horizon volume of the *AdS* Schwarzschild solution to field theory entropy was first made, in the *AdS*$_5$ case, in Ref. 3, using a somewhat different language. As in some other string-theoretic studies of Schwarzschild black holes,[53,54] and some earlier studies of BPS-saturated black holes,[55] but unlike some microscopic studies of BPS black holes,[56] in our discussion we are not able to determine the constant of proportionality between area and entropy.) This way of understanding black hole entropy is in keeping with the notion of "holography".[59-61] The result holds for black holes with Schwarzschild radius much greater than the radius of curvature of the *AdS* space, and so does not immediately imply the corresponding result for Schwarzschild black holes in Minkowski space.

In Section 3, we demonstrate, on the basis of the CFT/*AdS* correspondence, that the $\mathcal{N} = 4$ theory at nonzero temperature has the claimed properties, especially the breaking of the center of the gauge group, magnetic confinement, and the mass gap.

In Section 4, we present, using similar ideas, a proposal for studying ordinary large N gauge theory in four dimensions (without supersymmetry or matter fields) via string theory. In this proposal, we can exhibit confinement and the mass gap, precisely by the same arguments used in Section 3, along with the expected large N scaling, but we are not able to effectively compute hadron masses or show that the model is asymptotically free.

2. High Temperatures and *AdS* Black Holes

2.1. $\mathbf{R}^3$ *and* $\mathbf{S}^3$

We will study the $\mathcal{N} = 4$ theory at finite temperature on a spatial manifold $\mathbf{S}^3$ or $\mathbf{R}^3$. $\mathbf{R}^3$ will be obtained by taking an infinite volume limit starting with $\mathbf{S}^3$.

To study the theory at finite temperature on $\mathbf{S}^3$, we must compute the partition function on $\mathbf{S}^3 \times \mathbf{S}^1$ — with supersymmetry-breaking boundary conditions in the $\mathbf{S}^1$ directions. We denote the circumferences of $\mathbf{S}^1$ and $\mathbf{S}^3$ as β and β', respectively. By conformal invariance, only the ratio β/β' matters. To study the finite temperature theory on $\mathbf{R}^3$, we take the large β' limit, reducing to $\mathbf{R}^3 \times \mathbf{S}^1$.

Once we are on $\mathbf{R}^3 \times \mathbf{S}^1$, with circumference β for $\mathbf{S}^1$, the value of β can be scaled out via conformal invariance. Thus, the $\mathcal{N} = 4$ theory on $\mathbf{R}^3$ cannot have a phase transition at any nonzero temperature. Even if one breaks conformal invariance by formulating the theory on $\mathbf{S}^3$ with some circumference β', there can be no phase transition as a function of temperature, since theories with finitely many local fields have in general no phase transitions as a function of temperature.

However, in the large N limit, it is possible to have phase transitions even in finite volume.[62] In Section 3.2 of Ref. 49, it was shown that the $\mathcal{N} = 4$ theory on $\mathbf{S}^3 \times \mathbf{S}^1$ has in the large N limit a phase transition as a function of β/β'. The large β/β' phase has some properties in common with the usual large β (or small temperature) phase of confining gauge theories, while the small β/β' phase is analogous to a deconfining phase.

When we go to $\mathbf{R}^3 \times \mathbf{S}^1$ by taking $\beta' \to \infty$ for fixed β, we get $\beta/\beta' \to 0$. So the unique nonzero temperature phase of the $\mathcal{N} = 4$ theory on $\mathbf{R}^3$ is on the high temperature side of the phase transition and should be compared to the deconfining phase of gauge theories. Making this comparison will be the primary goal of Section 3. We will also make some remarks in Section 3 comparing the low temperature phase on $\mathbf{S}^3 \times \mathbf{S}^1$ to the confining phase of ordinary gauge theories. Here one can make some suggestive observations, but the scope is limited because in the particular $\mathcal{N} = 4$ gauge theory under investigation, the low temperature phase on $\mathbf{S}^3$ arises only in finite volume, while most of the deep questions of statistical mechanics and quantum dynamics refer to the infinite volume limit.

2.2. *Review of gauge theories*

We will now review the relevant expectations concerning finite temperature gauge theories in four dimensions.

Deconfinement at high temperatures can be usefully described, in a certain sense, in terms of spontaneous breaking of the center of the gauge group (or more precisely of the subgroup of the center under which all fields transform trivially). For our purposes, the gauge group will be $G = SU(N)$, and the center is $\Gamma = \mathbf{Z}_N$; it acts trivially on all fields, making possible the following standard construction.

Consider $SU(N)$ gauge theory on $Y \times \mathbf{S}^1$, with Y any spatial manifold. A conventional gauge transformation is specified by the choice of a map $g : Y \times \mathbf{S}^1 \to G$ which we write explicitly as $g(y, z)$, with y and z denoting respectively points in Y and in $\mathbf{S}^1$. (In describing a gauge transformation in this way, we are assuming that the G-bundle has been trivialized at least locally along Y; global properties along Y are irrelevant in the present discussion.) Such a map has $g(y, z + \beta) = g(y, z)$. However, as all fields transform trivially under the center of G, we can more generally consider gauge transformations by gauge functions $g(y, z)$ that obey

$$g(y, z + \beta) = g(y, z)h, \tag{1}$$

with h an arbitrary element of the center. Let us call the group of such extended gauge transformations (with arbitrary dependence on z and y and any h) $\overline{G}$ and the group of ordinary gauge transformations (with $h = 1$ but otherwise unrestricted) G'. The quotient $\overline{G}/G'$ is isomorphic to the center Γ of G, and we will denote it simply as Γ. Factoring out G' is natural because it acts trivially on all local observables and physical states (for physical states, G'-invariance is the statement of Gauss's law), while Γ can act nontrivially on such observables.

An order parameter for spontaneous breaking of Γ is the expectation value of a temporal Wilson line. Thus, let C be any oriented closed path of the form $y \times \mathbf{S}^1$ (with again y a fixed point in Y), and consider the operator

$$W(C) = \operatorname{Tr} P \exp \int_C A, \tag{2}$$

with A the gauge field and the trace taken in the N-dimensional fundamental representation of $SU(N)$. Consider a generalized gauge transformation of the form (1), with h an Nth root of unity representing an element of the center of $SU(N)$. Action by such a gauge transformation multiplies the holonomy of A around C by h, so one has

$$W(C) \to hW(C). \tag{3}$$

Hence, the expectation value $\langle W(C) \rangle$ is an order parameter for the spontaneous breaking of the Γ symmetry.

Of course, such spontaneous symmetry breaking will not occur (for finite N) in finite volume. But a nonzero expectation value $\langle W(C) \rangle$ in the infinite volume limit, that is with Y replaced by $\mathbf{R}^3$, is an important order parameter for deconfinement. Including the Wilson line $W(C)$ in the system means including an external static quark (in the fundamental representation of $SU(N)$), so an expectation value for $W(C)$ means intuitively that the cost in free energy of perturbing the system by such an external charge is finite. In a confining phase, this free energy cost is infinite and $\langle W(C) \rangle = 0$. The $\mathcal{N} = 4$ theory on $\mathbf{R}^3$ corresponds to a high temperature or deconfining phase; we will confirm in Section 3, using the CFT/*AdS* correspondence, that it has spontaneous breaking of the center.

Other important questions arise if we take the infinite volume limit, replacing X by $\mathbf{R}^3$. The theory at long distances along $\mathbf{R}^3$ is expected to behave like a pure $SU(N)$ gauge theory in three dimensions. At nonzero temperature, at least for weak coupling, the fermions get a mass at tree level from the thermal boundary conditions in the $\mathbf{S}^1$ direction, and the scalars (those present in four dimensions, as well as an extra scalar that arises from the component of the gauge field in the $\mathbf{S}^1$ direction) get a mass at one loop level; so the long distance dynamics is very plausibly that of three-dimensional gauge fields. The main expected features of three-dimensional pure Yang-Mills theory are confinement and a mass gap. The mass gap means simply that correlation functions $\langle \mathcal{O}(y, z)\mathcal{O}'(y', z) \rangle$ vanish exponentially for $|y - y'| \to \infty$. Confinement is expected to show up in an area law for the expectation value of a spatial Wilson loop. The area law means the following. Let C be now an oriented closed loop encircling an area A in $\mathbf{R}^3$, at a fixed point on $\mathbf{S}^1$. The area law means that if C is scaled up, keeping its shape fixed and increasing A, then the expectation value of $W(C)$ vanishes exponentially with A.

"Confinement" In Finite Volume

Finally, there is one more issue that we will address here. In the large N limit, a criterion for confinement is whether (after subtracting a constant from the ground state energy) the free energy is of order one — reflecting the contributions of color singlet hadrons — or of order N^2 — reflecting the contributions of gluons. (This criterion has been discussed in Ref. 63.) In Ref. 49, it was shown that in the $\mathcal{N} = 4$ theory on $\mathbf{S}^3 \times \mathbf{S}^1$, the large N theory has a low temperature phase with a free energy of order $1 - a$ "confining" phase — and a high temperature phase with a free energy of order N^2 — an "unconfining" phase.

Unconfinement at high temperatures comes as no surprise, of course, in this theory, and since the theory on $\mathbf{R}^3 \times \mathbf{S}^1$, at any temperature, is in the high temperature phase, we recover the expected result that the infinite volume theory is not confining. However, it seems strange that the finite volume theory on $\mathbf{S}^3$, at low temperatures, is "confining" according to this particular criterion.

This, however, is a general property of large N gauge theories on $\mathbf{S}^3$, at least for weak coupling (and presumably also for strong coupling). On a round three-sphere, the classical solution of lowest energy is unique up to gauge transformations (flat directions in the scalar potential are eliminated by the $R\phi^2$ coupling to scalars, R being the Ricci scalar), and is given by setting the gauge field A, fermions ψ, and scalars ϕ all to zero. This configuration is invariant under global $SU(N)$ gauge transformations. The Gauss law constraint in finite volume says that physical states must be invariant under the global $SU(N)$. There are no zero modes for any fields (for scalars this statement depends on the $R\phi^2$ coupling). Low-lying excitations are obtained by acting on the ground state with a finite number of A, ψ, and ϕ creation operators, and then imposing the constraint of global $SU(N)$ gauge invariance. The creation operators all transform in the adjoint representation, and so are represented in color space by matrices $M_1, M_2, \ldots, M_s$. $SU(N)$ invariants are constructed as traces, say $\mathrm{Tr}\ M_1 M_2 \ldots M_s$. The number of such traces is given by the number of ways to order the factors and is independent of N. So the multiplicity of low energy states is independent of N, as is therefore the low temperature free energy.

This result, in particular, is kinematic, and has nothing to do with confinement. To see confinement from the N dependence of the free energy, we must go to infinite volume. On $\mathbf{R}^3$, the Gauss law constraint does not say that the physical states are invariant under global $SU(N)$ transformations, but only that their global charge is related to the electric field at spatial infinity. If the free energy on $\mathbf{R}^3$ is of order 1 (and not proportional to N^2), this actually is an order parameter for confinement.

Now let us go back to finite volume and consider the behavior at high temperatures. At high temperatures, one cannot effectively compute the free energy by counting elementary excitations. It is more efficient to work in the "crossed channel." In $\mathbf{S}^3 \times \mathbf{S}^1$ with circumferences β' and β, if we take $\beta' \to \infty$ with fixed β, the free energy is proportional to the volume of $\mathbf{S}^3$ times the ground state energy density of the $2 + 1$-dimensional theory that is obtained by compactification on $\mathbf{S}^1$ (with circumference β and supersymmetry-breaking boundary conditions). That free energy is of order N^2 (the supersymmetry breaking spoils the cancellation between bosons and fermions already at the one-loop level, and the one-loop contribution is proportional to N^2). The volume of $\mathbf{S}^3$ is of order $(\beta')^3$. So the free energy on $\mathbf{S}^3 \times \mathbf{S}^1$ scales as $N^2(\beta')^3$ if one takes $\beta' \to \infty$ at fixed β, or in other words as $N^2\beta^{-3}$ if one takes $\beta \to 0$ at fixed β'. Presently we will recover this dependence on β by comparing to black holes.

2.3. *AdS correspondence and Schwarzschild black holes*

The version of the CFT/*AdS* correspondence that we will use asserts that conformal field theory on an n-manifold M is to be studied by summing over contributions of Einstein manifolds B of dimension $n + 1$ which (in a sense explained in Refs. 29, 49) have M at infinity.

We will be mainly interested in the case that $M = \mathbf{S}^{n-1} \times \mathbf{S}^1$, or $\mathbf{R}^{n-1} \times \mathbf{S}^1$. For $\mathbf{S}^{n-1} \times \mathbf{S}^1$, there are two known B's, identified by Hawking and Page[64] in the context of quantum gravity on *AdS* space. One manifold, X_1, is the quotient of *AdS* space by a subgroup of $SO(1, n + 1)$ that is isomorphic to $\mathbf{Z}$. The metric (with Euclidean signature) can be written

$$ds^2 = \left(\frac{r^2}{b^2} + 1\right) dt^2 + \frac{dr^2}{\left(\frac{r^2}{b^2}\right) + 1} + r^2 d\Omega^2, \tag{4}$$

with $d\Omega^2$ the metric of a round sphere $\mathbf{S}^{n-1}$ of unit radius. Here t is a periodic variable of arbitrary period. We have normalized (4) so that the Einstein equations read

$$R_{ij} = -nb^{-2} g_{ij}; \tag{5}$$

here b is the radius of curvature of the anti-de Sitter space. With this choice, n does not appear explicitly in the metric. This manifold can contribute to either the standard thermal ensemble $\mathrm{Tr}e^{-\beta H}$ or to $\mathrm{Tr}(-1)^F e^{-\beta H}$, depending on the boundary conditions one uses for fermions in the t direction. The topology of X_1 is $\mathbf{R}^n \times \mathbf{S}^1$, or $\mathbf{B}^n \times \mathbf{S}^1$ ($\mathbf{B}^n$ denoting an n-ball) if we compactify it by including the boundary points at $r = \infty$.

The second solution, X_2, is the Schwarzschild black hole, in AdS space. The metric is

$$ds^2 = \left(\frac{r^2}{b^2} + 1 - \frac{w_n M}{r^{n-2}}\right) dt^2 + \frac{dr^2}{\left(\frac{r^2}{b^2} + 1 - \frac{w_n M}{r^{n-2}}\right)} + r^2 d\Omega^2. \quad (6)$$

Here w_n is the constant

$$w_n = \frac{16\pi G_N}{(n-1)\mathrm{Vol}(\mathbf{S}^{n-1})}. \quad (7)$$

Here G_N is the $n+1$-dimensional Newton's constant and $\mathrm{Vol}(\mathbf{S}^{n-1})$ is the volume of a unit $n-1$-sphere; the factor w_n is included so that M is the mass of the black hole (as we will compute later). Also, the spacetime is restricted to the region $r \geq r_+$, with r_+ the largest solution of the equation

$$\frac{r^2}{b^2} + 1 - \frac{w_n M}{r^{n-2}} = 0. \quad (8)$$

The metric (6) is smooth and complete if and only if the period of t is

$$\beta_0 = \frac{4\pi b^2 r_+}{nr_+^2 + (n-2)b^2}. \quad (9)$$

For future use, note that in the limit of large M one has

$$\beta_0 \sim \frac{4\pi b^2}{n(w_n b^2)^{1/n} M^{1/n}}. \quad (10)$$

As in the $n = 3$ case considered in Ref. 64, β_0 has a maximum as a function of r_+, so the Schwarzschild black hole only contributes to the thermodynamics if β is small enough, that is if the temperature is high enough. Moreover, X_2 makes the dominant contribution at sufficiently high temperature, while X_1 dominates at low temperature. The topology of X_2 is $\mathbf{R}^2 \times \mathbf{S}^{n-1}$, or $\mathbf{B}^2 \times \mathbf{S}^{n-1}$ if we compactify it to include boundary points. In particular, X_2 is simply-connected, has a unique spin structure, and contributes to the standard thermal ensemble but not to $\mathrm{Tr}(-1)^F e^{-\beta H}$.

With either (4) or (6), the geometry of the $\mathbf{S}^{n-1} \times \mathbf{S}^1$ factor at larger r can be simply explained: the $\mathbf{S}^1$ has radius approximately $\beta = (r/b)\beta_0$, and the $\mathbf{S}^{n-1}$ has radius $\beta' = r/b$. The ratio is thus $\beta/\beta' = \beta_0$. If we wish to go to $\mathbf{S}^1 \times \mathbf{R}^{n-1}$, we must take $\beta/\beta' \to 0$, that is $\beta_0 \to 0$; this is the limit of large temperatures. (9) seems to show that this can be done with either $r_+ \to 0$ or $r_+ \to \infty$, but the $r_+ \to 0$ branch is thermodynamically unfavored[64] (having larger action), so we must take the large r_+ branch, corresponding to large M.

A scaling that reduces (9) to a solution with boundary $\mathbf{R}^{n-1} \times \mathbf{S}^1$ may be made as follows. If we set $r = (w_n M/b^{n-2})^{1/n}\rho$, $t = (w_n M/b^{n-2})^{-1/n}\tau$, then for large M we can reduce $r^2/b^2 + 1 - w_n M/r^{n-2}$ to $(w_n M/b^{n-2})^{2/n}(\rho^2/b^2 - b^{n-2}/\rho^{n-2})$. The period of τ become $\beta_1 = (w_n M/b^{n-2})^{1/n}\beta_0$ or (from (10)) for large M

$$\beta_1 = \frac{4\pi b}{n}. \tag{11}$$

The metric becomes

$$ds^2 = \left(\frac{\rho^2}{b^2} - \frac{b^{n-2}}{\rho^{n-2}}\right) d\tau^2 + \frac{d\rho^2}{\frac{\rho^2}{b^2} - \frac{b^{n-2}}{\rho^{n-2}}} + (w_n M/b^{n-2})^{2/n}\rho^2 d\Omega^2. \tag{12}$$

The $M^{2/n}$ multiplying the last term means that the radius of $\mathbf{S}^{n-1}$ is of order $M^{1/n}$ and so diverges for $M \to \infty$. Hence, the $\mathbf{S}^{n-1}$ is becoming flat and looks for $M \to \infty$ locally like $\mathbf{R}^{n-1}$. If we introduce near a point $P \in \mathbf{S}^{n-1}$ coordinates y_i such that at P, $d\Omega^2 = \sum_i dy_i^2$, and then set $y_i = (w_n M/b^{n-2})^{-1/n}x_i$, then the metric becomes

$$ds^2 = \left(\frac{\rho^2}{b^2} - \frac{b^{n-2}}{\rho^{n-2}}\right) d\tau^2 + \frac{d\rho^2}{\left(\frac{\rho^2}{b^2} - \frac{b^{n-2}}{\rho^{n-2}}\right)} + \rho^2 \sum_{i=1}^{n-1} dx_i^2. \tag{13}$$

This is the desired solution $\tilde{X}$ that is asymptotic at infinity to $\mathbf{R}^{n-1} \times \mathbf{S}^1$ instead of $\mathbf{S}^{n-1} \times \mathbf{S}^1$. Its topology, if we include boundary points, is $\mathbf{R}^{n-1} \times \mathbf{B}^2$. The same solution was found recently by scaling of a near-extremal brane solution.[45]

2.4. *Entropy of Schwarzschild black holes*

Following Hawking and Page[64] (who considered the case $n = 3$), we will now describe the thermodynamics of Schwarzschild black holes in AdS_{n+1}. Our normalization of the cosmological constant is stated in (5). The bulk Einstein action with this value of the cosmological constant is

$$I = -\frac{1}{16\pi G_N} \int d^{n+1}x \sqrt{g} \left(R + \frac{\frac{1}{2}n(n-1)}{b^2} \right). \tag{14}$$

For a solution of the equations of motion, one has $R = -\frac{1}{2}n(n+1)/b^2$, and the action becomes

$$I = \frac{n}{8\pi G_N} \int d^{n+1}x \sqrt{g}, \tag{15}$$

that is, the volume of spacetime times $n/8\pi G_N$. The action additionally has a surface term,[65,66] but the surface term vanishes for the AdS Schwarzschild black hole, as noted in,[64] because the black hole correction to the AdS metric vanishes too rapidly at infinity.

Actually, both the AdS spacetime (4) and the black hole spacetime (6) have infinite volume. As in Ref. 64, one subtracts the two volumes to get a finite result. Putting an upper cutoff R on the radial integrations, the regularized volume of the AdS spacetime is

$$V_1(R) = \int_0^{\beta'} dt \int_0^R dr \int_{S^{n-1}} d\Omega \, r^{n-1}, \tag{16}$$

and the regularized volume of the black hole spacetime is

$$V_2(R) = \int_0^{\beta_0} dt \int_{r_+}^R dr \int_{S^{n-1}} d\Omega \, r^{n-1}. \tag{17}$$

One difference between the two integrals is obvious here: in the black hole spacetime $r \geq r_+$, while in the AdS spacetime $r \geq 0$. A second and slightly more subtle difference is that one must use different periodicities β' and β_0 for the t integrals in the two cases. The black hole spacetime is smooth only if β_0 has the value given in (9), but for the AdS spacetime, any value of β' is possible. One must adjust β' so that the geometry of the hypersurface $r = R$ is the same in the two cases; this is done by setting $\beta'\sqrt{(r^2/b^2)+1} = \beta_0\sqrt{(r^2/b^2)+1} - w_n M/r^{n-2}$. After doing so, one finds that the action difference is

$$I = \frac{n}{8\pi G_N} \lim_{R\to\infty} (V_2(R) - V_1(R)) = \frac{\text{Vol}(S^{n-1})(b^2 r_+^{n-1} - r_+^{n+1})}{4G_N(nr_+^2 + (n-2)b^2)}.$$

(18)

This is positive for small r_+ and negative for large r_+, showing that the phase transition found in Ref. 64 occurs for all n.

Then, as in Ref. 64, one computes the energy

$$E = \frac{\partial I}{\partial \beta_0} = \frac{(n-1)\text{Vol}(S^{n-1})(r_+^n b^{-2} + r_+^{n-2})}{16\pi G_N} = M \qquad (19)$$

and the entropy

$$S = \beta_0 E - I = \frac{1}{4G_N} r_+^{n-1} \text{Vol}(S^{n-1}) \qquad (20)$$

of the black hole. The entropy can be written

$$S = \frac{A}{4G_N}, \qquad (21)$$

with A the volume of the horizon, which is the surface at $r = r_+$.

Comparison To Conformal Field Theory

Now we can compare this result for the entropy to the predictions of conformal field theory.

The black hole entropy should be compared to boundary conformal field theory on $S^{n-1} \times S^1$, where the two factors have circumference 1 and β_0/b, respectively. In the limit as $\beta_0 \to 0$, this can be regarded as a high temperature system on S^{n-1}. Conformal invariance implies that the entropy density on S^{n-1} scales, in the limit of small β_0, as $\beta_0^{-(n-1)}$. According to (9), $\beta_0 \to 0$ means $r_+ \to \infty$ with $\beta_0 \sim 1/r_+$. Hence, the boundary conformal field theory predicts that the entropy of this system is of order r_+^{n-1}, and thus asymptotically is a fixed multiple of the horizon volume which appears in (21). This is of course the classic result of Bekenstein and Hawking, for which microscopic explanations have begun to appear only recently. Note that this discussion assumes that $\beta_0 \ll 1$, which means that $r_+ \gg b$; so it applies only to black holes whose Schwarzschild radius is much greater than the radius of curvature of *AdS* space. However, in this limit, one does get a simple explanation of why the black hole entropy is proportional to area. The explanation is entirely "holographic" in spirit.[59,61]

To fix the constant of proportionality between entropy and horizon volume (even in the limit of large black holes), one needs some additional general insight, or some knowledge of the quantum field theory on the boundary. For $2 + 1$-dimensional black holes, in the context of an old framework[69] for a relation to boundary conformal field theory which actually is a special case of the general CFT/*AdS* correspondence, such additional information is provided by modular invariance of the boundary conformal field theory.[67,68]

3. High Temperature Behavior of the $\mathcal{N} = 4$ Theory

In this section we will address three questions about the high temperature behavior of the $\mathcal{N} = 4$ theory that were raised in Section 2: the behavior of temporal Wilson lines; the behavior of spatial Wilson lines; and the existence of a mass gap.

In discussing Wilson lines, we use a formalism proposed recently.[39,40] Suppose one is doing physics on a four-manifold M which is the boundary of a five-dimensional Einstein manifold B (of negative curvature). To compute a Wilson line associated with a contour $C \subset M$, we study elementary strings on B with the property that the string worldsheet D has C for its boundary. Such a D has an infinite area, but the divergence is proportional to the circumference of C. One can define therefore a regularized area $\alpha(D)$ by subtracting from the area of D an infinite multiple of the circumference of C. The expectation value of a Wilson loop $W(C)$ is then roughly

$$\langle W(C) \rangle = \int_{\mathcal{D}} d\mu \, e^{-\alpha(D)} \tag{22}$$

where $\mathcal{D}$ is the space of string worldsheets obeying the boundary conditions and $d\mu$ is the measure of the worldsheet path integral. Moreover, according to Refs. 39, 40, in the regime in which supergravity is valid (large N and large $g^2 N$), the integral can be evaluated approximately by setting D to the surface of smallest $\alpha(D)$ that obeys the boundary conditions.

The formula (22) is oversimplified for various reasons. For one thing, worldsheet fermions must be included in the path integral. Also, the description of the $\mathcal{N} = 4$ theory actually involves not strings on B but strings on the ten-manifold $B \times \mathbf{S}^5$. Accordingly, what are considered in Refs. 39, 40 are some generalized Wilson loop operators with scalar fields included in the definition; the boundary behavior of D in the $\mathbf{S}^5$ factor depends on which operator one uses. But if all scalars have masses, as they do in the $\mathcal{N} = 4$ theory at positive temperature, the generalized Wilson loop operators are equivalent at long distances to conventional ones. An important conclusion from (22) nonetheless stands: Wilson loops on $\mathbf{R}^4$ will obey an area law if, when C is scaled up, the minimum value of $\alpha(D)$ scales like a positive multiple of the area enclosed by C. (22) also implies vanishing of $\langle W(C) \rangle$ if suitable D's do not exist, that is, if C is not a boundary in B.

3.1. *Temporal Wilson lines*

Our first goal will be to analyze temporal Wilson lines. That is, we take spacetime to be $\mathbf{S}^3 \times \mathbf{S}^1$ or $\mathbf{R}^3 \times \mathbf{S}^1$, and we take $C = P \times \mathbf{S}^1$, with P a point in $\mathbf{S}^3$ or in $\mathbf{R}^3$.

We begin on $\mathbf{S}^3$ in the low temperature phase. We recall that this is governed by a manifold X_1 with the topology of $\mathbf{B}^4 \times \mathbf{S}^1$. In particular, the contour C, which wraps around the $\mathbf{S}^1$, is not homotopic to zero in X_1 and is not the boundary of any D. Thus, the expectation value of a temporal Wilson line vanishes at low temperatures. This is the expected result, corresponding to the fact that the center Γ of the gauge group is unbroken at low temperatures.

Now we move on to the high temperature phase on $\mathbf{S}^3$. This phase is governed by a manifold X_2 that is topologically $\mathbf{S}^3 \times \mathbf{B}^2$. In this phase, $C = P \times \mathbf{S}^1$ is a boundary; in fact it is the boundary of $D = P \times \mathbf{B}^2$. Thus, it appears at first sight that the temporal Wilson line has a vacuum expectation value and that the center of the gauge group is spontaneously broken.

There is a problem here. Though we expect these results in the high temperature phase on $\mathbf{R}^3$, they cannot hold on $\mathbf{S}^3$, because the center (or any other bosonic symmetry) cannot be spontaneously broken in finite volume.

The resolution of the puzzle is instructive. The classical solution on X_2 is not unique. We must recall that Type IIB superstring theory has a two-form field B that couples to the elementary string world-sheet D by

$$i \int_D B. \tag{23}$$

The gauge-invariant field strength is $H = dB$. We can add to the solution a "world-sheet theta angle," that is a B field of $H = 0$ with an arbitrary value of $\psi = \int_D B$ (here D is any surface obeying the boundary conditions, for instance $D = P \times B^2$). Since discrete gauge transformations that shift the flux of B by a multiple of 2π are present in the theory, ψ is an angular variable with period 2π.

If this term is included, the path integrand in (22) receives an extra factor $e^{i\psi}$. Upon integrating over the space of all classical solutions — that is integrating over the value of ψ — the expectation value of the temporal Wilson line on S^3 vanishes.

Now, let us go to $\mathbf{R}^3 \times \mathbf{S}^1$, which is the boundary of $\mathbf{R}^3 \times \mathbf{B}^2$. In infinite volume, ψ is best understood as a massless scalar field in the low energy theory on $\mathbf{R}^3$. One still integrates over local fluctuations in ψ, but not over the vacuum expectation value of ψ, which is set by the value at spatial infinity. The expectation value of $W(C)$ is nonzero and is proportional to $e^{i\psi}$.

What we have seen is thus spontaneous symmetry breaking: in infinite volume, the expectation value of $\langle W(C) \rangle$ is nonzero, and depends on the choice of vacuum, that is on the value of ψ. The field theory analysis that we reviewed in Section 2 indicates that the symmetry that is spontaneously broken by the choice of ψ is the center, Γ, of the gauge group. Since ψ is a continuous angular variable, it seems that the center is $U(1)$. This seems to imply that the gauge group is not $SU(N)$, with center $\mathbf{Z}_N$, but $U(N)$. However, a variety of arguments[49] show that the AdS theory encodes a $SU(N)$ gauge group, not $U(N)$. Perhaps the apparent $U(1)$ center should be understood as a large N limit of $\mathbf{Z}_N$.

't Hooft Loops

We would also like to consider in a similar way 't Hooft loops. These are obtained from Wilson loops by electric-magnetic duality. Electric-magnetic duality of $\mathcal{N} = 4$ arises[70,71] directly from the $\tau \to -1/\tau$ symmetry of

Type IIB. That symmetry exchanges elementary strings with D-strings. So
to study the 't Hooft loops we need only, as in Ref. 48, replace elementary
strings by D-strings in the above discussion.

The $\tau \to -1/\tau$ symmetry exchanges the Neveu-Schwarz two-form B
which entered the above discussion with its Ramond-Ramond counterpart
B'; the D-brane theta angle $\psi' = \int_D B'$ thus plays the role of ψ in the
previous discussion. In the thermal physics on $\mathbf{R}^3 \times \mathbf{S}^1$, the center of the
"magnetic gauge group" is spontaneously broken, and the temporal 't Hooft
loops have an expectation value, just as we described for Wilson loops. The
remarks that we make presently about spatial Wilson loops similarly carry
over for spatial 't Hooft loops.

3.2. Spatial Wilson loops

Now we will investigate the question of whether at nonzero temperature the
spatial Wilson loops obey an area law. The main point is to first understand
why there is *not* an area law at zero temperature. At zero temperature, one
works with the AdS metric

$$ds^2 = \frac{1}{x_0^2} \left(dx_0^2 + \sum_{i=1}^{4} dx_i^2 \right). \tag{24}$$

We identify the spacetime M of the $\mathcal{N} = 4$ theory with the boundary at
$x_0 = 0$, parametrized by the Euclidean coordinates x_i, $i = 1, \ldots, 4$. M
has a metric $d\tilde{s}^2 = \sum_i dx_i^2$ obtained by multiplying ds^2 by x_0^2 and setting
$x_0 = 0$. (If we use a function other than x_0^2, the metric on M changes
by a conformal transformation.) We take a closed oriented curve $C \subset M$
and regard it as the boundary of an oriented compact surface D in AdS
space. The area of D is infinite, but after subtracting an infinite counterterm
proportional to the circumference of C, we get a regularized area $\alpha(D)$.
In the framework of Refs. 39, 40, the expectation value of the Wilson line
$W(C)$ is proportional to $\exp(-\alpha(D))$, with D chosen to minimize $\alpha(D)$.

Now the question arises: why does not this formalism *always* give an
area law? As the area enclosed by C on the boundary is scaled up, why
is not the area of D scaled up proportionately? The answer to this is clear
from conformal invariance. If we scale up C via $x_i \to t x_i$, with large
positive t, then by conformal invariance we can scale up D, with $x_i \to t x_i$,

$x_0 \to tx_0$, without changing its area (except for a boundary term involving the regularization). Thus the area of D need not be proportional to the area enclosed by C on the boundary. Since, however, in this process we had to scale $x_0 \to tx_0$ with very large t, the surface D which is bounded by a very large circle C "bends" very far away from the boundary of AdS space. If such a bending of D were prevented — if D were limited to a region with $x_0 \le L$ for some cutoff L — then one would get an area law for $W(C)$. This is precisely what will happen at nonzero temperature.

At nonzero temperature, we have in fact the metric (13) obtained earlier, with $n = 4$:

$$ds^2 = \left(\frac{\rho^2}{b^2} - \frac{b^2}{\rho^2} \right) d\tau^2 + \frac{d\rho^2}{\left(\frac{\rho^2}{b^2} - \frac{b^2}{\rho^2} \right)} + \rho^2 \sum_{i=1}^{3} dx_i^2. \qquad (25)$$

The range of ρ is $b \le \rho \le \infty$. Spacetime — a copy of $\mathbf{R}^3 \times \mathbf{S}^1$ — is the boundary at $\rho = \infty$. We define a metric on spacetime by dividing by ρ^2 and setting $\rho = \infty$. In this way we obtain the spacetime metric

$$d\tilde{s}^2 = \frac{d\tau^2}{b^2} + \sum_{i=1}^{3} dx_i^2. \qquad (26)$$

As the period of τ is β_1, the circumference of the $\mathbf{S}^1$ factor in $\mathbf{R}^3 \times \mathbf{S}^1$ is β_1/b and the temperature is

$$T = \frac{b}{\beta_1} = \frac{1}{\pi}. \qquad (27)$$

Because of conformal invariance, the numerical value of course does not matter.

Now, let C be a Wilson loop in $\mathbf{R}^3$, at a fixed value of τ, enclosing an area A in $\mathbf{R}^3$. A bounding surface D in the spacetime (25) is limited to $\rho \ge b$, so the coefficient of $\sum_i dx_i^2$ is always at least b^2. Apart from a surface term that depends on the regularization and the detailed solution of the equation for a minimal surface, the regularized area of D is at least $\alpha(D) = b^2 A$ (and need be no larger than this). The Wilson loops therefore obey an area law, with string tension b^2 times the elementary Type IIB string tension.

We could of course have used a function other than ρ^2 in defining the spacetime metric, giving a conformally equivalent metric on spacetime. For instance, picking a constant s and using $s^2\rho^2$ instead of ρ^2 would scale the temperature as $T \rightarrow T/s$ and would multiply all lengths on $\mathbf{R}^3$ by s. The area enclosed by C would thus become $A' = As^2$. As $\alpha(D)$ is unaffected, the relation between $\alpha(D)$ and A' becomes $\alpha(D) = (b^2/s^2)$. A'. The string tension in the Wilson loop area law thus scales like s^{-2}, that is, like T^2, as expected from conformal invariance.

3.3. *The mass gap*

The last issue concerning the $\mathcal{N} = 4$ theory at high temperature that we will discuss here is the question of whether there is a mass gap. We could do this by analyzing correlation functions, using the formulation of Refs. 29, 49, but it is more direct to use a Hamiltonian approach (discussed at the end of Ref. 49) in which one identifies the quantum states of the supergravity theory with those of the quantum field theory on the boundary.

So we will demonstrate a mass gap by showing that there is a gap, in the three-dimensional sense, for quantum fields propagating on the five-dimensional spacetime

$$ds^2 = \left(\frac{\rho^2}{b^2} - \frac{b^2}{\rho^2} \right) d\tau^2 + \frac{d\rho^2}{\left(\frac{\rho^2}{b^2} - \frac{b^2}{\rho^2} \right)} + \rho^2 \sum_{i=1}^{3} dx_i^2. \qquad (28)$$

This spacetime is the product of a three-space $\mathbf{R}^3$, parametrized by the x_i, with a two-dimensional "internal space" $\mathbf{W}$, parametrized by ρ and τ. We want to show that a quantum free field propagating on this five-dimensional spacetime gives rise, in the three-dimensional sense, to a discrete spectrum of particle masses, all of which are positive. When such a spectrum is perturbed by interactions, the discreteness of the spectrum is lost (as the very massive particles become unstable), but the mass gap persists.

If $\mathbf{W}$ were compact, then discreteness of the mass spectrum would be clear: particle masses on $\mathbf{R}^3$ would arise from eigenvalues of the Laplacian (and other wave operators) on $\mathbf{W}$. Since $\mathbf{W}$ is not compact, it is at first sight surprising that a discrete mass spectrum will emerge. However, this does occur, by essentially the same mechanism that leads to discreteness of particle energy levels on AdS space[72,73] with a certain notion of energy.

For illustrative purposes, we will consider the propagation of a Type IIB dilaton field ϕ on this spacetime. Other cases are similar. The action for ϕ is

$$I(\phi) = \frac{1}{2} \int_b^\infty d\rho \int_0^{\beta_1/b} d\tau \int_{-\infty}^\infty d^3x \, \rho^3 \left(\left(\frac{\rho^2}{b^2} - \frac{b^2}{\rho^2} \right) \left(\frac{\partial \phi}{\partial \rho} \right)^2 \right.$$

$$\left. + \left(\frac{\rho^2}{b^2} - \frac{b^2}{\rho^2} \right)^{-1} \left(\frac{\partial \phi}{\partial \tau} \right)^2 + \rho^{-2} \sum_i \left(\frac{\partial \phi}{\partial x_i} \right)^2 \right). \qquad (29)$$

Since translation of τ is a symmetry, modes with different momentum in the τ direction are decoupled from one another. The spectrum of such momenta is discrete (as τ is a periodic variable). To simplify things slightly and illustrate the essential point, we will write the formulas for the modes that are independent of τ; others simply give, by the same argument, additional three-dimensional massive particles with larger masses.

We look for a solution of the form $\phi(\rho, x) = f(\rho)e^{i\bar{k}\cdot x}$, with $\bar{k}$ the momentum in $\mathbf{R}^3$. The effective Lagrangian becomes

$$I(f) = \frac{1}{2} \int_b^\infty d\rho \, \rho^3 \left((\rho^2/b^2 - b^2/\rho^2) \left(\frac{df}{d\rho} \right)^2 + \rho^{-2}k^2 f^2 \right). \qquad (30)$$

The equation of motion for f is

$$-\rho^{-1} \frac{d}{d\rho} \left(\rho^3 (\rho^2/b^2 - b^2/\rho^2) \frac{df}{d\rho} \right) + k^2 f = 0. \qquad (31)$$

A mode of momentum k has a mass m, in the three-dimensional sense, that is given by $m^2 = -k^2$. We want to show that the equation (31) has acceptable solutions only if m^2 is in a certain discrete set of positive numbers.

Acceptable solutions are those that obey the following boundary conditions:

(1) At the lower endpoint $\rho = b$, we require $df/d\rho = 0$. The reason for this is that ρ behaves near this endpoint as the origin in polar coordinates; hence f is not smooth at this endpoint unless $df/d\rho = 0$ there.

(2) For $\rho \to \infty$, the equation has two linearly independent solutions, which behave as $f \sim$ constant and $f \sim \rho^{-4}$. We want a normalizable solution, so we require that $f \sim \rho^{-4}$.

For given k^2, Eq. (31) has, up to a constant multiple, a unique solution that obeys the correct boundary condition near the lower endpoint. For generic k^2, this solution will approach a nonzero constant for $\rho \to \infty$. As in standard quantum mechanical problems, there is a normalizable solution only if k^2 is such that the solution that behaves correctly at the lower endpoint also vanishes for $\rho \to \infty$. This "eigenvalue" condition determines a discrete set of values of k^2.

The spectrum thus consists entirely of a discrete set of normalizable solutions. There are no such normalizable solutions for $k^2 \geq 0$. This can be proved by noting that, given a normalizable solution f of the equation of motion, a simple integration by parts shows that the action (30) vanishes. For $k^2 \geq 0$, vanishing of $I(f)$ implies that $df/d\rho = 0$, whence (given normalizability) $f = 0$. So the discrete set of values of k^2 at which there are normalizable solutions are all negative; the masses $m^2 = -k^2$ are hence strictly positive. This confirms the existence of the mass gap.

To understand the phenomenon better, let us compare to what usually happens in quantum mechanics. In typical quantum mechanical scattering problems, with potentials that vanish at infinity, the solutions with positive energy (analogous to $m^2 > 0$) are oscillatory at infinity and obey plane wave normalizability. When this is so, both solutions at infinity are physically acceptable (in some situations, for example, they are interpreted as incoming and outgoing waves), and one gets a continuous spectrum that starts at zero energy. The special property of the problem we have just examined is that even for negative k^2, there are no oscillatory solutions at infinity, and instead one of the two solutions must be rejected as being unnormalizable near $\rho = \infty$. This feature leads to the discrete spectrum.

If instead of the spacetime (28), we work on AdS spacetime (4), there is a continuous spectrum of solutions with plane wave normalizability for all $k^2 < 0$; this happens because for $k^2 < 0$ one gets oscillatory solutions near the lower endpoint, which for the AdS case is at $r = 0$. Like confinement, the mass gap of the thermal $\mathcal{N} = 4$ theory depends on the cutoff at small r.

4. Approach to QCD

One interesting way to study four-dimensional gauge theory is by compactification from a certain exotic six-dimensional theory with (0, 2) supersymmetry. This theory can be realized in Type IIB compactification on K3[74] or in the dynamics of parallel M-theory fivebranes[75] and can apparently be interpreted[11] in terms of M-theory on $AdS_7 \times S^4$. This interpretation is effective in the large N limit — as the M-theory radius of curvature is of order $N^{1/3}$. Since compactification from six to four dimensions has been an effective approach to gauge theory dynamics (for instance, in deducing Montonen-Olive duality[74] using a strategy proposed in Ref. 76), it is natural to think of using the solution for the large N limit of the six-dimensional theory as a starting point to understand the four-dimensional theory.

Our basic approach will be as follows. If we compactify the six-dimensional (0, 2) theory on a circle C_1 of radius R_1, with a supersymmetry-preserving spin structure (fermions are periodic in going around the circle), we get a theory that at low energies looks like five-dimensional $SU(N)$ supersymmetric Yang-Mills theory, with maximal supersymmetry and five-dimensional gauge coupling constant $g_5^2 = R_1$. Now compactify on a second circle C_2, orthogonal to the first, with radius R_2. If we take $R_2 \gg R_1$, we can determine what the resulting four-dimensional theory is in a two-step process, compactifying to five dimensions on C_1 to get five-dimensional supersymmetric Yang-Mills theory and then compactifying to four-dimensions on C_2.

No matter what spin structure we use on C_2, we will get a four-dimensional $SU(N)$ gauge theory with gauge coupling $g_4^2 = R_1/R_2$. If we take on C_2 (and more precisely, on $C_1 \times C_2$) the supersymmetry-preserving spin structure, then the low energy theory will be the four-dimensional $\mathcal{N} = 4$ theory some of whose properties we have examined in the present paper. We wish instead to break supersymmetry by taking the fermions to be antiperiodic in going around C_2. Then the fermions get masses (of order $1/R_2$) at tree level, and the spin zero bosons very plausibly get masses (of order $g_4^2 N/R_2$) at one-loop level. If this is so, the low energy theory will be the pure $SU(N)$ theory without supersymmetry. If $g_4^2 \ll 1$, the theory will flow at very long distances to strong coupling; at such long distances the spin one-half and spin one fields that receive tree level or one-loop masses will be irrelevant. So this is a possible framework for studying the pure Yang-Mills theory without supersymmetry.

We want to take the large N limit with $g_4 \to 0$ in such a way that $\eta = g_4^2 N$ has a limit. So we need $g_4^2 = \eta/N$, or in other words

$$R_1 = \frac{\eta R_2}{N}. \tag{32}$$

We actually want η fixed and small, so that the four-dimensional Yang-Mills theory is weakly coupled at the compactification scale, and flows to strong coupling only at very long distances at which the detailed six-dimensional setup is irrelevant.

To implement this approach, we first look for an Einstein manifold that is asymptotic at infinity to $\mathbf{R}^5 \times C_2$. Though it may seem to reverse the logic of the construction, starting with C_2 first in constructing the solution turns out to be more convenient. The supersymmetry-breaking boundary conditions on C_2 are the right ones for using the spacetime (13) that is constructed by scaling of the seven-dimensional AdS Schwarzschild solution:

$$ds^2 = \left(\frac{\rho^2}{b^2} - \frac{b^4}{\rho^4} \right) d\tau^2 + \frac{d\rho^2}{\left(\frac{\rho^2}{b^2} - \frac{b^4}{\rho^4} \right)} + \rho^2 \sum_{i=1}^{5} dx_i^2. \tag{33}$$

According to Ref. 11, we want here

$$b = 2 G_N^{1/9} (\pi N)^{1/3}. \tag{34}$$

(Here G_N is the eleven-dimensional Newton constant, so $G_N^{1/9}$ has dimensions of length. We henceforth set $G_N = 1$.)

To make the scaling with N clearer, we also set $\rho = 2(\pi N)^{1/3} \lambda$. And — noting from (11) that τ has period $4\pi b/n = (4/3)\pi^{4/3} N^{1/3}$ — we set

$$\tau = \theta \cdot \left(\frac{2\pi N}{3} \right)^{1/3}, \tag{35}$$

where θ is an ordinary angle, of period 2π. After also a rescaling of the x_i, the metric becomes

$$ds^2 = \frac{4}{9}\pi^{2/3}N^{2/3}\left(\lambda^2 - \frac{1}{\lambda^4}\right)d\theta^2 + 4\pi^{2/3}N^{2/3}\frac{d\lambda^2}{\left(\lambda^2 - \frac{1}{\lambda^4}\right)}$$

$$+ 4\pi^{2/3}N^{2/3}\lambda^2\sum_{i=1}^{5}dx_i^2. \tag{36}$$

Now we want to compactify one of the x_i, say x_5, on a second circle whose radius as measured at $\lambda = \infty$ should according to (32) should be η/N times the radius of the circle parametrized by θ. To do this, we write $x_5 = (\eta/N)\psi$ with ψ of period 2π. We also now restore the $\mathbf{S}^4$ factor that was present in the original M-theory on $AdS_7 \times \mathbf{S}^4$ and has so far been suppressed. The metric is now

$$ds^2 = \frac{4}{9}\pi^{2/3}N^{2/3}\left(\lambda^2 - \frac{1}{\lambda^4}\right)d\theta^2 + \frac{4}{9}\eta^2\pi^{2/3}N^{-4/3}\lambda^2 d\psi^2$$

$$+ 4N^{2/3}\frac{d\lambda^2}{\left(\lambda^2 - \frac{1}{\lambda^4}\right)} + 4\pi^{2/3}N^{2/3}\rho^2\sum_{i=1}^{4}dx_i^2 + \pi^{2/3}N^{2/3}d\Omega_4^2. \tag{37}$$

At this stage, θ and ψ are both ordinary angular variables of radius 2π, and $d\Omega_4^2$ is the metric on a unit four-sphere.

Now, we want to try to take the limit as $N \to \infty$. The metric becomes large in all directions except that one circle factor — the circle C_1, parametrized by ψ — shrinks. Thus we should try to use the equivalence between M-theory compactified on a small circle and weakly coupled Type IIA superstrings. We see that the radius $R(\lambda)$ of the circle parametrized by C is in fact

$$R(\lambda) = \frac{2}{3}\eta\lambda\pi^{1/3}N^{-2/3}. \tag{38}$$

To relate an M-theory compactification on a circle to a Type IIA compactification, we must[77] multiply the metric by R. All factors of N felicitously disappear from the metric, which becomes

$$ds^2 = \frac{8}{27}\eta\lambda\pi\left(\lambda^2 - \frac{1}{\lambda^4}\right)d\theta^2 + \frac{8\pi}{3}\eta\lambda\frac{d\lambda^2}{\left(\lambda^2 - \frac{1}{\lambda^4}\right)}$$

$$+ \frac{8\pi}{3}\eta\lambda^3 \sum_{i=1}^{4} dx_i^2 + \frac{2\pi}{3}\eta\lambda d\Omega_4^2. \tag{39}$$

The string coupling constant is meanwhile

$$g_{st}^2 = R^{3/2} = \frac{(2/3)^{3/2}\eta^{3/2}\lambda^{3/2}\pi^{1/2}}{N}. \tag{40}$$

This result clearly has some of the suspected properties of large N gauge theories. The metric (39) is independent of N, so in the weak coupling limit, the spectrum of the string theory will be independent of N. Meanwhile, the string coupling constant (40) is of order $1/N$, as expected[1] for the residual interactions between color singlet states in the large N limit. The very ability to get a description such as this one in which $1/N$ only enters as a coupling constant (and not explicitly in the multiplicity of states) is a reflection of confinement. Confinement in the form of an area law for Wilson loops can be demonstrated along the lines of our discussion in Section 3: it follows from the fact that the coefficient in the metric of $\sum_{i=1}^{4} dx_i^2$ is bounded strictly above zero. A mass gap likewise can be demonstrated, as in Section 3, by using the large λ behavior of the metric.

On the other hand, it is not obvious how one could hope to compute the spectrum or even show asymptotic freedom. Asymptotic freedom should say that as $\eta \to 0$, the particle masses become exponentially small (with an exponent determined by the gauge theory beta function). It is not at all clear how to demonstrate this. A clue comes from the fact that the coupling of the physical hadrons should be independent of η (and of order $1/N$) as $\eta \to 0$. In view of the formula (40), this means that we should take $\eta\lambda$ of order one as $\eta \to 0$. If we set $\tilde{\lambda} = \eta\lambda$, and write the metric in terms of $\tilde{\lambda}$, then the small η limit becomes somewhat clearer: a singularity develops at small $\tilde{\lambda}$ for $\eta \to 0$. Apparently, in this approach, the mysteries of four-dimensional quantum gauge theory are encoded in the behavior of string theory near this singularity.

This singularity actually has a very simple and intuitive interpretation which makes it clearer why four-dimensional gauge theory can be described by string theory in the spacetime (39). The Euclidean signature Type IIA nonextremal fourbrane solution is described by the metric[78]

$$ds^2 = \left(1 - \left(\frac{r_+}{r}\right)^3\right)\left(1 - \left(\frac{r_-}{r}\right)^3\right)^{-1/2} dt^2$$

$$+ \left(1 - \left(\frac{r_+}{r}\right)^3\right)^{-1}\left(1 - \left(\frac{r_-}{r}\right)^3\right)^{-5/6} dr^2$$

$$+ \left(1 - \left(\frac{r_-}{r}\right)^3\right)^{1/2} \sum_{i=1}^{4} dx_i^2 + r^2 \left(1 - \left(\frac{r_-}{r}\right)^3\right)^{1/6} d\Omega_4^2, \quad (41)$$

with $r_+ > r_- > 0$. The string coupling constant is

$$g_{st}^2 = \left(1 - \left(\frac{r_-}{r}\right)^3\right)^{1/2}. \quad (42)$$

The horizon is at $r = r_+$, and the spacetime is bounded by $r \geq r_+$. This spacetime is complete and smooth if t has period

$$T = 12\pi \left(1 - \left(\frac{r_-}{r_+}\right)^3\right)^{-1/6}. \quad (43)$$

If one continues (via Lorentzian or complex values of the coordinates) past $r = r_+$, there is a singularity at $r = r_-$. The extremal fourbrane solution is obtained by setting $r_+ = r_-$ and is singular. But this singularity is exactly the singularity that arises in (39) upon taking $\eta \to 0$, with $\eta\lambda \sim 1$! In fact, if we set $\lambda^6 = (r^3 - r_-^3)/(r_+^3 - r_-^3)$, identify η with $(1 - (r_-/r_+)^3)^{1/6}$, and take the limit of $r_+ \to r_-$, then (41) reduces to (39), up to some obvious rescaling. Moreover, according to (43), $r_+ \to r_-$ is the limit that T is large, which (as $1/g_4^2 = T/g_5^2$) makes the four-dimensional coupling small.

So in hindsight we could discuss four-dimensional gauge theories in the following way, without passing through the CFT/*AdS* correspondence. In a spacetime $\mathbf{R}^9 \times \mathbf{S}^1$, consider N Type IIA fourbranes wrapped on $\mathbf{R}^4 \times \mathbf{S}^1$. Pick a spin structure on the $\mathbf{S}^1$ that breaks supersymmetry. This system looks at low energies like four-dimensional $U(N)$ gauge theory, with Yang-Mills coupling $g_4^2 = g_5^2/T = g_{st}/T$. Take $N \to \infty$ with $g_{st}N$ fixed. The D-brane system has both open and closed strings. The dominant string diagrams for large N with fixed $g_5^2 N$ and fixed T are the planar diagrams of 't Hooft[1] — diagrams of genus zero with any number of holes. (This fact was exploited recently[31] in analyzing the beta function of certain field theories.) Summing them up is precisely the long-intractable problem of the $1/N$ expansion.

Now, at least if η is large, supergravity effectively describes the sum of planar diagrams in terms of the metric (41) which is produced by the D-branes. This is a smooth metric, with no singularity and no D-branes. So we get a description with closed Type IIA strings only. Thus the old prophecy[1] is borne out: nonperturbative effects close up the holes in the Feynman diagrams, giving a confining theory with a mass gap, and with $1/N$ as a coupling constant, at least for large η. To understand large N gauge theories, one would want, from this point of view, to show that there is no singularity as a function of η, except at $\eta = 0$, and to exhibit asymptotic freedom and compute the masses for small η. (This looks like a tall order, given our limited knowledge of worldsheet field theory with Ramond-Ramond fields in the Lagrangian.) The singularity at $\eta = 0$ is simply the singularity of the fourbrane metric at $r_+ = r_-$; it reflects the classical $U(N)$ gauge symmetry of N parallel fourbranes, which disappears quantum mechanically when $\eta \neq 0$ and the singularity is smoothed out.

Acknowledgment

I have benefited from comments by N. Seiberg. This work was supported in part by NSF Grant PHY-9513835.

References

1. G. 't Hooft, "A Planar Diagram Theory For Strong Interactions," *Nucl. Phys.* **B72** (1974) 461.
2. A. M. Polyakov, "String Theory and Quark Confinement," hep-th/9711002.
3. S. S. Gubser, I. R. Klebanov, and A. W. Peet, "Entropy and Temperature of Black 3-Branes," *Phys. Rev.* **D54** (1996) 3915.

4. I. R. Klebanov, "World Volume Approach To Absorption By Nondilatonic Branes," *Nucl. Phys.* **B496** (1997) 231.
5. S. S. Gubser, I. R. Klebanov, and A. A. Tseytlin, "String Theory and Classical Absorption By Threebranes," *Nucl. Phys.* **B499** (1997) 217.
6. S. S. Gubser and I. R. Klebanov, "Absorption By Branes and Schwinger Terms in the World Volume Theory," *Phys. Lett.* **B413** (1997) 41.
7. J. Maldacena and A. Strominger, "Semiclassical Decay of Near Extremal Fivebranes," hep-th/9710014.
8. G. Gibbons and P. Townsend, "Vacuum Interpolation in Supergravity Via Super p-Branes," *Phys. Rev. Lett.* **71** (1993) 5223.
9. M. J. Duff, G. W. Gibbons, and P. K. Townsend, "Macroscopic Superstrings as Inter-polating Solitons," *Phys. Lett.* **B332** (1994) 321.
10. S. Ferrara, G. W. Gibbons, and R. Kallosh, "Black Holes and Critical Points in Moduli Space," *Nucl. Phys.* **B500** (1997) 75, hep-th/9702103; A. Chamseddine, S. Ferrara, G. W. Gibbons, and R. Kallosh, "Enhancement of Supersymmetry Near 5 — D Black Hole Horizon," *Phys. Rev.* **D55** (1997) 3647, hep-th/9610155.
11. J. Maldacena, "The Large N Limit of Superconformal Field Theories and Supergravity," hep-th/9711200.
12. M. J. Duff, B. E. W. Nilsson, and C. N. Pope, "Kaluza-Klein Supergravity," *Phys. Rep.* **130** (1986) 1.
13. M. J. Duff, H. Lu, and C. N. Pope, "$AdS_5 \times S^5$ Untwisted," hep-th/9803061.
14. M. Flato and C. Fronsdal, "Quantum Field Theory of Singletons: The Race," *J. Math. Phys.* **22** (1981) 1278; C. Fronsdal, "The Dirac Supermultiplet," *Phys. Rev.* **D23** (1988) 1982; E. Angelopoulos, M. Flato, C. Fronsdal, and D. Sternheimer, "Massless Particles, Conformal Group, and De Sitter Universe," *Phys. Rev.* **D23** (1981) 1278.
15. E. Bergshoeff, M. J. Duff, E. Sezgin, and C. N. Pope, "Supersymmetric Supermembrane Vacua and Singletons," *Phys. Lett.* **B199** (1988) 69; M. P. Blencowe and M. J. Duff, "Supersingletons," *Phys. Lett.* **B203** (1988) 203.
16. S. Hyun, "U-Duality Between Three and Higher Dimensional Black Holes," hep-th/9704005; S. Hyun, Y. Kiem, and H. Shin, "Infinite Lorentz Boost Along the M Theory Circle and Nonasymptotically Flat Solutions in Supergravities," hep-th/9712021.
17. K. Sfetsos and K. Skenderis, "Microscopic Derivation of the Bekenstein-Hawking Formula For Nonextremal Black Holes," hep-th/9711138, H. J. Boonstra, B. Peeters, and K. Skenderis, "Branes and Anti-de Sitter Space-times," hep-th/9801206.

18. P. Claus, R. Kallosh, and A. van Proeyen, "*M* Five-brane and Superconformal (0, 2) Tensor Multiplet in Six-Dimensions," hep-th/9711161; P. Claus, R. Kallosh, J. Kumar, P. Townsend, and A. van Proeyen, "Conformal Field Theory of *M*2, *D*3, *M*5, and *D*1-Branes + *D*5-Branes," hep-th/9801206.

19. R. Kallosh, J. Kumar, and A. Rajaraman, "Special Conformal Symmetry of World-volume Actions," hep-th/9712073.

20. S. Ferrara and C. Fronsdal, "Conformal Maxwell Theory As A Singleton Field Theory On *AdS*(5), IIB Three-branes and Duality," hep-th/9712239.

21. S. Hyun, "The Background Geometry of DLCQ Supergravity," hep-th/9802026.

22. M. Gunaydin and D. Minic, "Singletons, Doubletons, and *M* Theory," hep-th/9802047.

23. S. Ferrara and C. Fronsdal, "Gauge Fields As Composite Boundary Excitations," hep-th/9802126.

24. G. T. Horowitz and H. Ooguri, "Spectrum of Large *N* Gauge Theory From Super-gravity," hep-th/9802116.

25. N. Itzhaki, J.M. Maldacena, J. Sonnenschein, and S. Yankielowicz, "Super-gravity and the Large *N* Limite of Theories With Sixteen Supercharges," hep-th/9802042.

26. S. Kachru and E. Silverstein, "4d Conformal Field Theories and Strings On Orbifolds," hep-th/9802183.

27. M. Berkooz, "A Supergravity Dual of A (1, 0) Field Theory in Six Dimensions," hep-th/9802195.

28. V. Balasumramanian and F. Larsen, "Near Horizon Geometry and Black Holes in Four Dimensions," hep-th/9802198.

29. S. S. Gubser, I. R. Klebanov, and A. M. Polyakov, "Gauge Theory Correlators From Noncritical String Theory," hep-th/9802109.

30. M. Flato and C. Fronsdal, "Interacting Singletons," hep-th/9803013.

31. A. Lawrence, N. Nekrasov, and C. Vafa, "On Conformal Theories in Four Dimensions," hep-th/9803015.

32. M. Bershadsky, Z. Kakushadze, and C. Vafa, "String Expansion as Large *N* Expansion of Gauge Theories," hep-th/9803076.

33. S. S. Gubser, A. Hashimoto, I. R. Klebanov, and M. Krasnitz, "Scalar Absorption and the Breaking of the World Volume Conformal Invariance," hep-th/9803023.

34. I. Ya. Aref'eva and I. V. Volovich, "On Large *N* Conformal Theories, Field Theories on Anti-de Sitter Space, and Singletons," hep-th/9803028.

35. L. Castellani, A. Ceresole, R. D'Auria, S. Ferrara, P. Fré, and M. Trigiante, "*G*/*H* *M*-Branes and *AdS*$_{p+2}$ Geometries," hep-th/9803039.

36. S. Ferrara, C. Fronsdal, and A. Zaffaroni, "On $N = 8$ Supergravity On AdS_5 and $N = 4$ Superconformal Yang-Mills Theory," hep-th/9802203.

37. O. Aharony, Y. Oz, and Z. yin, hep-th/9803051.

38. S. Minwalla, "Particles on $AdS_{4/7}$ and Primary Operators On $M_{2/5}$ Brane World volumes," hep-th/9803053.

39. J. Maldacena, "Wilson Loops in Large N Field Theories," hep-th/9803002.

40. S.-J. Rey and J. Yee, "Macroscopic Strings As Heavy Quarks in Large N Gauge Theory and Anti de Sitter Supergravity," hep-th/9803001.

41. E. Halyo, "Supergravity On $AdS_{4/7} \times S^{7/4}$ and M Branes," hep-th/9803077.

42. R. G. Leigh and M. Rozali, "The Large N Limit of the (2, 0) Superconformal Field Theory," hep-th/9803068.

43. M. Bershadsky, Z. Kakushadze, and C. Vafa, "String Expansion As Large N Expansion of Gauge Theories," hep-th/9803076.

44. A. Rajaraman, "Two-Form Fields and the Gauge Theory Description of Black Holes," hep-th/9803082.

45. G. T. Horowitz and S. F. Ross, "Possible Resolution of Black Hole Singularities From Large N Gauge Theory," hep-th/9803085.

46. J. Gomis, "Anti de Sitter Geometry and Strongly Coupled Gauge Theories," hep-th/9803119.

47. S. Ferrara, A. Kehagias, H. Partouche, and A. Zaffaroni, "Membranes and Fivebranes With Lower Supersymmetry and Their AdS Supergravity Duals," hep-th/9803109.

48. J. A. Minahan, "Quark-Monopole Potentials in Large N Super Yang-Mills," hep-th/9803111.

49. E. Witten, "Anti-de Sitter Space and Holography," hep-th/9802150.

50. A. M. Polyakov, "Thermal Properties of Gauge Fields and Quark Liberation," *Phys. Lett.* **72B** (1978) 477.

51. L. Susskind, "Lattice Models of Quark Confinement At High Temperature," *Phys. Rev.* **D20** (1979) 2610.

52. D. J. Gross, R. D. Pisarski, and L. G. Yaffe, "QCD and Instantons At Finite Temperature," *Rev. Mod. Phys.* **53** (1981) 43.

53. G. Horowitz and J. Polchinski, "Corresondence Principle For Black Holes and Strings," *Phys. Rev.* **D55** (1997) 6189.

54. T. Banks, W. Fischler, I. R. Klebanov, and L. Susskind, "Schwarzschild Black Holes From Matrix Theory," *Phys. Rev. Lett.* **80** (1998) 226, hep-th/9709091; I.R. Klebanov and L. Susskind, "Schwarzschild Black Holes in Various Dimensions From Matrix Theory," *Phys. Lett.* **B416** (1998) 62, hep-th/ 9719108, "Schwarzschild Black Holes in Matrix Theory 2," hep-th/971 1005.

55. A. Sen, "Extremal Black Holes and Elementary String States," *Mod. Phys. Lett.* **A10** (1995), hep-th/9504147.
56. A. Strominger and C. Vafa, "Microscopic Origin of the Bekenstein-Hawking Entropy," *Phys. Lett.* **B379** (1996) 99, hep-th/9601029.
57. J. Bekenstein, "Black Holes and Entropy," *Phys. Rev.* **D7** (1973) 2333.
58. S. W. Hawking, "Particle Creation By Black Holes," *Commun. Math. Phys.* **43** (1975) 199.
59. G. 't Hooft, "Dimensional Reduction in Quantum Gravity," in *Salamfest 93*, p. 284, gr-qc/9310026.
60. C. Thorn, "Reformulating String Theory with the $1/N$ Expansion," lecture at First A. D. Sakharov Conference on Physics, hep-th/9405069.
61. L. Susskind, "The World as a Hologram," *J. Math. Phys.* **36** (1995) 6377.
62. D. J. Gross and E. Witten, "Possible Third Order Phase Transition in the Large N Lattice Gauge Theory," *Phys. Rev.* **D21** (1980) 446.
63. C. Thorn, "Infinite N (C) QCD At Finite Temperature: Is There An Ultimate Temperature?" *Phys. Lett.* **B** (1981) 458.
64. S. W. Hawking and D. Page, "Thermodynamics of Black Holes in Anti-de Sitter Space," *Commun. Math. Phys.* **87** (1983) 577.
65. J. W. York, "Role of Conformal Three-Geometry in the Dynamics of Gravitation," *Phys. Rev. Lett.* **28** (1972) 1082.
66. G. W. Gibbons and S. W. Hawking, "Action Integrals and Partition Functions in Quantum Gravity," *Phys. Rev.* **D15** (1977) 2752.
67. J. Cardy, "Operator Content of Two-Dimensional Conformally-Invariant Theories," *Nucl. Phys.* **B270** (1986) 186.
68. A. Strominger, "Black Hole Entropy From Near-Norizon Microstates," hep-th/9712251.
69. J. D. Brown and M. Henneaux, "Central Charges in the Canonical Realization of Asymptotic Symmetries: An Example From Three-Dimensional Gravity," *Commun. Math. Phys.* **104** (1986) 207.
70. M. B. Green and M. Gutperle, "Comments on Three-Branes," *Phys. Lett.* **B377** (1996) 28.
71. A. A. Tseytlin, "Self-duality of Born-Infeld Action and Dirichlet 3-Brane of Type IIB Superstring," *Nucl. Phys.* **B469** (1996) 51.
72. S. J. Avis, C. Isham, and D. Storey, "Quantum Field Theory in Anti-de Sitter Space-time," *Phys. Rev.* **D18** (1978) 3565.
73. P. Breitenlohner and D. Z. Freedman, "Positive Energy in Anti-de Sitter Backgrounds and Gauged Extended Supergravity," *Phys. Lett.* **115B** (1982) 197, "Stability in Gauged Extended Supergravity," *Ann. Phys.* **144** (1982) 197.

74. E. Witten, "Some Comments On String Dynamics," in *Strings '95*, ed. I. Bars *et al.* (World Scientific, 1997), hep-th/9507121.

75. A. Strominger, "Open *p*-Branes," *Phys. Lett.* **B383** (1996) 44, hep-th/9512059.

76. M. Duff, "Strong/Weak Coupling Duality From the Dual String," *Nucl. Phys.* **B442** (1995) 47, hep-th/9501030.

77. E. Witten, "String Theory Dynamics in Various Dimensions," *Nucl. Phys.* **B443** (1995) 85, hep-th/9503124.

78. G. Horowitz and A. Strominger, "Black Strings and *p*-Branes," *Nucl. Phys.* **B360** (1991) 197.

CAN THERE BE PHYSICS WITHOUT EXPERIMENTS? CHALLENGES AND PITFALLS

Gerard 't Hooft*

Institute for Theoretical Physics
University of Utrecht, Princetonplein 5
3584 CC Utrecht, the Netherlands
and
Spinoza Institute
Postbox 80.195
3508 TD Utrecht, the Netherlands

Physicists investigating space, time and matter at the Planck scale will probably have to work with much less guidance from experimental input than has ever happened before in the history of Physics. This may imply that we should insist on much higher demands of logical and mathematical rigour than before. Working with long chains of arguments linking theories to experiment, we must be able to rely on logical precision when and where experimental checks cannot be provided.

Introduction

During the last few decades our view upon matter, the fundamental forces and the geometry of space-time have undergone a metamorphosis, enabling us to extrapolate our knowledge towards time- and distance scales that in fact cannot be directly probed by experiment. This extrapolation leads us towards the so-called Planck scale, where lengths are of the order of 10^{-33} cm, time intervals of the order of 10^{-44} sec, and masses of the order of 20 μg. Here, gravitational effects cause curvature in space and time, and today's physical models, the "superstring theories" and related ideas, tend to become extremely complex. In general, even our book-keeping procedures,

---

*E-mail: g.thooft@phys.uu.nl; internet: http://www.phys.uu.nl/~thooft/.

needed to formulate the physical degrees of freedom, their coordinates in space and time, and the disciplines for their evolution, present fundamental difficulties.

Physicists will not easily accept any new theory unless it has been extensively tested experimentally and all conceivable alternatives have been ruled out. The extremely successful Standard Model is a perfect example. Being the first complete synthesis of the theories of Special Relativity and Quantum Mechanics, it came into being by a close collaboration between theoreticians and experimentalists.

But with String Theory,[1] the situation will be vastly different. Experimentally, this theory will be extremely difficult to assess. Most of our evidence will be extremely indirect at best: the theory shows a slight preference for predicting the emergence of super symmetry at the TeV scale, although this prediction actually was made long before String Theory took its present form. Some cosmological models inspired by String Theory will be experimentally testable but here also the input from these theories is indirect. The most direct predictions refer to the nature of space, time, and the particle spectrum at the Planck scale, where no experiment will ever be able to check any of their claims.

Yet, String Theory is of extreme importance. It is considered by many to be as yet the only detailed avenue towards a better understanding of the synthesis of general Relativity with Quantum Mechanics. One day, we hope to be able to understand the spectral and numerical details of the Standard Model itself. Presently, there are about 26 fundamental constants of nature, whose numerical values we are unable to derive from present theories. It is hoped that String Theory or any other similar construction will do this for us. The question to be discussed now is: Will there be any chance to succeed, if experimental support will be as indirect as we all expect?

To address this question, let us look at the development of important theories in the past, and estimate in what way they owe their existence to the numerous and essential experimental checks.

Theory and Experiment in Physics

Physics is an experimental science. One studies phenomena observed in our world, ranging from the mundane ones to the extremely esoteric. Obvious, but nevertheless of essential importance are observations such as the fact that our space is three dimensional, and that there exists such a force as the gravitational one, pulling us towards the ground; we see that our universe has matter in it, divided into little particles. It seems that the properties of matter can be reduced to the properties of these particles, which appear to be point like, and their properties, such as the forces acting between them, are universal. Our entire universe appears to be built out of the same material.

The most esoteric observations, requiring the most sophisticated techniques to be realized at all, are for instance the fact that our universe has curvature in it, and that some particles long thought to be massless, actually turn out to have mass after all.[2] Much of the progress in modern physics results from such eminent achievements of the scientific method.

It is the theoretician's task to try to find order in these phenomena. We have experienced that the phenomena obey rules, which we call "laws of physics". It appears that the laws are obeyed with extreme precision, and, most importantly, it seems that some rules are obeyed everywhere, under the most extreme circumstances. They appear to be universal.

An essential complication however is that, at atomic distance and time scales, a new kind of 'fuzziness' arises, allowing us only to apply certain rules of statistics, rather than deterministic rules. This is what we call Quantum Mechanics. But apart from this important side remark, law and order are everywhere.

I am not the first to be surprised by humanity's continuing progress in unraveling these laws, in spite of the fact that these laws do not seem to have been made by humans. Indeed, they appear to be far superior to the man-made laws. History nevertheless demonstrated that we can get into grips with them.

How does this process work? Philosophers of science sometimes give answers of the following sort:

(1) At a given stage of our understanding, there may be competing theories for the explanation of an observed phenomenon, theory #1 and theory #2, say.
(2) Theory #1 leads to prediction #1, and theory #2 leads to prediction #2.

(3) And now an experiment is done. If it agrees with one of the two predictions this helps us to select the right theory, until new phenomena are observed that require further refinements. Go back to (1).

According to this picture, the role of theoreticians is to come with a number of alternative scenarios. They are tested by the experiments. Experiment has the last word. But this picture, often taught at high schools, is not quite the way physicists themselves experience what is really happening. It is far too simplistic to suggest that theoreticians are just sitting there producing theories, preparing them just such that they can be confronted with experiment. The point is that in most cases, the different theories are not "equivalent" at all. They are not equally probable or equally credible. Very often, the choice is between

1) a theory,   and   2) no theory.

Or else, the theories are of very different quality. I am talking now of highly 'fundamental' branches of physics, where the laws at the very end of the scales are searched for: either at extremely tiny distance, at the highest conceivable energies, or, conversely, at the largest distance scale. Nowadays we have a very detailed theoretical picture of what is going on, but there are uncertainties, in regions of physics that have been inaccessible in the past.

We try to produce a theoretical scheme that is more advanced, more refined or more ambitious than the more established versions. To find out whether the new version makes sense or not, one naturally first asks whether experiments or new astronomical observations can be envisioned that can give us further guidance. If this is not the case, at least in the foreseeable future, should we then abandon such theoretical research, or do we have guidelines other than experimental verifications?

It turns out that we do. New theories for space, time, matter and forces are being proposed, which, in practice, appear to be such that they may be subjected to tests of the following kinds:

(i) A theory should be *logically coherent*. This demand is far from trivial or obvious. Often, theories are formulated in terms of abstract mathematical equations that ought to control infinitely many degrees of freedom. If these equations display a clear causal structure, and if wild oscillations can be seen to be subject to constraints derived from a bounded quantity such as energy, one might conclude that the theory makes sense logically. However, many complications can arise that imperil logical consistency. If point like particles are replaced by line like objects, locality of the equations is no longer obvious, so that the causal structure may get lost;[a] if gravitational effects remove the lower bound to the energy, then energy can no longer be used to show the absence of run-away solutions, and if the theory is only formulated in perturbative terms, the existence and internal consistency of a corresponding non-perturbative system may be questioned. And so on.

(ii) The theory should *be capable of* making predictions. Even if the predictions cannot be tested experimentally, the mere fact that meaningful statements are made concerning the outcome of some *Gedanken* experiment could distinguish a theory from less viable competitors. In short: a theory should at least be capable of predicting coherently the outcome of experiments done in our imagination. Requiring such outcomes to be informative and coherent is an important criterion for our esoterical theory.

(iii) The theory should agree with older theories that are well-established. Thus, most advanced particle theories such as String Theory, M-theory and the like are demanded to agree at least with Quantum Mechanics, and Special and General Relativity. They should allow for the existence of fermions, and a four-dimensional spacetime evolving into something resembling our present universe.

(iv) There should be *agreement* with already observed phenomena. This means that, even if no *new* experiments can be thought of to test our theory, we can still make use of experiments done long ago. For instance, the fact that a gravitatonal force exists, the fact that there are three generations of quarks and leptons, that the effective gauge forces at low energy exhibit the mathematical structure of $SU(3) \otimes SU(2) \otimes U(1)$, all these should be consistent with our theory,

---

[a]In the *perturbative* formulation of String Theory, causality appears to be obeyed perfectly, including its coupling to gravity. It is in the circular nature of any attempt at a non-perturbative formulation where reference to time ordering is often lost, so that causality may be questionable.

and in practice these are very important constraints. Some string theorists proclaim that their theory "predicts" gravity. This is an absurd statement, of course, but it indicates how stringent this requirement appears to be in the eyes of these physicists.

(v) Eventually, a successful theory must be able to make *real* predictions. For string theory, or other considerations concerning physics at the Planck scale, this may mean that we hope to be able, one day, to 'predict' the value of fundamental constants of nature that as yet have eluded any genuine prediction, such as the fine structure constant. Even if this constant could be reproduced with less accuracy than its presently known experimental value, this would be a tremendous achievement. It would imply that, at least in principle, such numbers could be computed with arbitrary precision, so that definitely predictions of the conventional type can be expected. This is surely what we are aiming at eventually. But long before this situation is arrived at, a theory may have become acceptable and respected by a fairly large community, simply because it may have provided more and deeper understanding.

Examples from the Past

It is difficult to compare this situation with our experiences in the past, because physicists always have been so fortunate that clever experiments could be devised to check *directly* whether the equations described in a theory worked the way they were expected to.

Consider the Maxwell equations. James Clark Maxwell discovered around 1864 that the descriptions existing in his time were inconsistent. His equations for electricity and magnetism were a mathematical completion of existing knowledge. He quickly realized that light propagation had to be closely related to electromagnetism, and as the theory agreed with everything known before, it was soon accepted as obeying the first few of the criteria written down above.

Needless to say however, that the theory was tested at numerous occasions when electromagnetic phenomena were further studied. Would we have accepted the theory if such tests had not been there? Clearly, the theory would not have enjoyed the same status as it does today, but the mere fact that, at one stroke, the theory of optics was unified together with electricity and magnetism would have made it highly respectable.

Einstein's theory of relativity was just such a case. The theory was not primarily intended to "explain" observed phenomena. Although Einstein presumably knew about the Michelson-Morley experiments, his primary concern was to give a more satisfactory description of the geometry of space and time in its relation to electro-magnetism. His analysis was a purely mathematical one. His successful attempts to include the gravitational force, led to the generalized theory of relativity, and this happened to provide for an unexpected explanation of an observation that had already been made but was not understood until then: the anomalous shift of the orbit of the planet Mercury. It was also very fortunate that Einstein's theory gave rise to an opportunity for further experimental tests, notably the famous Eddington expeditions.

In this case, the role of the experiment was not to discriminate between Einstein's theory and some would-be alternative; there hardly were alternatives,[b] but the theory could have been wrong or incomplete for other reasons. The test was made in order to find new confirmations of the fact that Physics here was on the right track. this is a situation that we see occur quite often in our field of science. Historically, these tests were generally thought to be of extreme importance. What this means, is that theories that are so novel as relativity was at that time, cannot flourish without as much support as possible from independent sources of information.

Many of our present theories are to be compared to relativity *without* such checks. The builders of these theories are facing hard times; yet I believe that we *can* make progress, provided our techniques, as well as our critical attitude regarding the required logical consistence, improve.

A milestone in physics was Paul Adrien Maurice Dirac's improved equation for the electron. Previously, Wolfgang Pauli had the following

---

[b]There *do* exist various kinds of *modifications* of Einstein's theory, such as Dicke's scalar-tensor theory, which could be tested by these experiments — they were all but ruled out — but even these were based on Einstein's principle of invariance under general coordinate transformations. There actually also existed an alternative explanation for Mercury's perihelion shift: an unknown planet very near to the sun ...

(non-relativistic) equation:

$$H\psi = \left(\frac{1}{2m}(\hat{\mathbf{p}} - e\mathbf{A})^2 - \frac{ge}{2m}(\mathbf{B} \cdot \boldsymbol{\sigma}) \right) \psi,$$

where the last term describes the interaction between the electron's magnetic moment with the magnetic field **B**. It features an unknown constant, the electron's gyro-magnetic ratio, g. Now Dirac's equation[4] was

$$H\psi = (\boldsymbol{\alpha} \cdot (\hat{\mathbf{p}} - e\mathbf{A}) + \beta m)\psi,$$

where α and β are the Dirac matrices.

In the non-relativistic limit, Dirac's equation turns into Pauli's equation, but it offers a bonus: the gyromagnetic ratio g is predicted to be exactly 2, which was very close to the experimentally observed value. However, Dirac realized very well that there also was a difficulty with his equation. His electron is described by 4 components, not 2, as in Pauli's equation. The two extra degrees of freedom could be interpreted as describing fermions with electric charge $+e$ instead of $-e$, as in the electron. About this episode, A. Pais writes[5]:

> While in Leipzig, June 1928, Dirac visited his friend and colleague Werner Heisenberg, who had recently been appointed there. Heisenberg was also well aware of the difficulties. In May he had written to Pauli: "In order not to be forever irritated with Dirac, I have done something else for a change" (the something else was his quantum theory of ferromagnetism). Dirac and Heisenberg discussed several aspects of the new theory. Shortly thereafter, Heisenberg wrote again to Pauli: "The saddest chapter of modern physics is and remains the Dirac theory." He mentioned some of his own work which demonstrated the difficulties, and added that the magnetic electron had made Jordan "trübsinnig" (melancholic). At about the same time, Dirac, not feeling so good either, wrote to Oskar Klein: "I have not met with any success in my attempts to solve the $\pm e$ difficulty. Heisenberg (whom I met in Leipzig) thinks the problem will not be solved until one has a theory of the proton and the electron together."

Indeed, during some time, Dirac and his colleagues suspected that the new components had to be describing the proton, but this gave rise to even more difficulties: Oppenheimer[6] and Tamm[7] correctly objected that, if this were true, it would make the hydrogen atom unstable against decay into two or more photons. Finally, in 1931, Dirac had to admit, much against his will, that his equation predicted a particle not observed until then. Just before the year's end, Carl Anderson made the first announcement of experimental evidence for the anti-electron.[8]

What this history tells us is that the relation between theory and experiment is quite a lot more subtle than a game of differentiation and rejection of false theories. Theories are more often than not rejected solely on the grounds that they do not seem to be reconcilable with known facts.

Another case of interest is Pauli's theory of the existence of a "neutrino" to account for the missing energy in β decay. In February 1929, Pauli complained in a letter to Oskar Klein that Bohr was serving him up all kinds of ideas about beta decay

"… by appealing to the Cambridge authorities but without reference to the literature", and he said: "With his considerations about a violation of the energy law Bohr is on a *completely wrong* track".[5]

The existence of neutrinos was confirmed by Frederick Reines' experiment[9] at the Savannah River reactor, in 1956. When he Announced the discovery in a telegram to Pauli, the latter reacted not the least surprised: "I knew already that they exist".

Again a different situation occurred when the new theories for the weak interaction were proposed. In the late '60s the problem was that the existing theories for weak interactions were simply too diffuse to be believable.

When C.N. Yang and R.L. Mills did their work on the generalization of QED equations, in 1954, the paper they published[10] contained a theory that *did not seem to apply to any known situation* in particle physics. Yet, the theory had such a high degree of elegance that they believed it to contain some deep truth. Although this new theory was ignored by many physicists, it did provide inspiration to some of the deeper thinkers. Richard Feynman used it at various occasions: the Feynman-Gell-Mann expression[11] for an effective weak interaction theory (a refinement of an earlier proposal by Enrico Fermi), carried some elements of the Yang-Mills idea, and somewhat later when Feynman made an attempt to understand the quantization of the gravitational force, he was made aware by Gell-Mann of the deep mathematical relations between gravity and Yang-Mills.[12] So a theory without any experimental verification nevertheless played a role.

Insignificant as it seemed to be at the time, this role would later turn out to be an essential one.

In 1961, C.N. Yang was asked to lecture about "The Future of Physics",[13] at a panel discussion at the Celebration of the MIT Centennial, April 8. At that time, this is what Yang dared to say about his expectations:

> "[What I know now, does] enable us to say with some certainty that great clarification will come in the field of the weak interactions in the next few years. With luck on our side, we might even hope to see some integration of the various manifestations of the weak interactions."

Yang hoped, *with luck on his side*, to bring all *weak* interactions under one denominator. The fact that, within ten years, one single theory, based on his own ideas, would not only achieve that, but also include electromagnetism, apparently was way beyond his hopes.

In the early 70's it was discovered that *theories of this type were indeed logically fully consistent*, and superior to any of their alternatives. But it was also discovered that a whole class of such theories could be constructed, all having the Yang-Mills equations as a back bone. The earliest, already existing model was due to Steven Weinberg[14] and Abdus Salam,[15] and it was based on $SU(2) \otimes U(1)$. But among the alternatives was a model proposed by Pati and Salam,[16] whose version was based on $SU(3)$, and a more interesting alternative theory was proposed by H. Georgi and S.L. Glashow,[17] based on the group $SO(3)$. See Fig. 1, where some of the models that were proposed are illustrated.

Most of the models predicted not only the charged vector bosons that intermediate the observed weak interactions, but also one or more neutral components, called Z^0. Such particles would cause interactions between neutral components of the currents in hadrons and leptons. Such interactions had not yet been detected. What made the Georgi-Glashow model interesting was that, having no Z^0 boson, it would predict the *absence* of any such effects. In the '70's, a combined effort by several experimental teams (CERN,[19] Fermilab,[20] Argonne,[21] Brookhaven[22]) reinvestigated the spurious weak interaction events associated to the neutral current interactions. Notably, according to most of the new theories, a neutrino should be able to hit against hadrons as well as leptons without itself changing into a charged lepton. Actually, such events had been searched

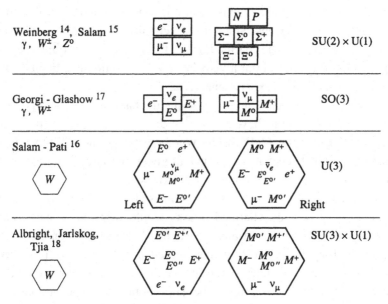

Fig. 1.   A few weak interaction models of the early '70's, showing some of the proposed multiplet structures. *E* and *M* are proposed heavy leptons. Also indicated are the predicted vector bosons. The hexagons labeled "*W*" indicate nonets of charged and neutral vector bosons. For the last three models only the leptonic sectors are displayed.

for earlier, with apparently negative results: common wisdom had been that these events do not occur. But it was also agreed upon that these investigations were not totally conclusive, as the events searched for could easily have been mistaken for background noise.

Around 1974 the results of these experimentations became clear: the events were there, more or less as they were predicted by the simplest of all models: the Weinberg-Salam scheme. For theoreticians and experimentalists alike, this was decisive for selecting the original Weinberg-Salam scheme as the most likely theory for the electro-weak interactions. Could we have *derived* from purely theoretical arguments that the model was superior?

The Weinberg-Salam scheme did require a special mechanism, called the GIM mechanism,[23] to cancel out all those weak interaction effects that are associated to a change of strangeness, without the exchange of electric charge ('strangeness changing neutral current events'). It was well-known that such interactions, if they existed at all, were very much weaker than the other weak interactions. Since this mechanism required the existence of a new quark species, the charmed quark, the search for this object began, leading to its discovery about a year later. What would have happened if we had not had all these experimental clues?

The alternative models did not describe the hadronic sector very well. In particular, they would have been very difficult to apply to the quark constituents of hadrons. Only for the Weinberg-Salam model did we have a scheme, the GIM mechanism. This had given the model an edge, even before the results of the experiments were known.

One crucial prediction was essentially missed: the new hadronic objects that can consist of only charmed quarks would turn out to have quite conspicuous properties. Experimenters justifiably sometimes ignore theory, by just searching for interesting things even if not explicitly predicted. Some theoreticians did indeed urge to search for narrow resonances, but these were not really considered to be so pronounced or special that they would have striking properties, that would stand out so much that a new field of physics would emerge. But we could have known. When these objects were discovered, called J (by S. Ting[24]) and ψ (by B. Richter[25]), in 1974, they were striking indeed. Most remarkable was the fact that their strong interaction properties were far weaker than normal. It was only during a short time that several "alternative theories" were considered to explain J/ψ For instance, it was suggested that they might be the bosons intermediating the weak interactions, or maybe they were totally new objects of physics. Several alternative theories were concocted, completely in Popperian style. Yet the charm theory stood out from the beginning. After half a year it was the only theory.

Since that time, theoretical physics has enormously gained in importance. What we have presently is an extremely detailed description of all particles and forces known to us. It is called the "Standard Model". There is actually only a limited amount of information that is still waiting to be provided by experiments in the near future: more details about the neutrino masses and their mixings,[2] more precise determination of the mixing angles between quarks, responsible for the violation of PC invariance and top and bottom decays, and the mass (and possibly the spectrum) of the Higgs particle.

Of course we hope that accelerators in the near future will allow us to find further extensions of the Standard Model. Indeed, extensions are expected: supersymmetric versions of the theory (the Minimally Supersymmetric Standard Model, or MSSM) require the existence of much more elementary particle types than the ones known presently: the "super partners". Again, this is the presently preferred scenario, to be confirmed, or to be falsified, in which latter case we have not much to put in its place, other than "new and unexpected physics". Since I do have certain objections against the super symmetric scenario, I would love to see a more complicated physical scene than the one predicted by the MSSM, but detailed predictions cannot be provided.

The reason for giving this talk, and for emphasizing the important shifts in the relations between theory and experiment, is my concern over the situation my field of research will be in, after these facts may or may not have been checked by experiments. We are rapidly proceeding with our theories. Our theoretical view of the particle scenery is presently so clear that we can project much further than the supersymmetry domain. Most importantly, we want to understand the gravitational force.

It is here that possibilities for experiment are severely limited. The Planck scale is determined by the combinations of the constants c (the velocity of light), h (Planck's constant divided by 2π), and G (Newton's constant) that are listed in Table 1.

Several theoretical approaches are addressing the fundamental questions we are confronted with here, but of these, Superstring Theory, and its successors (D-brane theory and M-theory), are taking the lead. String Theory started as an attempt to understand the strong interactions, but its mathematical elegance became as striking as that of Yang-Mills theory,

Table 1. The Planck units.

$$\sqrt{\frac{G\hbar}{c^3}} = 1.616 \times 10^{-33}\,\text{cm}$$

$$\sqrt{\frac{G\hbar}{c^5}} = 5.39 \times 10^{-44}\,\text{sec}$$

$$\sqrt{\frac{\hbar c}{G}} = 21.77\mu\text{g} \rightarrow 1.22 \times 10^{28}\,\text{eV}$$

20 years earlier. Here again, theoreticians were dealing with a general scheme that inspires us towards finding deeper connections between space, time and matter. The theories are gaining in power now that various extremely essential difficulties are being confronted and new resolutions are being discovered. The theoretical scheme now being devised is generally believed to be the "only credible theory that unifies the gravitational forces with other forms of matter in a quantum theory".

Most important here is the fact that the gravitational force is a *cumulative force*, which means that, as it acts, it may draw large amounts of matter closer together, and in so doing create regions of space-time where the field becomes increasingly stronger. This makes gravity itself *fundamentally unstable*: it may give rise to "gravitational collapse", or in more popular terms: a black hole.

All theoretical scenarios agree that, at least in principle, initial state conditions can be realized that inevitably lead to gravitational collapse: black holes, or objects very closely related to them, must be part of the matter spectrum at the Planck scale.

The black hole is a prototype of a physical system that can only be theorized about, but not be studied experimentally; nobody ever saw a black hole up close. Here in particular, one must ask the question whether the existing theories describing their behaviour are fully self-consistent, and whether these descriptions give rise to any kind of logical conflict. Indeed, attempts to arrive at a consistent description using well-known and accepted ingredients of physics, have led us to quite peculiar results.

A black hole appears to behave as if it is a "ball of material" that carries one "bit of information" in every 0.724×10^{-65} cm^2 of its surface area. At the same time, however, a region of space-time that undergoes gravitational collapse into a black hole must also be able to describe ordinary particles living there, and, at first sight, it seems as if these particles are not constrained by a limit on the information they carry. This at least is what we appear to arrive at when applying Einstein's theory of general relativity. On the other hand, black holes appear to drain away information from space-time, whereas in general such behaviour would be in conflict with standard formulations of quantum mechanics.[26]

How can one incorporate these apparently conflicting features correctly in a theory? Remarkably, string theoreticians recently succeeded in describing black holes (of certain types) in such a way that their results agree completely with everything just said above, so that their theory appears to be in good shape.[27]

Yet there are still numerous difficulties. Not the least of these is that theories of this nature do not provide simple answers to the question as to how they would predict the observed features of the Standard Model, or how in general any detailed calculation should be done to confront the theory with experiment at low energy scales.

There is no *direct* experiment to test any of the more esoteric ideas. This is not to say that there would not be *indirect* clues. There are still quite a few possibilities for future experiments and observations. The determination of the neutrino mass spectrum is just one of these. Clarification of the supersymmetric domain (if it exists!) and possible detection of superweakly interacting massive particles (if they exist!) are other possibilities. Furthermore, we may hope that astrophysical observations such as the spectrum and fluctuations of the cosmic microwave radiation will yield further data. But, to extrapolate from any of such observations all the way to a string theoretical description at the Planck scale, will continue to be a task of gargantuan proportions.

The only instrument we can use to study the Planck length in detail, is our minds.

In previous years we had experiment to guide us and to protect us against our numerous mistakes, so the correct theories could be identified and confirmed. This, as it turns out, was a luxury that we may have to do without in the future. Does such physics make sense?

Well, we have seen that, more often than not, experiment was not the only way to check a theory. It was logical consistence, general esthetics, agreement with previously known facts, and general rigour, that we could use as criterea in addition.

Unfortunately, history has also shown how easily we can be misled into wrong theories, and how important it was, throughout the ages, that we could ask Nature to settle our disputes. Numerous publications on the new esoteric theories show quite clearly that the moderating control of experiment has been absent for some time. The mathematics is often not as rigorous as one should wish. What disturbs me most is that main progress seems to be coming from theories that take the form of "conjectures". Failing to falsify the conjectures is often presented as a proof of their validity. In reality, these conjectures are essentially impossible to prove (or to disprove) with any kind of rigor since the theories are not sufficiently well defined. How sure are we that our logic is correct? In the past, such a situation has often showed to be detrimental for a theory.

On the other hand, if the logic in an argument appears to be less than perfect, this is not at all a sufficient reason to abandon further pursuit along the lines given. Remember that renormalization theory appeared to have serious shortcomings, yet it worked, just because the difficulties could be kept under control.

Do we have the difficulties of string theory under control? My presently most favored approach is that the most celebrated edifice of Physics, the theory of Quantum Mechanics, must undergo revision.[28] I am not the only one saying this, but I do think that I have a better way to write improved theories and to show why they are superior, and furthermore that I have indications as to *why* Nature chose to run in accordance with laws that feature the remarkable principles of statistics that characterize Quantum Mechanics.

It was a momentous achievement of a century of Physics to find how the dynamics of molecules, atoms and subatomic particles could be described in detail using the principles of Quantum Mechanics. It is conceivable that the next century will tell us how to reconcile these findings with a deterministic world view. All that is needed is a description of a *primordial basis* of states,

$$|\psi_1\rangle, |\psi_2\rangle, \ldots$$

in terms of which the evolution operators U behave as pure permutation operators, not just any set of unitary operators. A primordial basis may be seen to be generated by operators $\mathcal{O}(\mathbf{x}, t)$ which commute at all times:

$$[\mathcal{O}(\mathbf{x}, t), \mathcal{O}(\mathbf{x}', t')] = 0, \quad \forall \mathbf{x}, \mathbf{x}', t, t',$$

in contrast with more conventional fields, which do not commute within the light cones. Physicists are presently used to the idea that such operators $\mathcal{O}$ cannot possibly exist, but this is not totally obvious. When gravitational forces are taken into account, any 'no-go' theorem would be extremely difficult to prove.

Naturally, my critical audience asks what, in such a theory, would be the cause and the meaning of interference effects. A possible answer is that interference is only recognized as such *after* having assumed some law of physics describing particles in semi classical terms as objects that can be detected in an experiment. These laws of physics may in reality be extremely simplified versions of a statistical analysis using only a very small subspace of Hilbert space generated by the primordial states: the states with very low energies. This subspace is built from superpositions of the primordial basis elements, being the low-energy eigenstates of the hamiltonian (defined to be the true operator generating time shifts). Since these are *not* primordial states, physicists experience interference effects all the time.

The theory, of which the above is only a very brief outline, leads to several difficulties that are as yet far from being resolved. A primary difficulty is the simple observation that, in the real world, the hamiltonian happens to have a lower bound. If I define the hamiltonian to be the generator of time shifts, then the presence of a lower bound is difficult to reconcile with determinism. A possible way out of this dilemma was recently identified: information loss.

If ideas of this sort could be combined with string theory, this might lead to the insights still needed to produce the necessary logical coherence in our theories. But will it be enough?

Without much guidance from experiment, obtaining better understanding of Nature may perhaps be prohibitively difficult. But my confidence in human ingenuity is tremendous. Even if experiments will be difficult, smart new ideas for experiments will still be proposed, but they will require the utmost skill for a correct interpretation. To do our job, to reach the ultimate equations of physics, with the given limitations for experimental evidence, will probably take much more time than the 20 years that were S.W. Hawking's estimate.[29] During this time, probably many new generations of young physicists will enjoy the fine taste of new discoveries. We are all full of optimism for the new century.

References

1. M.B. Green, J.H. Schwarz and E. Witten, *Superstring Theory*, Cambridge Univ. Press; J. Polchinski, *String Theory*, Cambridge Univ. Press, 1998.

2. Y. Fukuda *et al.* (Super-Kamiokande Collaboration), *Phys. Rev. Lett.* **81** 1562 (1998); H. Terazawa, KEK Prepr. 98–226, *Proc. Int. Conf. on Modern Developments in Elementary Particle Physics*, Cairo Helwan and Assyut, Jan and Feb 1999.

3. A. Pais, *Subtle is the Lord...*, Oxford University Press, Oxford, 1982, pp. 115–119.

4. P.A.M. Dirac, *Proc. Roy. Soc.* **A117** (1928) 610; *ibid.* **A118** (1928) 315.

5. A. Pais, *Inward Bound, of Matter and Forces in the Physical World*, Clarendon, Oxford University Press, New York, 1986, p. 309, 346, . . .

6. R. Oppenheimer, *Phys. Rev.* **35** (1930) 562.

7. I. Tamm, *Zeitschr. f. Phys.* **62** (1930) 545.

8. C.D. Anderson, *Science.* **76** (1932) 238.

9. C.L. Cowan *et al.*, *Science.* **124** (1956) 103.

10. C.N. Yang and R.L. Mills, *Phys. Rev.* **96** (1954) 191, R. Shaw, Cambridge Ph.D. Thesis, unpublished.

11. R.P. Feynman and M. Gell-Mann, *Phys. Rev.* **109** (1958) 193.

12. R.P. Feynman, *Acta Phys. Polonica.* **24** (1963) 697.

13. C.N. Yang, *The Future of Physics*, in *Selected Papers 1945–1980 With Commentary*, C.N. Yang, W.H. Freeman and Co, 1983, ISBN 0-7167-1406-X, p. 319.

14. S. Weinberg, *Phys. Rev. Lett.* **19** (1967) 1264; *id.*, *Sci. Am.* **231** (1974) 50.

15. A. Salam and J.C. Ward, *Phys. Lett.* **13** (1964) 168.

16. J.C. Pati and A. Salam, *Phys. Rev.* **D8** (1973) 1240.

17. H. Georgi and S.L. Glashow, *Phys. Rev. Lett.* **28** (1972) 1494.

18. C.H. Albright, C. Jarskog and M.O. Tjia, *Nucl. Phys.* **B86** (1975) 535.

19. C. Rubbia, *Proc. of the XVIIth Int. Conf. on High Energy Physics*, London 1974, p. IV–114, 117; J. Sacton, *ibid.*, p. IV–121; A. Rousset, *ibid.* p. 128.

20. B.C. Barish *et al.*, *Proc. of the XVIIth Int. Conf. on High Energy Physics*, London 1974, p. IV–111; *Proc. Int. School of Subnuclear Physics, Erice, 1975*, A. Zichichi, ed., Plenum 1977, p. 897.

21. P. Schreiner, *Proc. of the XVIIth Int. Conf. on High Energy Physics*, London 1974, p. IV–123.

22. W. Lee, *Proc. of the XVIIth Int. Conf. on High Energy Physics*, London 1974, p. IV–127.

23. S.L. Glashow, J. Iliopoulos and L. Maiani, *Phys. Rev.* **D2** (1970) 1285.

24. C.C. Ting, *Proc. Int. School of Subnuclear Physics, Erice, 1975*, A. Zichichi, ed., Plenum 1977, p. 559; J.J. Aubert *et al.*, *Phys. Rev. Lett.* **33** (1974) 1404.

25. J.-E. Augustin *et al.*, *Phys. Rev. Lett.* **33** (1974) 1406.

26. G. 't Hooft, *J. Mod. Phys.* **A11** (1996) 4623; gr-qc/9607022.

27. J. Maldacena, *Black Holes in String Theory*, hep-th/9607235; A.W. Peet, *Class. Quant. Grav.* **15** (1998) 3291.
28. G. 't Hooft, *Quantum Gravity as a dissipative deterministic system*, gr-qc/9903084, *Class. Quant. Grav.*, to be publ.
29. S.W. Hawking, in several of his recent public lectures.

AUTHOR INDEX

Author Index

SUBJECT INDEX

Subject Index

Printed in the United States
By Bookmasters